T0331144

A CHEMIST'S GUIDE TO VALENCE BOND THEORY

BICENTENNIAL

1807

WILEY

2007

BICENTENNIAL

THE WILEY BICENTENNIAL—KNOWLEDGE FOR GENERATIONS

*E*ach generation has its unique needs and aspirations. When Charles Wiley first opened his small printing shop in lower Manhattan in 1807, it was a generation of boundless potential searching for an identity. And we were there, helping to define a new American literary tradition. Over half a century later, in the midst of the Second Industrial Revolution, it was a generation focused on building the future. Once again, we were there, supplying the critical scientific, technical, and engineering knowledge that helped frame the world. Throughout the 20th Century, and into the new millennium, nations began to reach out beyond their own borders and a new international community was born. Wiley was there, expanding its operations around the world to enable a global exchange of ideas, opinions, and know-how.

For 200 years, Wiley has been an integral part of each generation's journey, enabling the flow of information and understanding necessary to meet their needs and fulfill their aspirations. Today, bold new technologies are changing the way we live and learn. Wiley will be there, providing you the must-have knowledge you need to imagine new worlds, new possibilities, and new opportunities.

Generations come and go, but you can always count on Wiley to provide you the knowledge you need, when and where you need it!

WILLIAM J. PESCE
PRESIDENT AND CHIEF EXECUTIVE OFFICER

PETER BOOTH WILEY
CHAIRMAN OF THE BOARD

A CHEMIST'S GUIDE TO VALENCE BOND THEORY

Sason Shaik
The Hebrew University
Jerusalem, Israel

Philippe C. Hiberty
Université de Paris-Sud
Orsay, France

WILEY-INTERSCIENCE

A JOHN WILEY & SONS, INC., PUBLICATION

Published by John Wiley & Sons, Inc., Hoboken, New Jersey
Published simultaneously in Canada

Limit of Liability/Disclaimer of Warranty; While the publisher and author have used their best efforts in preparing this book, they make no representations or warranties with respect to the accuracy or completeness of the contents of this book and specifically disclaim any implied warranties of merchantability or fitness for a particular purpose. No warranty may be created or extended by sales representatives or written sales materials. The advice and strategies contained herein may not be suitable for your situation. You should consult with a professional where appropriate. Neither the publisher nor author shall be liable for any loss of profit or any other commercial damages, including but not limited to special, incidental, consequential, or other damages.

For general information on our other products and services or for technical support, please contact our Customer Care Department within the United States at (800) 762-2974, outside the United States at (317) 572-3993 or fax (317) 572-4002.

Wiley also publishes its books in a variety of electronic formats. Some content that appears in print may not be available in electronic formats. For more information about Wiley products, visit our web site at www.wiley.com.

Wiley Bicentennial Logo: Richard J. Pacifico

Library of Congress Cataloging-in-Publication Data:

Shaik, Sason S., 1943-
 A chemist's guide to valence bond theory / by Sason Shaik and Philippe C. Hiberty.
  p. cm.
 Includes index.
 ISBN 978-0-470-03735-5 (cloth)
1. Valence (Theoretical chemistry) I. Hiberty, Philippe C. II. Title.
 QD469.S53 2008
 541'.224–dc22   2007040432

Printed in the United States of America

10 9 8 7 6 5 4 3 2 1

Dedicated to our great teachers:
Roald Hoffman,
Nicolaos D. Epiotis,
Lionel Salem,
and the late Edgar Heilbronner.

CONTENTS

PREFACE

This text was written to fill the missing niche of a textbook that teaches Valence Bond (VB) theory. The theory that once charted the mental map of chemists had been abandoned since the mid-1960s for reasons that are discussed in Chapter 1. Consequently, the knowledge of VB theory and its teaching became gradually more scarce, and was effectively eliminated from the teaching curriculum in much of the chemical community. Nevertheless, a few elements of the theory somehow survived as the *Lingua Franca* of chemists, mostly due to the use of the Lewis bonding paradigm and the post-Lewis concepts of hybridization and resonance. But there is much more to VB theory than these concepts and ideas. Since its revival in the 1980s, VB theory has been enjoying a renaissance that is characterized by the development of a growing number of *ab initio* methods that can be applied to chemical problems of bonding and reactivity. Alongside these methodology developments, there has been a surge of new post-Pauling models and concepts that have rendered VB theory useful again as a central theory in chemistry; especially productive concepts arose by importing insights from molecular orbital (MO) theory and making the VB approach more portable and easier to apply. Following a recent review article by us (1) and two essays on VB theory and its relation to MO theory (2,3), we felt that the time had come to write a textbook dedicated to VB theory, its applications, and special insights.

This text is aimed at a nonexpert audience and designed as a tutorial material for teachers and students who would like to teach and use VB theory, but who otherwise have basic knowledge of quantum chemistry. As such, the primary focus of this textbook is a qualitative insight of the theory and ways to apply this theory to the problems of bonding and reactivity in the ground and excited states of molecules. Almost every chapter contains problem sets followed by answers. These problems provide the teachers, students, and interested readers with an opportunity to practice the art of VB theory. We will be indebted to readers—teachers—students for comments and more suggestions, which can be incorporated into subsequent editions of this book that we hope, will follow.

Another features in this book is the description of the main methods and programs available today for *ab initio* VB calculations, and how actually one may plan and run VB calculations. In this sense, the book provides a snapshot of the current VB capabilities in 2007. Regrettably, much important work had to be left out. The readers interested in technical and theoretical development aspects of VB theory may wish to consult two other monographs (4,5).

The two authors owe a debt of gratitude to colleagues and friends who read the chapters and provided useful comments and insights. In particular, we acknowledge Dr. Benoît Braïda, Professor Narahary Sastry, Professor Hendrik Zipse, Professor François Volatron, and Dr. Hajime Hirao for the many comments and careful reading of earlier drafts. François Volatron actually solved all the problem sets and checked the equations of Chapter 3: The equations are in much better shape thanks to his careful screening. Hajime Hirao went over the entire book and its galley proofs in search for glitches. Sebastian Kozuch (S. Shaik's PhD student) helped in the design of the book cover. Needless to say, none of these gentlemen should be held responsible for the content of the book.

In addition, we are thankful to all our co-workers and students during the years of collaboration (1981–present). Especially intense collaborations with Professor Addy Pross and Professor Wei Wu are acknowledged. Professor Wei Wu and Professor Joop van Lenthe are especially thanked for making their programs (XMVB and TURTLE) available to us. In fact, Professor Wei Wu has been kind enough to give us unlimited access to his XMVB code during the work on this book. Dr. David Danovich is thanked for producing all the inputs and outputs in this book, for helping with the final proofing, and for keeping alive the VB computational know-how at the Hebrew University throughout the years from 1992 onward.

Finally, any readers, teachers, or students who wish to comment on aspects of the content, problems-, and/or answer-sets, as well as on errors, are welcome to do so by contacting the authors directly by e-mail (sason@yfaat.ch.huji.ac.il and philippe.hiberty@lcp.u-psud.fr).

REFERENCES

1. S. Shaik, P. C. Hiberty, *Rev. Comput. Chem.*, **20**, 1 (2004). Valence Bond, Its History, Fundamentals and Applications: A Primer.
2. S. Shaik, P. C. Hiberty, *Helv. Chem. Acta* **86**, 1063 (2003). Myth and Reality in the Attitude Toward Valence-Bond (VB) Theory: Are Its 'Failures' Real?
3. R. Hoffmann, S. Shaik, P. C. Hiberty, *Acc. Chem. Res.* **36**, 750 (2003). A Conversation on VB vs. MO Theory: A Never Ending Rivalry?
4. G. A. Gallup, *Valence Bond Methods*, Cambridge University Press: Cambridge, 2002.
5. R. McWeeny, *Methods of Molecular Quantum Mechanics,* 2nd ed., Academic Press, New York, 1992.

SASON SHAIK
Jerusalem, Israel

PHILIPPE C. HIBERTY
Orsay, France

1 A Brief Story of Valence Bond Theory, Its Rivalry with Molecular Orbital Theory, Its Demise, and Resurgence

The new quantum mechanics of Heisenberg and Schrödinger provided chemistry with two general theories, one called valence bond (VB) theory and the other molecular orbital (MO) theory. The two theories were developed at about the same time, but have quickly diverged into rival schools that have competed, sometimes fervently, on charting the mental map and epistemology of chemistry. In brief, until the mid-1950s VB theory had dominated chemistry, then MO theory took over while VB theory fell into disrepute and was almost completely abandoned. The more recent period from the 1980s onward marked a comeback of VB theory, which has since then been enjoying a renaissance both in the qualitative application of the theory and in the development of new methods for its computer implementation (1). One of the great merits of VB theory is its pictorially intuitive wave function that is expressed as a linear combination of chemically meaningful structures. It is this feature that has made VB theory so popular in the 1930s−1950s, and it is the same feature that underlies its temporary demise and ultimate resurgence. This monograph therefore constitutes an attempt to guide the chemist in the use of VB theory, to highlight its insight into chemical problems, and some of its state-of-the-art methodologies.

Since VB is considered, as an obsolete theory, we thought it would be instructive to begin with a short historical account of VB theory, its rivalry against the alternative MO theory, its downfall, and the reasons for the past victory of MO and the current resurgence of VB theory. Part of this review is based on material from the fascinating historical accounts of Servos (2) and Brush (3,4). Other parts are not official historical accounts, but rational analyses of historical events; in some sense, we are reconstructing history in a manner that reflects our own opinions and the comments we received from colleagues, as well as ideas formed during the writing of the recent "conversation" the two authors have published with Roald Hoffmann (5).

1.1   ROOTS OF VB THEORY

The roots of VB theory in chemistry can be traced back to the famous paper of
Lewis *The Atom and The Molecule* (6), which introduces the notions of electron-
pair bonding and the octet rule (initially called the rule of eight) (6). Lewis was
seeking an understanding of weak and strong electrolytes in solution (2). This
interest led him to formulate the concept of the chemical bond as an intrinsic
property of the molecule that varies between the covalent (shared-pair) and
ionic extremes. In this article, Lewis uses his recognition that almost all known
stable compounds had an even number of electrons as the rationale that led
him to the notion of electron pairing as a mechanism of bonding. This and the
fact that helium was found by Mosely to possess only two electrons made it
clear to Lewis that electron pairing was more fundamental than the octet rule;
the latter rule was an upper bound for the number of electron pairs that can
surround an atom (6). In the same paper, Lewis invents an ingenious symbol
for electron pairing, the colon (e.g., H:H), which enabled him to draw
electronic structures for a great variety of molecules involving single, double,
and triple bonds. This article predated new quantum mechanics by 11 years
and constitutes the first effective formulation of bonding in terms of the
covalent–ionic classification, which is still taught today. This theory has
formed the basis for the subsequent construction and generalization of VB
theory. This work eventually had its greatest impact through the work of
Langmuir, who articulated the Lewis model, applied it across the periodic
table, and invented catchy terms like the octet rule and the covalent bond (7).
From then onward, the notion of electron pairing as a mechanism of bonding
became widespread and initiated the "electronic structure revolution" in
chemistry (8).

   The overwhelming chemical support of Lewis's idea presented an exciting
agenda for research directed at understanding the mechanism by which an
electron pair could constitute a bond. This, however, remained a mystery until
1927 when Heitler and London went to Zurich to work with Schrödinger. In
the summer of the same year, they published their seminal paper, *Interaction
Between Neutral Atoms and Homopolar Binding* (9,10). Here they showed that
the bonding in dihydrogen (H_2) originates in the quantum mechanical
"resonance" interaction that is contributed as the two electrons are allowed
to exchange their positions between the two atoms. This wave function and the
notion of resonance were based on the work of Heisenberg (11), who showed
earlier that, since electrons are indistinguishable particles, then for a two
electron systems, with two quantum numbers n and m, there exist two wave
functions that are linear combinations of the two possibilities of arranging
these electrons, as shown Equation 1.1.

$$\Psi_A = (1/\sqrt{2})[\varphi_n(1)\varphi_m(2) + \varphi_n(2)\varphi_m(1)] \tag{1.1a}$$

$$\Psi_B = (1/\sqrt{2})[\varphi_n(1)\varphi_m(2) - \varphi_n(2)\varphi_m(1)] \tag{1.1b}$$

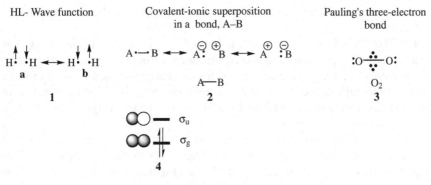

HL- Wave function   Covalent-ionic superposition   Pauling's three-electron
                                 in a bond, A–B   bond

Scheme 1.1

As demonstrated by Heisenberg, the mixing of $[\varphi_n(1)\varphi_m(2)]$ and $[\varphi_n(2)\varphi_m(1)]$ led to a new energy term that caused a splitting between the two wave functions Ψ_A and Ψ_B. He called this term "resonance" using a classical analogy of two oscillators that, by virtue of possessing the same frequency, form a resonating situation with characteristic exchange energy.

In modern terms, the bonding in H_2 can be accounted for by the wave function drawn in **1**, in Scheme 1.1. This wave function is a superposition of two covalent situations in which, in the first form (**a**) one electron has a spin-up (α spin), while the other has spin-down (β spin), and vice versa in the second form (**b**). Thus, the bonding in H_2 arises due to the quantum mechanical "resonance" interaction between the two patterns of spin arrangement that are required in order to form a singlet electron pair. This "resonance energy" accounted for ~75% of the total bonding of the molecule, and thereby projected that the wave function in **1**, which is referred to henceforth as the HL-wave function, can describe the chemical bonding in a satisfactory manner. This "resonance origin" of the bonding was a remarkable feat of the new quantum theory, since until then it was not obvious how two neutral species could be at all bonded.

In the winter of 1928, London extended the HL-wave function and drew the general principles of the covalent bonding in terms of the resonance interaction between the forms that allow interchange of the spin-paired electrons between the two atoms (10,12). In both treatments (9,12) the authors considered ionic structures for homopolar bonds, but discarded their mixing as being too small. In London's paper, there is also a consideration of ionic (so-called polar) bonding. In essence, the HL theory was a quantum mechanical version of Lewis's electron-pair theory. Thus, even though Heitler and London did their work independently and perhaps unaware of the Lewis model, the HL-wave function still precisely described the shared-pair bond of Lewis. In fact, in his letter to Lewis (8), and in his landmark paper (13), Pauling points out that the HL and London treatments are 'entirely equivalent to G.N. Lewis's successful theory of shared electron pair . . .". Thus, although the final formulation of the chemical bond has a physicists' dress, the origin is clearly the chemical theory of Lewis.

The HL-wave function formed the basis for the version of VB theory that later became very popular, and was behind some of the failings to be attributed to VB theory. In 1929, Slater presented his determinant-based method (14). In 1931, he generalized the HL model to n-electrons by expressing the total wave function as a product of $n/2$ bond wave functions of the HL type (15). In 1932, Rumer (16) showed how to write down all the possible bond pairing schemes for n-electrons and avoid linear dependencies among the forms in order to obtain canonical structures. We will refer hereafter to the kind of theory that considers only covalent structures as HLVB. Further refinements of the new theory of bonding (17) between 1928–1933 were mostly quantitative, focusing on improvement of the exponents of the atomic orbitals by Wang (18), and on the inclusion of polarization functions and ionic terms by Rosen and Weinbaum (19,20).

The success of the HL model and its relation to Lewis's model posed a wonderful opportunity for the young Pauling and Slater to construct a general quantum chemical theory for polyatomic molecules. In the same year (1931), they both published a few seminal papers in which they developed the notion of hybridization, the covalent–ionic superposition, and the resonating benzene picture (15,21–24). Especially effective were those Pauling's papers that linked the new theory to the chemical theory of Lewis, and rested on an encyclopedic command of chemical facts, much like the knowledge applied by Lewis to find his ingenious concept 15 years before (6). In the first paper (23), Pauling presented the electron-pair bond as a superposition of the covalent HL form and the two possible ionic forms of the bond, as shown in **2** in Scheme 1.1. He discussed the transition from covalent to ionic bonding. He then developed the notion of hybridization and discussed molecular geometries and bond angles in a variety of molecules, ranging from organic to transition metal compounds. For the latter compounds, he also discussed the magnetic moments in terms of the unpaired spins. In the following article (24), Pauling addressed bonding in molecules like diborane, and odd-electron bonds as in the ion molecule H_2^+, and in dioxygen, O_2, which Pauling represented as having two three-electron bonds, **3** in Scheme 1.1. These papers were followed by a stream of five papers, published from 1931 to 1933 in the *Journal of the American Chemical Society*, and entitled *The Nature of the Chemical Bond*. This series of papers enabled the description of any bond in any molecule, and culminated in the famous monograph in which all the structural chemistry of the time was treated in terms of the covalent–ionic superposition, resonance, and hybridization theory (25). The book, which was published in 1939, is dedicated to G.N. Lewis, and the 1916 paper of Lewis is the only reference cited in the preface to the first edition. Valence bond theory in Pauling's view is a quantum chemical version of Lewis's theory of valence. In Pauling's work, the long sought for basis for the Allgemeine Chemie (unified chemistry) of Ostwald, the father of physical chemistry, was finally found (2).

1.2 ORIGINS OF MO THEORY AND THE ROOTS OF VB–MO RIVALRY

At the same time that Slater and Pauling were developing their VB theory (17), Mulliken (25–29) and Hund (30,31) were developing an alternative approach called MO theory that has a spectroscopic origin. The term MO theory appeared in 1932, but the roots of the method can be traced back to earlier papers from 1928 (26), in which both Hund and Mulliken made spectral and quantum number assignments of electrons in molecules, based on correlation diagrams tracing the energies from separated to united atoms. According to Brush (3), the first person to write a wave function for a MO was Lennard-Jones in 1929, in his treatment of diatomic molecules (32). In this paper, Lennard-Jones easily shows that ·the O_2 molecule is paramagnetic, and mentions that the HLVB method runs into difficulties with this molecule (32). This molecule would eventually become a symbol for the failings of VB theory, although as we wrote above there was no obvious reason for this branding, since VB theory always described this molecule as a diradical with two three-electron bonds 3.

In MO theory, the electrons in a molecule occupy delocalized orbitals made from linear combination of atomic orbitals. Drawing 4 (Scheme 1.1) shows the MOs of the H_2 molecule, and the delocalized σ_g MO can be contrasted with the localized HL description in 1. Eventually, it would be the work of Hückel that would usher MO theory into the mainstream chemistry. The work of Hückel in the early 1930s initially had a chilly reception (33), but eventually it gave MO theory an impetus and formed a successful and widely applicable tool. In 1930, Hückel used Lennard-Jones's MO ideas on O_2, applied it to C=X (X=C, N, O) double bonds, and suggested the $\sigma-\pi$ separation (34). With this novel treatment, Hückel ascribed the restricted rotation in ethylene to the π-type orbital. Equipped with this facility of $\sigma-\pi$ separability, Hückel turned to solve the electronic structure of benzene using both HLVB theory and his new Hückel—MO (HMO) approach; the latter giving better "quantitative" results, and hence being preferred (35). The π-MO picture, 5 (Scheme 1.2), was quite unique in the sense that it viewed the molecule as a whole, with a σ-frame dressed by π-electrons that occupy three completely delocalized π-orbitals. The HMO picture also allowed Hückel to understand the special stability of benzene. Thus, the molecule was found to have a closed-shell π-component and its energy was calculated to be lower relative to that of three isolated π-bonds as in ethylene. In the same paper, Hückel treated the ion molecules of C_5H_5 and C_7H_7, as well as the molecules C_4H_4 (CBD) and C_8H_8 (COT). This treatment allowed him to understand why molecules with six π-electrons had special stability, and why molecules like COT or CBD either did not possess this stability (i.e., COT) or had not yet been made (i.e., CBD) at his time. In this and a subsequent paper (36), Hückel lays the foundations for what will become later known as the Hückel rule, regarding the special stability of aromatic molecules with $4n + 2$ π-electrons (3). This rule, its extension to

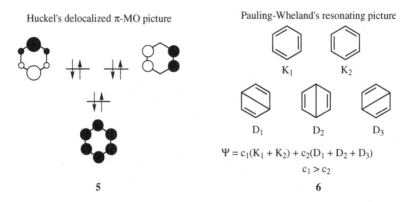

Scheme 1.2

antiaromaticity, and its articulation by organic chemists in the 1950s–1970s will constitute a major cause for the acceptance of MO theory and the rejection of VB theory (4).

The description of benzene in terms of a superposition (resonance) of two Kekulé structures appeared for the first time in the work of Slater, as a case belonging to a class of species in which each atom possesses more neighbors than electrons it can share, much like in metals (21). Two years later, Pauling and Wheland (37) applied HLVB theory to benzene. They developed a less cumbersome computational approach, compared with Hückel's previous HLVB treatment, using the five canonical structures in **6**, and approximated the matrix elements between the structures by retaining only close neighbor resonance interactions. Their approach allowed them to extend the treatment to naphthalene and to a great variety of other species. Thus, in the HLVB approach, benzene is described as a "resonance hybrid" of the two Kekulé structures and the three Dewar structures; the latter had already appeared before in Ingold's idea of mesomerism, which itself is rooted in Lewis's concept of electronic tautomerism (6). In his book, published for the first time in 1944, Wheland explains the resonance hybrid with the biological analogy of mule = donkey + horse (38). The pictorial representation of the wave function, the link to Kekulé's oscillation hypothesis, and to Ingold's mesomerism, which were known to chemists, made the HLVB representation very popular among practicing chemists.

With these two seemingly different treatments of benzene, the chemical community was faced with two alternative descriptions of one of its molecular icons, and this began the VB–MO rivalry that seems to accompany chemistry to the Twenty-first Century (5). This rivalry involved most of the prominent chemists of various periods (e.g., Mulliken, Hückel, J. Mayer, Robinson, Lapworth, Ingold, Sidgwick, Lucas, Bartlett, Dewar, Longuet-Higgins, Coulson, Roberts, Winstein, Brown). A detailed and interesting account of the nature of this rivalry and the major players can be found in the treatment of Brush (3,4). Interestingly, back in the 1930s, Slater (22) and van Vleck and

Sherman (39) stated that since the two methods ultimately converge, it is senseless to quibble on the issue of which one is better. Unfortunately, however, this rational attitude does not seem to have made much of an impression on this religious war-like rivalry.

1.3   ONE THEORY IS UP THE OTHER IS DOWN

By the end of World War II, Pauling's resonance theory was widely accepted, while most practicing chemists ignored HMO and MO theories. The reasons for this situation are analyzed by Brush (3). Mulliken suggested that the success of VB theory was due to Pauling's skill as a propagandist. According to Hager (a biographer of Pauling) VB won out in the 1930s because of Pauling's communication skills. However, the most important reason for this dominance is the direct lineage of VB-resonance theory to the structural concepts of chemistry dating from the days of Kekulé, Couper, and others through the electron-pair notion and electron-dot structures of Lewis. Pauling himself emphasized that his VB theory is a natural evolution of chemical experience, and that it emerges directly from the chemical conception of the chemical bond. This has made VB-resonance theory appear intuitive and chemically meaningful. Ingold was a great promoter of VB-resonance theory who saw in it a quantum chemical version of his own mesomerism concept (according to Brush, the terms resonance and mesomerism entered chemical vocabulary at the same time, due to Ingold's assimilation of VB-resonance theory; Reference 3, p. 57). Another very important reason is the facile qualitative application of this theory to all known structural chemistry of the time in Pauling's book (25), and to a variety of problems in organic chemistry in Wheland's book (38). The combination of an easily applicable general theory, and its ability to fit experiment so well, created a rare credibility nexus. In contrast, MO theory seemed alien to everything chemists had thought about the nature of the chemical bond. Even Mulliken admitted that MO theory departs from the chemical ideology (Reference 3, p. 51). To top it all, back at that period, MO theory offered no visual representation to compete with the resonance hybrid representation of VB-resonance theory with its direct lineage to the structure of molecules, the heartland of chemistry. At the end of World War II, VB-resonance theory dominated the epistemology of chemists.

By the mid-1950s, the tide had started shifting slowly in favor of MO theory, gaining momentum through the mid-1960s. What had caused the shift is a combination of factors, of which the following two may be decisive. First, there were many successes of MO theory, for example the experimental verification of the Hückel rules (33), the construction of intuitive MO theories, and their wide applicability for rationalization of structures (e.g., Walsh diagrams) and spectra [electronic and electron spin resonance (ESR)], the highly successful predictive application of MO theory in chemical reactivity, the instant rationalization of the bonding in newly discovered exotic molecules like ferrocene (40), for which the VB theory description was cumbersome, and the development of widely applicable MO-based computational techniques (e.g., extended Hückel and semiempirical programs). Last, but not least, is the publication of influential books, which taught MO theory to chemists, like the books of Dewar and Coulson, on MO theory, and the books of Roberts and Streitwieser on Hückel theory and its usage (41–43). On the other side, VB theory, in chemistry, suffered a detrimental conceptual arrest that has crippled the predictive ability of the theory, which, in addition, has started to accumulate "failures". Unlike its fresh exciting beginning, in the period of 1950s–1960s VB theory ceased to guide experimental chemists to new experiments. This process ultimately ended in the complete victory of MO theory. However, the MO victory was over resonance theory and other simplified versions of VB theory, but not over VB theory itself. In fact, the true VB theory was hardly being practiced anymore in the mainstream chemical community.

1.4   MYTHICAL FAILURES OF VB THEORY: MORE GROUND IS GAINED BY MO THEORY

One of the major registered failures is associated with the dioxygen molecule. Application of the simple Pauling–Lewis recipe of hybridization and bond pairing to rationalize and predict the electronic structure of molecules fails to predict the paramagneticity of O_2. In contrast, using MO theory reveals this paramagneticity instantaneously (32). Even though VB theory does not really fail with O_2, and Pauling himself preferred, without reasoning why, to describe it in terms of three-electron bonds (3) in his early papers (24) [see also Wheland's description on p. 39 of his book (38)], this "failure" of Lewis's recipe sticks to VB theory and becomes a fixture of the common chemical wisdom (Reference 3, p. 49, footnote 112).

A second sore spot concerned the VB treatments of CBD and COT. Thus, using HLVB theory leads to a an incorrect prediction that the resonance energy of CBD should be as large or even larger than that of benzene. The facts that CBD had not yet been made and that COT exhibited no special stability were in favor of HMO theory. Another impressive success of HMO theory was the prediction that due to the degenerate set of singly occupied MOs, square CBD should distort to a rectangular structure, which made a connection to the ubiquitous phenomena of Jahn-Teller and pseudo-Jahn-Teller effects amply observed by the spectroscopic community. Wheland analyzed the CBD problem early on, and his analysis pointed out that inclusion of ionic structures would probably change the VB predictions and make them identical to MO (38,44,45). Craig showed that HLVB theory in fact correctly assigns the ground state of CBD, in contrast to HMO theory (46,47). Despite this demonstration and the fact that modern VB theory has subsequently demonstrated unique and novel insight into the problems of benzene, CBD, and their isoelectronic species, nevertheless the early stamp of the CBD story as a failure of VB theory still persists.

The increasing interest of chemists in large molecules, as of the late 1940s, has started making VB theory impractical, compared with the emerging semiempirical MO methods that allowed the treatment of larger and larger molecules. A great advantage of semiempirical MO calculations was the ability to calculate bond lengths and angles rather than assume them as in VB theory (4). Skillful communicators like Longuet-Higgins, Coulson, and Dewar, were among the leading MO proponents. They handled MO theory in a visualizable manner, which was sorely missing before. In 1951, Coulson addressed the Royal Society meeting and expressed his opinion that despite the great success of VB theory, it has no good theoretical basis; it is just a semiempirical method, of little use for more accurate calculations (48). In 1949, Dewar's monograph, *Electronic Theory of Organic Chemistry* (49), summarized the faults of resonance theory, as being cumbersome, inaccurate, and too loose ("it can be played happily by almost anyone without any knowledge of the underlying principles involved").

In 1952, Coulson published his book *Valence* (50) which did for MO theory, at least in part, what Pauling's book (25) had done much earlier for VB theory. It is interesting that the great pedagogy of Coulson relied on combined insights of MO and VB theory, and the creation of a portable MO theory (43), using localized bond orbitals instead of delocalized MOs. As analyzed by Park (43), the famous pictures for ethylene and benzene using the sp^2 hybridization and π-bonding were Coulson's and not Pauling's, who was still using the tetrahedral carbon to describe ethylene with two bent "banana" bonds. At the same time, Coulson stressed that this localized picture could be converted to the delocalized one (43). Thus, Coulson has provided a lucid qualitative account of the mathematics of quantum mechanical theories of valence and reoriented MO theory from spectroscopic concerns to chemical applications. Pauling strongly objected to Coulson's simpler pictures of, for example, ethylene, and chose to cling to his use of sp^3 hybridization to describe the bonding in ethylene. Only in 1960, in the third edition of his book (25), page 137 did Pauling give the two alternative descriptions with sp^3 and sp^2 hybridization; by that time VB theory was losing grounds, at least in part, because its founder was reluctant to change it and perhaps to infuse it with insights from MO theory. In 1960 Mulliken won the Nobel Prize and Platt wrote, "MO is now used far more widely, and simplified versions of it are being taught to college freshmen and even to high school students" (51). Indeed, many communities took to MO theory due to its proven portability and successful predictions.

A decisive victory was won by MO theory when organic chemists were finally able to synthesize transient molecules and establish the stability patterns of $C_8H_8^{2-}$, $C_5H_5^{-,+}$, $C_3H_3^{+,-}$, and $C_7H_7^{+,-}$ during the 1950s–1960s (3,4,33). The results, which followed the Hückel rules, convinced most of the organic chemists that MO theory was correct, while HLVB and resonance theories were wrong. During the 1960s–1978, C_4H_4 was made, and its structure and properties were determined by MO theory, which challenged initial experimental determination of a square structure (3,4). The syntheses of nonbenzenoid aromatic compounds such as azulene, tropone, etc., further established the Hückel rules, and highlighted the failure of resonance theory (33). This era in organic chemistry marked a decisive downfall of VB theory.

By 1960, the 3rd edition of Pauling's book was published (25), and although it was still spellbinding for chemists, it contained errors and omissions. For example, the discussion of electron deficient boranes, where Pauling describes the molecule $B_{12}H_{12}$ instead of $B_{12}H_{12}^{2-}$ (Reference 25, p. 378), and a very cumbersome description of ferrocene and analogous compounds (on pp. 385–392), for which MO theory presented simple and appealing descriptions. These and other problems in the book, as well as the neglect of the then known species $C_5H_5^{-,+}$, $C_3H_3^{+,-}$, and $C_7H_7^{+,-}$, reflected the situation that unlike MO theory, VB theory did not have a useful Aufbau principle that could reliably predict the dependence of molecular stability on the number of electrons and project magic numbers as $4n/4n + 2$, and so on. As we have already pointed out, the conceptual development of VB theory was arrested since the 1950s, in part due to the insistence of Pauling himself that resonance theory was sufficient to deal with most problems (see, e.g., Reference 4, p. 283). Sadly, the creator himself contributed to the downfall of his own brainchild.

In 1952, Fukui published his Frontier MO Theory (52), which went initially unnoticed. In 1965, Woodward and Hoffmann published their principle of conservation of orbital symmetry, and applied it to all pericyclic chemical reactions. The immense success of these rules (53) renewed the interest in Fukui's approach and together they formed a new MO-based framework of thought for chemical reactivity (called, e.g., "giant steps forward in chemical theory" in Morrison and Boyd, pp. 934, 939, 1201, 1203). This success of MO theory resulted in its increased dissemination among chemists and in the effective decimation of the alternative VB theory. In this area, despite the early calculations of the Diels–Alder and 2 + 2 cycloaddition reactions by Evans (54), VB theory did not make an impact, in part at least, because of the blind adherence of its practitioners to simple resonance theory (33). Further, the reluctance of its proponents to infuse it with insights from its rival MO theory and thereby to derive the dependence of reactivity phenomenon on magic numbers led to the further decline of VB theory. All the subsequent VB derivations of the rules (e.g., by Oosterhoff and by Goddard) were "after the fact" and failed to reestablish the status of VB theory. In Hoffmann, MO theory found another great teacher who, in 1965, started his long march of teaching MO theory by applying it to almost any branch of chemistry, and by demonstrating how portable MO ideas were and how useful they could be for chemists. One of his key contributions, the "isolobal analogy", in fact relied on the localized bond orbital picture, which created a bridge between organic and organometallic chemistries (55).

The development of photoelectron spectroscopy (PES) and its application to molecules in the 1970s, in the hands of Heilbronner, showed that the spectra could be easily interpreted if one assumes that electrons occupy delocalized MO (56,57). This further strengthened the case for MO theory. Moreover, this has served to dismiss VB theory, because it describes electron pairs that occupy localized bond orbitals. A frequent example of this "failure" of VB theory is the photoelectron spectroscopy (PES) of methane, which shows two different ionization peaks. These peaks correspond to the a_1 and t_2 MOs, but not to the four C–H bond orbitals in Pauling's hybridization theory [see recent paper on a similar issue (58)]. With these and similar types of arguments, VB theory has eventually fell into a state of disrepute and became known, at least when the present authors were students, either as a "wrong theory" or simply as a "dead theory".

The late 1960s and early 1970s mark the era of mainframe computing. In contrast to VB theory, which is very difficult to implement computationally ("the $N!$ problem", which is a misnomer since no one really calculates $N!$ terms anymore), MO theory easily could be implemented (even GVB was implemented through an MO-based formalism—see later). In the early 1970s, Pople and co-workers developed the GAUSSIAN70 package that uses *ab initio* MO theory with no approximations other than the choice of basis set. Sometime later density functional theory made a spectacular entry into chemistry. Suddenly, it has become possible to calculate real molecules, and to probe their properties with increasing accuracy. The *lingua franca* of all these methods was MO theory, and even when density function theory (DFT) entered into chemistry it used Kohn–Sham orbitals that look almost identical to MOs. This theory further cemented the role of MO theory as the primary conceptual tool acceptable in chemistry.

The new and user-friendly tool created a subdiscipline of computational chemists who explored the molecular world with the GAUSSIAN series and many of the other packages, which sprouted alongside the dominant one. Today leading textbooks hardly include VB theory anymore, and when they do, the theory is misrepresented (59,60). Advanced quantum chemistry courses regularly teach MO theory, but books that teach VB theory are rare. This development of user-friendly *ab initio* MO-based software and the lack of similar VB software put the "last nail in the coffin of VB theory" and substantiated MO theory as the only legitimate chemical theory.

Nevertheless, despite this seemingly final judgment and the obituaries showered on VB theory in textbooks and in the public opinion of chemists, the theory never really died. Due to its close affinity to chemistry and its utmost clarity, it has remained an integral part of the thought process of many chemists, even among proponents of MO theory (see comment by Hoffmann on page 284 in Reference 4). Within the chemical dynamics community, the usage of the theory has never been arrested, and it lived in terms of computational methods called LEPS, BEBO, DIM, an so on, which were (and still are) used for generation of potential energy surfaces. Moreover, around the 1970s, but especially from 1980s onward, VB theory began to rise from the ashes, to dispel many myths about its "failures" and to offer a sound and attractive alternative to MO theory. Before some of these developments are described, it is important to go over some of the mythical "failures" of VB theory and inspect them a bit more closely.

1.5   ARE THE FAILURES OF VB THEORY REAL?

All the so-called failures of VB theory are due to misuse and failures of very simplified versions of the theory. Simple resonance theory enumerates structures without proper consideration of their interaction matrix elements (or overlaps). It will fail whenever the matrix element is important, as in the case of aromatic viz. antiaromatic molecules, and so on (61,62). The hybridization-bond pairing theory (modern day Lewis theory) assumes that the most important energetic effect for a molecule is the bonding, and hence, one should hybridize the atoms and make the maximum number of bonds, henceforth, "perfect pairing". The perfect-pairing approach will fail whenever other factors (see below) become equally or more important than bond pairing (62–64). The HLVB theory is based on covalent structures only, which become insufficient and require inclusion of ionic structures explicitly or implicitly (through delocalization tails of the atomic orbitals, as in the GVB method described later). In certain cases such as antiaromatic molecules, this deficiency of HLVB makes incorrect predictions (63,64). In the space below we consider four iconic "failures" and show that some of them stuck to VB in unexplained ways:

1.5.1   The O_2 Failure

It is doubtful whether this so-called failure can be attributed to Pauling himself, because in his landmark paper (23), Pauling was careful enough to state that the molecule does not possess a normal state, but rather one with two three-electron bonds (3), (also see Reference 38 where Wheland made the same statement on page 39). In 1934, Heitler and Pöschl (65) published a *Nature* paper describing the O_2 molecule with VB principles and concluded that "the $^3\Sigma_g^-$ term ... giving the fundamental state of the molecule". It is not clear how the myth of this "failure" grew, spread so widely, and was accepted so unanimously. Curiously, while Wheland acknowledged the prediction of MO theory by a proper citation of Lennard-Jones's paper (32), Pauling did not, at least not in his landmark papers (23,24), nor in his book (25). In these works, the Lennard-Jones paper is either not cited (24,25), or is mentioned only as a source of the state symbols (23) that Pauling used to characterize the states of CO, CN, and so on. One wonders what role the animosity between the MO and VB camps played in propagating the notion of the "failures" of VB to predict the ground state of O_2. Sadly, scientific history is determined also by human weaknesses. As we repeatedly stated, it is true that a naïve application of hybridization and perfect pairing approach (simple Lewis pairing) without consideration of the important effect played by the four-electron repulsion, would fail and predict a $^1\Delta_g$ ground state. As we will see later, in the case of O_2, perfect pairing in the $^1\Delta_g$ state leads to four-electron repulsion, which more than cancels the π-bond. To avoid the repulsion, we can form two three-electron π-bonds, and by keeping the two odd-electrons in a high spin situation, the ground state becomes $^3\Sigma_g^-$, which is further lowered by exchange energy due to the two triplet electrons (62).

1.5.2 The C_4H_4 Failure

This finding is a failure of the HLVB approach that does not involve ionic structures. Their inclusion in an all-electron VB theory, either explicitly (64,66), or implicitly through delocalization tails of the atomic orbitals (67), correctly predicts the geometry and resonance energy. In fact, even HLVB theory makes a correct assignment of the ground state of CBD as the $^1B_{1g}$ state. In contrast, monodeterminantal MO theory makes an incorrect assignment of the ground state as the triplet $^3A_{2g}$ state (46,47). Moreover, HMO theory was successful for the wrong reason, since the Hückel MO determinant for the singlet state corresponds to a single Kekulé structure and for this reason, CBD exhibits zero resonance energy in HMO (44). This idea is of course incorrect, but is reinforced by the idea from experimental facts that the species is highly unstable.

1.5.3 The $C_5H_5^+$ Failure

This idea is a failure of simple resonance theory, not of VB theory. Taking into account the sign of the matrix element (overlap) between the five VB structures shows that singlet $C_5H_5^+$ is Jahn—Teller unstable, and the ground state is in fact the triplet state. As shown later in Chapter 5, this is generally the case for all of the antiaromatic ionic species having 4n electrons over $4n + 1$ or $4n - 1$ centers (61).

1.5.4 The Failure Associated with the Photoelectron Spectroscopy of CH_4

Starting from a naïve application of the VB picture of methane (CH_4), it follows that since methane has four equivalent localized bond orbitals (LBOs), ergo the molecule should exhibit only one ionization peak in PES. However, since the PES of methane shows two peaks, ergo VB theory "fails"! This argument is false for two reasons: first, as known since the 1930s, LBOs for methane or any molecule, can be obtained by a unitary transformation of the delocalized MOs (68). Thus, both MO and VB descriptions of methane can be cast in terms of LBOs. Second, if one starts from the LBO picture of methane, the electron can come out of any one of the LBOs. A physically correct representation of the CH_4^+ cation would be a linear combination of the four forms that ascribe electron ejection to each of the four bonds. One can achieve the correct physical description, either by combining the LBOs back to canonical MOs (57), or by taking a linear combination of the four VB configurations that correspond to one bond ionization (69,70). As seen later, correct linear combinations are $^2A_1$ and $^2T_2$, the later in a triply degenerate VB state.

1.6   VALENCE BOND IS A LEGITIMATE THEORY ALONGSIDE MOLECULAR ORBITAL THEORY

Obviously, the rejection of VB theory cannot continue to invoke failures, because a properly executed VB theory does not fail, much as a properly executed MO-based calculation. This notion of VB failure, which is traced back to the VB–MO rivalry in the early days of quantum chemistry, should now be considered obsolete, unwarranted, and counterproductive. A modern chemist should know that there are two ways of describing electronic structure that are not two contrasting theories, but rather two representations or two guises of the same reality. Their capabilities and insights into chemical problems are complementary and the exclusion of any one of them undermines the intellectual heritage of chemistry. Indeed, theoretical chemists in the community of chemical dynamics continue to use VB theory and maintain an uninterrupted chain of VB usage from London, through Eyring, Polanyi, to Wyatt, Truhlar, and others today. Physicists also, continue to use VB theory, and one of the main proponents is the Nobel Laureate P.W. Anderson, who developed a resonating VB theory of superconductivity. In terms of the focus of this book, in mainstream chemistry too, VB theory begins to enjoy a slow but steady Renaissance in the form of modern VB theory.

1.7   MODERN VB THEORY: VALENCE BOND THEORY IS COMING OF AGE

The Renaissance of VB theory is marked by a surge in the following two-fold activity: (1) creation of general qualitative models based on VB theory; and (2) development of new methods and program packages that enable applications to moderate-sized molecules. Below we briefly mention some of these developments without pretence of creating exhaustive lists. We apologize for any omissions.

A few general qualitative models based on VB theory began to appear in the late 1970s and early 1980s. Among these models semiempirical approaches are also included based, for example, on the Heisenberg and Hubbard Hamiltonians (71–79), as well as Hückeloid VB methods (61,80–82), which can handle with clarity ground and excited states of molecules. Methods that map MO-based wave functions to VB wave functions offer a good deal of interpretative insight. Among these mapping procedures, we note the half-determinant method of Hiberty and Leforestier (83), and the CASVB methods of Thorsteinsson et al (84,85) and Hirao et al (86,87). General qualitative VB models for chemical bonding were proposed in the early 1980s and the late

1990s by Epiotis (88,89). A general model for the origins of barriers in chemical reactions was proposed in 1981 by one of the present authors, in a manner that incorporates the role of orbital symmetry (61,90). Subsequently, in collaboration with Pross (91,92) and Hiberty (93), the model has been generalized for a variety of reaction mechanisms (94), and used to shed new light on the problems of aromaticity and antiaromaticity in isoelectronic series (66). Following Linnett's reformulation of three-electron bonding in the 1960s (95), Harcourt (96,97) developed a VB model that describes electron-rich bonding in terms of increased valence structures, and showed its occurrence in bonds of main elements and transition metals.

Valence bond ideas also contributed to the revival of theories for photochemical reactivity. Early VB calculations by Oosterhoff et al (98,99). revealed a potentially general mechanism for the course of photochemical reactions. Michl (100,101) articulated this VB-based mechanism and highlighted the importance of "funnels" as the potential energy features that mediate the excited-state species back into the ground state. Subsequently, Robb and co-workers (102–105) showed that these "funnels" are conical intersections that can be predicted by simple VB arguments, and computed at a high level of sophistication. Similar applications of VB theory to deduce the structure of conical intersections in photoreactions were done by Shaik and Reddy (106) and recently by Haas and Zilberg (107).

Valence bond theory enables a very straightforward account of environmental effects, such as those imparted by solvents and/or protein pockets. A major contribution to the field was made by Warshel, who has created his empirical VB (EVB) method, and, by incorporating van der Waals and London interactions by molecular mechanical (MM) methods, created the QM(VB)/MM method for the study of enzymatic reaction mechanisms (108–110). His pioneering work ushered the now emerging QM/MM methodologies for studying enzymatic processes (111). Hynes et al. showed how to couple solvent models into VB and create a simple and powerful model for understanding and predicting chemical processes in solution (112–114). One of us has shown how solvent effect can be incorporated in an effective manner to the reactivity factors that are based on VB diagrams (115,116).

Generally, VB theory is seen to offer a widely applicable framework for thinking and predicting chemical trends. Some of these qualitative models and their predictions are discussed in the application sections of this book.

Sometime in the 1970s a stream of nonempirical VB methods began to appear and were followed by many applications of rather accurate calculations. All these programs divide the orbitals in a molecule into inactive and active subspaces, treating the former as a closed shell and the latter by some VB formalism. The programs optimize the orbitals, and the coefficients of the VB structures, but they differ in the manner by which the VB orbitals are defined. Goddard and co-workers developed the generalized VB (GVB) method (117–120) that uses semilocalized atomic orbitals (having small delocalization tails), employed originally by Coulson and Fisher for the H_2 molecule (121). The GVB method is incorporated now in GAUSSIAN and in most other MO-based packages. Somewhat later, Gerratt and co-workers developed their VB method called the spin coupled (SC) theory and its follow-up by configuration interaction using the SCVB method (122–124). Both the GVB and SC theories do not employ covalent and ionic structures explicitly, but instead use semilocalized atomic orbitals that effectively incorporate all the ionic structures, and thereby enable one to express the electronic structures in compact forms based on formally covalent pairing schemes. Balint-Kurti and Karplus (125) developed a multistructure VB method that utilizes covalent and ionic structures with localized atomic orbitals. In a later development by van Lenthe and Balint-Kurti (126,127) and by Verbeek and van Lenthe (128,129), the multistructure method is being referred to as a VB self-consistent field (VBSCF) method. In a subsequent development, van Lenthe (130) and Verbeek et al. (131) generated the multipurpose VB program called TURTLE, which has recently been incorporated into the MO-based package of programs GAMESS-UK. Matsen (132,133), McWeeny (134), and Zhang and co-workers (135,136) developed their spin-free VB approaches based on symmetric group methods. Subsequently, Wu et al. extended the spin-free approach, and produced a general purpose VB program called the XIAMEN-99 package most recently named XMVB (137,138). Soon after, Li and McWeeny announced their VB2000 software, which is also a general purpose program, including a variety of methods (139). Another software of multiconfigurational VB (MCVB), called CRUNCH and based on the symmetric group methods of Young was written by Gallup and co-workers (140,141). During the early 1990s, Hiberty and co-workers developed the breathing orbital VB (BOVB) method, which also utilizes covalent and ionic structures, but additionally allows them to have their own unique set of orbitals (142–147). The method is now incorporated into the programs TURTLE and XMVB. Very recently, Wu et al. (148) developed a VBCI method that is akin to BOVB, but can be applied to larger systems. In a more recent work, the same authors coupled VB theory with the solvent model, PCM, and produced the VBPCM program that enables one to study reactions in solution (149). The recent biorthogonal VB method of McDouall has the potential to carry out VB calculations up to 60 electrons outside the closed shell (150). Finally, Truhlar and co-workers (151) developed the VB-based multiconfiguration molecular mechanics method (MCMM) to treat dynamic aspects of chemical reactions, while Landis and co-workers (152), introduced the VAL-BOND method that is capable of predicting structures of transition metal complexes using Pauling's ideas of orbital hybridization. A recent monograph by Landis and Weinhold makes use of VAL-BOND as well as of natural resonance theory to discuss a variety of problems in inorganic and organometallic chemistry (153). In the section dedicated to VB methods, we mention the main program packages and methods, which we used, and outline their features, capabilities, and limitations.

This plethora of acronyms of the VB programs starts to resemble a similar development that had accompanied the ascent of MO theory. While this may sound like good news, recall the biblical admonition, "let us go down, and there confound their language that they may not understand one another's speech" (*Genesis* 11, 7). Certainly, the situation is also a call for systematization much like what Pople and co-workers enforced on computational MO terminology. Nonetheless, at the moment the important point is that the advent of so many good VB programs has caused a surge in applications of VB theory, to problems ranging from bonding in main group elements to transition metals, conjugated systems, aromatic and antiaromatic species, all the way to excited states and full pathways of chemical reactions, with moderate-to-very good accuracies. For example, a recent calculation of the barrier for the identity hydrogen-exchange reaction, H + H−H' → H−H + H', by Song et al. (154) shows that it is possible to calculate the reaction barrier accurately with just eight classical VB structures! Thus, in many respects, VB theory is coming of age, with the development of faster, and more accurate *ab initio* VB methods (155,156), and with generation of new post-Pauling concepts. As these activities further flourish, so will the usage of VB theory spread among practicing chemists (157). This book aims to ease this goal and to serve as a source for teachers in advanced classes.

REFERENCES

1. D. L. Cooper, Ed. *Valence Bond Theory*, Elsevier, Amsterdam, The Netherlands, 2002.
2. J. W. Servos, *Physical Chemistry from Ostwald to Pauling*, Princeton University Press, New Jersey, 1990.
3. S. G. Brush, *Stud. Hist. Philos. Sci.* **30**, 21 (1999). Dynamics of Theory Change in Chemistry: Part 1. The Benzene Problem 1865–1945.
4. S. G. Brush, *Stud. Hist. Philos. Sci.* **30**, 263 (1999). Dynamics of Theory Change in Chemistry: Part 2. Benzene and Molecular Orbitals, 1945–1980.
5. R. Hoffmann, S. Shaik, P. C. Hiberty, *Acc. Chem. Res.* **36**, 750 (2003). A Conversation on VB vs. MO Theory: A Never Ending Rivalry?
6. G. N. Lewis, *J. Am. Chem. Soc.* **38**, 762 (1916). The Atom and the Molecule.
7. I. Langmuir, *J. Am. Chem. Soc.* **41**, 868 (1919). The Arrangement of Electrons in Atoms and Molecules.
8. S. Shaik, *J. Comput. Chem.* **28**, 51 (2007). The Lewis Legacy: The Chemical Bond A Territory and Heartland of Chemistry.
9. W. Heitler, F. London, *Z. Phys.* **44**, 455 (1927). Wechselwirkung neutraler Atome und homöopolare Bindung nach der Quantenmechanik.
10. For an English translation, see H. Hettema, *Quantum Chemistry Classic Scientific Paper*, World Scientific, Singapore, 2000.
11. W. Heisenberg, *Z. Phys.* **38**, 411 (1926). Multi-body Problem and Resonance in the Quantum Mechanics.

12. F. London, *Z. Phys.* **46**, 455 (1928). On the Quantum Theory of Homo-polar Valence Numbers.

13. L. Pauling, *Proc. Natl. Acad. Sci. U.S.A.* **14**, 359 (1928). The Shared-Electron Chemical Bond.

14. J. C. Slater, *Phys. Rev.* **34**, 1293 (1929). The Theory of Complex Spectra.

15. J. C. Slater, *Phys. Rev.* **38**, 1109 (1931). Molecular Energy Levels and Valence Bonds.

16. G. Rumer, *Gottinger Nachr.* 337 (1932). Zum Theorie der Spinvalenz.

17. G. A. Gallup, in *Valence Bond Theory*, D. L. Cooper, Ed., Elsevier, Amsterdam, The Netherlands, 2002, pp 1–40. A Short History of VB Theory.

18. S. C. Wang, *Phys. Rev.* **31,** 579 (1928). The Problem of the Normal Hydrogen Molecule in the New Quantum Mechanics.

19. N. Rosen, *Phys. Rev.* **38**, 2099 (1931). The Normal State of the Hydrogen Molecule.

20. S. Weinbaum, *J. Chem. Phys.* **1**, 593 (1933). The Normal State of the Hydrogen Molecule.

21. J. C. Slater, *Phys. Rev.* **37**, 481 (1931). Directed Valence in Polyatomic Molecules.

22. J. C. Slater, *Phys. Rev.* **41**, 255 (1932). Note on Molecular Structure.

23. L. Pauling, *J. Am. Chem. Soc.* **53**, 1367 (1931). The Nature of the Chemical Bond. Application of Results Obtained from the Quantum Mechanics and from a Theory of Magnetic Susceptiblity to the Structure of Molecules.

24. L. Pauling, *J. Am. Chem. Soc.* **53**, 3225 (1931). The Nature of the Chemical Bond. II. The One-Electron Bond and the Three-Electron Bond.

25. L. Pauling, *The Nature of the Chemical Bond,* Cornell University Press, Ithaca, NY, 1939 (3rd ed., 1960).

26. R. S. Mullliken, *Phys. Rev.* **32**, 186 (1928). The Assignment of Quantum Numbers for Electrons in Molecules. I.

27. R. S. Mullliken, *Phys. Rev.* **32**, 761 (1928). The Assignment of Quantum Numbers for Electrons in Molecules. II. Correlation of Molecular and Atomic Electron States.

28. R. S. Mullliken, *Phys. Rev.* **33**, 730 (1929). The Assignment of Quantum Numbers for Electrons in Molecules. III. Diatomic Hydrides.

29. R. S. Mulliken, *Phys. Rev.* **41**, 49 (1932). Electronic Structures of Polyatomic Molecules and Valence. II. General Considerations.

30. F. Hund, *Z. Phys.* **73**, 1 (1931). Zur Frage der Chemischen Bindung.

31. F. Hund, *Z. Phys.* **51**, 759 (1928). Zur Deutung der Molekelspektren. IV.

32. J. E. Lennard-Jones, *Trans. Faraday Soc.* **25**, 668 (1929). The Electronic Structure of Some Diatomic Molecules.

33. J. A. Berson, *Chemical Creativity. Ideas from the Work of Woodward, Hückel, Meerwein, and Others*, Wiley–VCH, NewYork, 1999.

34. E. Hückel, *Z. Phys.* **60**, 423 (1930). Zur Quantentheorie der Doppelbindung.

35. E. Hückel, *Z. Phys.* **70**, 204 (1931). Quantentheoretische Beiträge zum Benzolproblem. I. Dies Elektronenkonfiguration des Benzols und verwandter Verbindungen.

36. E. Hückel, *Z. Phys.* **76**, 628 (1932). Quantentheoretische Beiträge zum Problem der aromatischen und ungesättigten Verbindungen. III.

37. L. Pauling, G. W. Wheland, *J. Chem. Phys.* **1**, 362 (1933). The Nature of the Chemical Bond. V. The Quantum-Mechanical Calculation of the Resonance Energy of Benzene and Naphthalene and the Hydrocarbon Free Radicals.

38. G. W. Wheland, *Resonance in Organic Chemistry*, John Wiley & Sons, Inc., New York, 1955, pp. 4, 39, 148.
39. J. H. van Vleck, A. Sherman, *Rev. Mod. Phys.* **7**, 167 (1935). The Quantum Theory of Valence.
40. P. Laszlo, R. Hoffmann, *Angew. Chem. Int. Ed. Engl.* **39**, 123 (2000). Ferrocene: Ironclad History or Rashomon Tale?
41. J. D. Roberts, *Acc. Chem. Res.* **37**, 417 (2004). The Conversation Continues I.
42. A. Streitwieser, *Acc. Chem. Res.* **37**, 419 (2004). The Conversation Continues II.
43. B. S. Park, in *Pedagogy and the Practice of Science*, D. Kaiser, Ed., MIT Press, Boston, 2005, pp. 287–319. In the Context of Padagogy: Teaching Strategy and Theory Change in Quantum Chemistry.
44. G. W. Wheland, *Proc. R. Soc. London, Ser. A* **159**, 397 (1938). The Electronic Structure of Some Polyenes and Aromatic Molecules. V—A Comparison of Molecular Orbital and Valence Bond Methods.
45. G. W. Wheland, *J. Chem. Phys.* **2**, 474 (1934). The Quantum Mechanics of Unsaturated and Aromatic Molecules: A Comparison of Two Methods of Treatment.
46. D. P. Craig, *J. Chem. Soc.* 3175 (1951). Cyclobutadiene and Some Other Pseudoaromatic Compounds.
47. D. P. Craig, *Proc. R. Soc. London, Ser. A* **200**, 498 (1950). Electronic Levels in Simple Conjugated Systems. I. Configuration Interaction in Cyclobutadiene.
48. C. A. Coulson, *Proc. R. Soc. London, Ser. A* **207**, 63 (1951). Critical Survey of the Method of Ionic-Homopolar Resonance.
49. M. J. S. Dewar, *Electronic Theory of Organic Chemistry*, Clarendon Press, Oxford, 1949, pp. 15–17.
50. C. A. Coulson, *Valence*, Oxford University Press, London, 1952.
51. J. R. Platt, *Science*, **154**, 745 (1966). Nobel Laureate in Chemistry: Robert S. Mulliken.
52. K. Fukui, T. Yonezawa, H. Shingu, *J. Chem. Phys.* **20**, 722 (1952). A Molecular Orbital Theory of Reactivity in Aromatic Hydrocarbons.
53. R. B. Woodward, R. Hoffmann, *The Conservation of Orbital Symmetry*, Verlag Chemie, Weinheim, 1971.
54. M. G. Evans, *Trans. Faraday Soc.* **35**, 824 (1939). The Activation Energies in Reactions Involving Conjugated Systems.
55. R. Hoffmann, *Angew. Chem. Int. Ed. Engl.* **21**, 711 (1982). Building Bridges Between Inorganic and Organic Chemistry (Nobel Lecture).
56. E. Heilbronner, H. Bock, *The HMO Model and its Applications*, John Wiley & Sons, Inc., New York, 1976.
57. E. Honegger, E. Heilbronner, in *Theoretical Models of Chemical Bonding*, Vol. 3, Z. B. Maksic, Ed., Springer Verlag, Berlin–Heidelberg, 1991, pp. 100–151. The Equivalent Orbital Model and the Interpretation of PE Spectra.
58. S. Wolfe, Z. Shi, *Isr. J. Chem.* **40**, 343 (2000). The S−C−O (O−C−S) Edward-Lemieux Effect is Controlled by p-Orbital on Oxygen. Evidence from Electron Momentum Spectroscopy, Photoelectron Spectroscopy, X-Ray Crystallography, and Density Functional Theory.
59. P. Atkins, L. Jones, *Chemical Principles. The Quest for Insight*, W. H. Freeman and Co., New York, 1999, pp. xxiii and 121–122.

60. L. Jones, P. Atkins, *Chemistry: Molecules, Matter and Change*, W. H. Freeman and Co., New York, 2000, pp. 400–401.

61. S. S. Shaik, in *New Theoretical Concepts for Understanding Organic Reactions*, J. Bertrán and I. G. Csizmadia, Eds., NATO ASI Series, **C267**, Kluwer Academic Publishers, Norwell, MA, 1989, pp. 165–217. A Qualitative Valence Bond Model for Organic Reactions.

62. S. Shaik, P. C. Hiberty, *Adv. Quant. Chem.* **26**, 100 (1995). Valence Bond Mixing and Curve Crossing Diagrams in Chemical Reactivity and Bonding.

63. P. C. Hiberty, *J. Mol. Struc. (Theochem)* **398**, 35 (1997). Thinking and Computing Valence Bond in Organic Chemistry.

64. A. Shurki, P. C. Hiberty, F. Dijkstra, S. Shaik, *J. Phys. Org. Chem.* **16**, 731 (2003). Aromaticity and Antiaromaticity: What Role Do Ionic Configurations Play in Delocalization and Induction of Magnetic Properties?

65. W. Heitler, G. Pöschl, *Nature (London)* **133**, 833 (1934). Ground State of C_2 and O_2 and the Theory of Valency.

66. S. Shaik, A. Shurki, D. Danovich, P. C. Hiberty, *Chem. Rev.* **101**, 1501 (2001). A Different Story of π-Delocalization—The Distortivity of the π-Electrons and Its Chemical Manifestations.

67. A. F. Voter, W. A. Goddard, III, *J. Am. Chem. Soc.* **108**, 2830 (1986). The Generalized Resonating Valence Bond Description of Cyclobutadiene.

68. J. H. van Vleck, *J. Chem. Phys.* **3**, 803 (1935). The Group Relation Between the Mulliken and Slater–Pauling Theories of Valence.

69. H. Zuilhof, J. P. Dinnocenzo, A. C. Reddy, S. Shaik, *J. Phys. Chem.* **100**, 15774 (1996). Comparative Study of Ethane and Propane Cation Radicals by B3LYP Density Functional and High-Level *Ab Initio* Methods.

70. S. Shaik, P. C. Hiberty, *Helv. Chim. Acta* **86**, 1063 (2003). Myth and Reality in the Attitude Toward Valence-Bond (VB) Theory: Are Its 'Failures' Real?"

71. D. J. Klein, in *Valence Bond Theory*, D. L. Cooper, Ed., Elsevier, Amsterdam, The Netherlands, 2002, pp. 447–502. Resonating Valence Bond Theories for Carbon π-Networks and Classical/Quantum Connections.

72. T. G. Schmalz, in *Valence Bond Theory*, D. L. Cooper, Ed., Elsevier, Amsterdam, The Netherlands, 2002, pp. 535–564. A Valence Bond View of Fullerenes.

73. J. P. Malrieu, in *Models of Theoretical Bonding*, Z. B. Maksic, Ed., Springer Verlag, New York, 1990, pp. 108–136. The Magnetic Description of Conjugated Hydrocarbons.

74. J. P. Malrieu, D. Maynau, *J. Am. Chem. Soc.* **104**, 3021 (1982). A Valence Bond Effective Hamiltonian for Neutral States of π-Systems. 1. Methods.

75. Y. Jiang, S. Li, in *Valence Bond Theory*, D. L. Cooper, Ed., Elsevier, Amsterdam, The Netherlands, 2002, pp. 565–602. Valence Bond Calculations and Theoretical Applications to Medium-Sized Conjugated Hydrocarbons.

76. F. A. Matsen, *Acc. Chem. Res.* **11**, 387 (1978). Correlation of Molecular Orbital and Valence Bond States in π Systems.

77. M. A. Fox, F. A. Matsen, *J. Chem. Educ.* **62**, 477 (1985). Electronic Structure in π-Systems. Part II. The Unification of Hückel and Valence Bond Theories.

78. M. A. Fox, F. A. Matsen, *J. Chem. Educ.* **62**, 551 (1985). Electronic Structure in π-Systems. Part III. Applications in Spectroscopy and Chemical Reactivity.

79. S. Ramasesha, Z. G. Soos, in *Valence Bond Theory*, D. L. Cooper, Ed., Elsevier, Amsterdam, The Netherlands, 2002, pp. 635–697. Valence Bond Theory of Quantum Cell Models.

80. W. Wu, S. J. Zhong, S. Shaik, *Chem. Phys. Lett.* **292**, 7 (1998). VBDFT(s): A Hückel-Type Semi-empirical Valence Bond Method Scaled to Density Functional Energies. Application to Linear Polyenes.

81. W. Wu, D. Danovich, A. Shurki, S. Shaik, *J. Phys. Chem. A* **104**, 8744 (2000). Using Valence Bond Theory to Understand Electronic Excited States. Applications to the Hidden Excited State (2^1A_g) of $C_{2n}H_{2n+2}$ (n = 2–14) Polyenes.

82. W. Wu, Y. Luo, L. Song, S. Shaik, *Phys. Chem. Chem. Phys.* **3**, 5459 (2001). VBDFT(s): A Semiempirical Valence Bond Method: Application to Linear Polyenes Containing Oxygen and Nitrogen Heteroatoms.

83. P. C. Hiberty, C. Leforestier, *J. Am. Chem. Soc.* **100**, 2012 (1978). Expansion of Molecular Orbital Wave Functions into Valence Bond Wave Functions. A Simplified Procedure.

84. T. Thorsteinsson, D. L. Cooper, J. Gerratt, M. Raimondi, *Molecular Engineering* **7**, 67 (1997). A New Approach to Valence Bond Calculations: CASVB.

85. D. L. Thorsteinsson, J. Cooper, Gerratt, P. B. Karadakov, M. Raimondi, *Theor. Chim Acta (Berlin)* **93**, 343 (1996). Modern Valence Bond Representations of CASSCF Wavefunctions.

86. H. Nakano, K. Sorakubo, K. Nakayama, K. Hirao, in *Valence Bond Theory*, D. L. Cooper, Ed. Elsevier, Amsterdam, The Netherlands, 2002, pp. 55–77. Complete Active Space Valence Bond (CASVB) Method and its Application in Chemical Reactions.

87. K. Hirao, H. Nakano, K. Nakayama, *J. Chem. Phys.* **107**, 9966 (1997). A Complete Active Space Valence Bond Method with Nonorthogonal Orbitals.

88. N. D. Epiotis, J. R. Larson, H. L. Eaton, *Lecture Notes Chem.* **29**, 1–305 (1982). Unified Valence Bond Theory of Electronic Structure.

89. N. D. Epiotis, *Deciphering the Chemical Code. Bonding Across the Periodic Table*, VCH Publ., New York, 1996, pp. 1–933.

90. S. S. Shaik, *J. Am. Chem. Soc.* **103**, 3692 (1981). What Happens to Molecules as They React? A Valence Bond Approach to Reactivity.

91. A. Pross, S. S. Shaik, *Acc. Chem. Res.* **16**, 363 (1983). A Qualitative Valence Bond Approach to Chemical Reactivity.

92. A. Pross, *Theoretical and Physical Principles of Organic Reactivity*, John Wiley & Sons, Inc., New York, 1995, pp. 83–124, 235–290.

93. S. Shaik, P. C. Hiberty, in *Theoretical Models of Chemical Bonding*, Vol. 4, Z. B. Maksic, Ed., Springer Verlag, Berlin–Heidelberg, 1991, pp. 269–322. Curve Crossing Diagrams as General Models for Chemical Structure and Reactivity.

94. S. Shaik, A. Shurki, *Angew. Chem. Int. Ed. Engl.* **38**, 586 (1999). Valence Bond Diagrams and Chemical Reactivity.

95. J. W. Linnett, *The Electronic Structure of Molecules. A New Approach*, Methuen & Co Ltd, London, 1964, pp. 1–162.

96. R. D. Harcourt, *Lecture Notes Chem.* **30**, 1–260 (1982). Qualitative Valence-Bond Descriptions of Electron-Rich Molecules: Pauling "3-Electron Bonds" and "Increased-Valence" Theory.

97. R. D. Harcourt, in *Valence Bond Theory*, D. L. Cooper, Ed., Elsevier, Amsterdam, The Netherlands, 2002, pp. 349–378. Valence Bond Structures for Some Molecules with Four Singly-Occupied Active-Space Orbitals: Electronic Structures, Reaction Mechanisms, Metallic Orbitals.

98. W. Th. A. M. Van der Lugt, L. J. Oosterhoff, *J. Am. Chem. Soc.* **91**, 6042 (1969). Symmetry Control and Photoinduced Reactions.

99. J. J. C. Mulder, L. J. Oosterhoff, *Chem. Commun.* 305 (1970). Permutation Symmetry Control in Concerted Reactions.

100. J. Michl, *Topics Curr. Chem.* **46**, 1 (1974). Physical Basis of Qualitative MO Arguments in Organic Photochemistry.

101. W. Gerhartz, R. D. Poshusta, J. Michl, *J. Am. Chem. Soc.* **99**, 4263 (1977). Excited Potential Energy Hypersurfaces for H_4. 2. "Triply Right" (C_{2v}) Tetrahedral Geometries. A Possible Relation to Photochemical "Cross Bonding" Processes.

102. F. Bernardi, M. Olivucci, M. Robb, *Isr. J. Chem.* **33**, 265 (1993). Modelling Photochemical Reactivity of Organic Systems. A New Challenge to Quantum Computational Chemistry.

103. M. Olivucci, I. N. Ragazos, F. Bernardi, M. A. Robb, *J. Am. Chem. Soc.* **115**, 3710 (1993). A Conical Intersection Mechanism for the Photochemistry of Butadiene. A MC-SCF Study.

104. M. Olivucci, F. Bernardi, P. Celani, I. Ragazos, M. A. Robb, *J. Am. Chem. Soc.* **116**, 1077 (1994). Excited-State Cis–Trans Isomerization of *cis*-Hexatriene. A CAS–SCF Computational Study.

105. M. A. Robb, M. Garavelli, M. Olivucci, F. Bernardi, *Rev. Comput. Chem.* **15**, 87 (2000). A Computational Strategy for Organic Photochemistry.

106. S. Shaik, A. C. Reddy, *J. Chem. Soc. Faraday Trans.* **90**, 1631 (1994). Transition States, Avoided Crossing States and Valence Bond Mixing: Fundamental Reactivity Paradigms.

107. S. Zilberg, Y. Haas, *Chem. Eur. J.* **5**, 1755 (1999). Molecular Photochemistry: A General Method for Localizing Conical Intersections Using the Phase-Change Rule.

108. A. Warshel, R. M. Weiss, *J. Am. Chem. Soc.* **102**, 6218 (1980). An Empirical Valence Bond Approach for Comparing Reactions in Solutions and in Enzymes.

109. A. Warshel, S. T. Russell, *Quart. Rev. Biophys.* **17**, 283 (1984). Calculations of Electrostatic Interactions in Biological Systems and in Solutions.

110. A. Warshel, *J. Phys. Chem.* **83**, 1640 (1979). Calculations of Chemical Processes in Solutions.

111. S. Braun-Sand, M. H. M. Olsson, A. Warshel, *Adv. Phys. Org. Chem.* **40**, 201 (2005). Computer Modeling of Enzyme Catalysis and Its Relationship to Concepts in Physical Organic Chemistry.

112. H. J. Kim, J. T. Hynes, *J. Am. Chem. Soc.* **114**, 10508 (1992). A Theoretical Model for S_N1 Ionic Dissociation in Solution. 1. Activation Free Energies and Transition-State Structure.

113. J. R. Mathias, R. Bianco, J. T. Hynes, *J. Mol. Liq.* **61**, 81 (1994). On the Activation Free Energy of the $Cl^- + CH_3Cl$ S_N2 Reaction in Solution.

114. J. I. Timoneda, J. T. Hynes, *J. Phys. Chem.* **95**, 10431 (1991). Nonequilibrium Free Energy Surfaces for Hydrogen-bonded Proton Transfer Compexes in Solution.

115. S. S. Shaik, *J. Am. Chem. Soc.* **106**, 1227 (1984). Solvent Effect on Reaction Barriers. The S_N2 Reactions. 1. Application to the Identity Exchange.

116. S. S. Shaik, *J. Org. Chem.* **52**, 1563 (1987). Nucleophilicity and Vertical Ionization Potentials in Cation–Anion Recombinations.

117. W. A. Goddard, III, T. H. Dunning, Jr., W. J. Hunt, P. J. Hay, *Acc. Chem. Res.* **6**, 368 (1973). Generalized Valence Bond Description of Bonding in Low-Lying States of Molecules.

118. W. A. Goddard, III, *Phys. Rev.* **157**, 81 (1967). Improved Quantum Theory of Many-Electron Systems. II. The Basic Method.

119. W. A. Goddard, III, L. B. Harding, *Annu. Rev. Phys. Chem.* **29**, 363 (1978). The Description of Chemical Bonding from *Ab-Initio* Calculations.

120. W. A. Goddard, III, *Int. J. Quant. Chem.* **IIIS**, 593 (1970). The Symmetric Group and the Spin Generalized SCF Method.

121. C. A. Coulson, I. Fischer, *Philos. Mag.* **40**, 386 (1949). Notes on the Molecular Orbital Treatment of the Hydrogen Molecule.

122. D. L. Cooper, J. Gerratt, M. Raimondi, *Chem. Rev.* **91**, 929 (1991). Applications of Spin-Coupled Valence Bond Theory.

123. D. L. Cooper, J. P. Gerratt, M. Raimondi, *Adv. Chem. Phys.* **69**, 319 (1987). Modern Valence Bond Theory.

124. M. Sironi, M. Raimondi, R. Martinazzo, F. A. Gianturco, in *Valence Bond Theory*, D. L. Cooper, Ed. Elsevier, Amsterdam, The Netherlands, 2002, pp. 261–277. Recent Developments of the SCVB Method.

125. G. G. Balint-Kurti, M. Karplus, *J. Chem. Phys.* **50**, 478 (1969). Multistructure Valence-Bond and Atoms-in-Molecules Calculations for LiF, F_2, and F_2^-.

126. J. H. van Lenthe, G. G. Balint-Kurti, *Chem. Phys. Lett.* **76**, 138 (1980). The Valence-Bond SCF (VB SCF) Method. Synopsis of Theory and Test Calculations of OH Potential Energy Curve.

127. J. H. van Lenthe, G. G. Balint-Kurti, *J. Chem. Phys.* **78**, 5699 (1983). The Valence-Bond Self-Consistent Field Method (VB–SCF): Theory and Test Calculations.

128. J. Verbeek, J. H. van Lenthe, *J. Mol. Struct. (THEOCHEM)* **229**, 115 (1991). On the Evaluation of Nonorthogonal Matrix Elements.

129. J. Verbeek, J. H. van Lenthe, *Int. J. Quant. Chem.* **XL**, 201 (1991). The Generalized Slater–Condon Rules.

130. J. H. van Lenthe, F. Dijkstra, W. A. Havenith, in *Valence Bond Theory*, D. L. Cooper, Ed., Elsevier, Amsterdam, The Netherlands, 2002, pp. 79–116. TURTLE—A Gradient VBSCF Program Theory and Studies of Aromaticity.

131. J. Verbeek, J. H. Langenberg, C. P. Byrman, F. Dijkstra, J. H. van Lenthe, *TURTLE: An Ab Initio VB/VBSCF Program (1998–2000)*.

132. F. A. Matsen, *Adv. Quant. Chem.* **1**, 59 (1964). Spin-Free Quantum Chemistry.

133. F. A. Matsen, *J. Phys. Chem.* **68**, 3282 (1964). Spin-Free Quantum Chemistry. II. Three-Electron Systems.

134. R. McWeeny, *Int. J. Quant. Chem.* **XXXIV**, 25 (1988). A Spin Free Form of Valence Bond Theory.

135. Z. Qianer, L. Xiangzhu, *J. Mol. Struct. (THEOCHEM)* **198**, 413 (1989). Bonded Tableau Method for Many Electron Systems.

136. X. Li, Q. Zhang, *Int. J. Quant. Chem.* **XXXVI**, 599 (1989). Bonded Tableau Unitary Group Approach to the Many-Electron Correlation Problem.

137. W. Wu, Y. Mo, Z. Cao, Q. Zhang, in *Valence Bond Theory*, D. L. Cooper, Ed., Elsevier, Amsterdam, The Netherlands, 2002, pp. 143–186. A Spin Free Approach for Valence Bond Theory and Its Application.

138. (a) W. Wu, L. Song, Y. Mo, Q. Zhang, *XIAMEN-99—An Ab Initio Spin-free Valence Bond Program*, Xiamen University, Xiamen, 1999. (b) L. Song, Y. Mo, Q. Zhang, W. Wu, *J. Comput. Chem.* **26**, 514 (2005). XMVB: A Program for *Ab Initio* Nonorthogonal Valence Bond Computations.

139. J. Li, R. McWeeny, *VB2000: An Ab Initio Valence Bond Program Based on Generalized Product Function Method and the Algebrant Algorithm, 2000.*

140. G. A. Gallup, R. L. Vance, J. R. Collins, J. M. Norbeck, *Adv. Quant. Chem.* **16**, 229 (1982). Practical Valence Bond Calculations.

141. G. A. Gallup, *The CRUNCH Suite of Atomic and Molecular Structure Programs*, 2001. ggallup@phy-ggallup.unl.edu.

142. P. C. Hiberty, J. P. Flament, E. Noizet, *Chem. Phys. Lett.* **189**, 259 (1992). Compact and Accurate Valence Bond Functions with Different Orbitals for Different Configurations: Application to the Two-Configuration Description of F_2.

143. P. C. Hiberty, S. Humbel, C. P. Byrman, J. H. van Lenthe, *J. Chem. Phys.* **101**, 5969 (1994). Compact Valence Bond Functions with Breathing Orbitals: Application to the Bond Dissociation Energies of F_2 and FH.

144. P. C. Hiberty, S. Humbel, P. Archirel, *J. Phys. Chem.* **98**, 11697 (1994). Nature of the Differential Electron Correlation in Three-Electron Bond Dissociation. Efficiency of a Simple Two-Configuration Valence Bond Method with Breathing Orbitals.

145. P. C. Hiberty, in *Modern Electronic Structure Theory and Applications in Organic Chemistry*, E. R. Davidson, Ed., World Scientific, River Edge, NJ, 1997, pp. 289–367. The Breathing Orbital Valence Bond Method.

146. P. C. Hiberty, S. Shaik, in *Valence Bond Theory*, D. L. Cooper, Ed., Elsevier, Amsterdam, The Netherlands, 2002, pp. 187–226. Breathing-Orbital Valence Bond—A Valence Bond Method Incorporating Static and Dynamic Electron Correlation Effects.

147. P. C. Hiberty, S. Shaik, *Theor. Chem. Acc.* **108**, 255 (2002). BOVB—A Modern Valence Bond Method That Includes Dynamic Correlation.

148. W. Wu, L. Song, Z. Cao, Q. Zhang, S. Shaik, *J. Phys. Chem. A* **106**, 2721 (2002). Valence Bond Configuration Interaction: A Practical Valence Bond Method Incorporating Dynamic Correlation.

149. L. Song, W. Wu, Q. Zhang, S. Shaik, *J. Phys. Chem. A* **108**, 6017 (2004). VBPCM: A Valence Bond Method that Incorporates a Polarizable Continuum Model.

150. J. J. W. McDouall, in *Valence Bond Theory*, D. L. Cooper, Ed., Elsevier, Amsterdam, The Netherlands, 2002, pp. 227–260. The Biorthogonal Valence Bond Method.

151. T. V. Albu, J. C. Corchado, D. G. Truhlar, *J. Phys. Chem. A* **105**, 8465 (2001). Molecular Mechanics for Chemical Reactions: A Standard Strategy for Using Multiconfiguration Molecular Mechanics for Variational Transition State Theory with Optimized Multidimensional Tunneling.

152. T. K. Firman, C. R. Landis, *J. Am. Chem. Soc.* **123**, 11728 (2001). Valence Bond Concepts Applied to the Molecular Mechanics Description of Molecular Shapes. 4. Transition Metals with π-Bonds.

153.   F. Weinhold, C. Landis, *Valency and Bonding: A Natural Bond Orbital Donor-Acceptor Perspective*, Cambridge University Press, Cambridge, 2005.

154. L. Song, W. Wu, P. C. Hiberty, S. Shaik, *Chem. Eur. J.* **9**, 4540 (2003). An Accurate Barrier for the Hydrogen Exchange Reaction from Valence Bond Theory: Is this Theory Coming of Age?

155. S. Shaik, P. C. Hiberty, *Rev. Comput. Chem.* **20**, 1 (2004). Valence Bond, Its History, Fundamentals and Applications: A Primer.

156. P. C. Hiberty, S. Shaik, *J. Comput. Chem.* **28**, 137 (2007). A Survey of Some Recent Developments of *Ab Initio* Theory.

157. D. G. Truhlar, *J. Comput. Chem.* **28**, 74 (2007). Valence Bond Theory for Chemical Dynamics.

2 A Brief Tour Through Some Valence Bond Outputs and Terminology

Since VB theory has not been taught in a systematic way for a few decades, it is a good idea to get a feeling for the method first through the inspection of some VB outputs of modern VB calculations. We are therefore going to look at two sets of calculations for the H_2 and HF molecules using a simple *ab initio* VB method, the so-called VBSCF (1), and with a minimal basis set, STO-3G. Subsequently, we will look at the HF molecule using a double zeta basis set, 3-21G. As in every VB calculation, here too the wave function is represented as a linear combination of VB structures, where the electrons are distributed in hybrid atomic orbitals (HAOs). Thus, for the two examples discussed in this chapter, we use one covalent structure, called also a Heitler–London (HL) structure, marking the names of those who pioneered the use of this wave function (2,3), and two ionic structures, as shown in Scheme 2.1. The VBSCF method optimizes the structural coefficients of the wave function as well as the orbitals of all the structures, as in any MCSCF procedure.

2.1 VALENCE BOND OUTPUT FOR THE H_2 MOLECULE

The first output for H_2 was calculated with the program TURTLE (4), and is presented in Output 2.1. The relevant output information begins with the section called "vb-symbolic", which specifies the multiplicity of the molecule, the number of electrons, and the number of VB configurations (i.e., structures). This process is followed by a symbolic representation of the VB structures, namely, the VB basis set of structures for the problem, or in short, the VB structure-set. Here, the numbers "1" and "2" designate the atomic orbitals (AOs) in which the VB program distributes the bonding electrons. The number "1" is the 1s AO of one of the hydrogen atoms, say H_1, and "2" is the AO of the other atom, H_2. Thus, structure **1**, represented as "1 2", is the covalent (HL) structure, where each atom possesses one electron, while structures **2** and

---

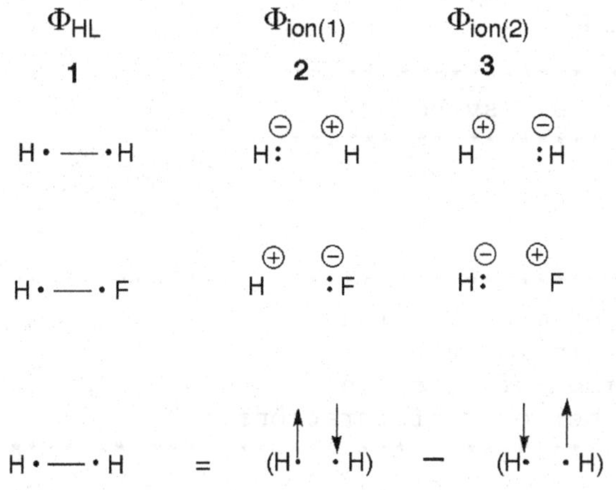

Φ_{HL}	$\Phi_{ion(1)}$	$\Phi_{ion(2)}$
1	**2**	**3**

H •———• H   H :   H   H   : H

H •———• F   H   : F   H :   F

H •———• H   =   (H   • H)   —   (H•   • H)

Scheme 2.1

3, written as "1 1" and "2 2", represent the ionic structures where one of the atoms (H_1 or H_2) possesses two electrons and the other none.

The structure representation is followed by a symbolic representation of the wave functions, along with a coarse normalization, in the section entitled "list of configurations". As can be seen, the wave function of the HL structure (structure **1**) involves two determinants. In the description of the determinants, the letters "a" and "b" stand for α and β spins, respectively. By convention, the orbitals are written in the same orders in both determinants, irrespective of their spins. If we represent our AOs by the symbols $1s_1$ and $1s_2$, these two determinants are the ones shown in Equation 2.1,

$$\Phi_{HL} = 0.70711|1s_1 \overline{1s_2}| - 0.70711|\overline{1s_1} 1s_2| \tag{2.1}$$

where the presence of a bar over the orbital indicates spin-down (β), while the absence of a bar indicates spin-up (α) (see Scheme 2.1). Since the HL structure is described by two determinants, it is coarsely (and temporarily) normalized, and as such, the coefficients of the determinants are given as -0.70711 and 0.70711, which correspond to the square root of 0.5. As shown in Chapter 3, determinants of opposite signs correspond to singlet spin coupling of the two electrons. The other structures are closed shell and each one is described by a single determinant.

The next section entitled, "final VBSCF results...", gives the total energy ("vb-energy") and the wave function ("vb-vector"), the latter is expressed in the usual manner in terms of the coefficient of the structure-set, as follows:

$$\Psi_{VB-full} = 0.787469\Phi_{HL} + 0.133870(\Phi_{ion(1)} + \Phi_{ion(2)}) \tag{2.2}$$

Thus, the wave function that describes the H—H bond is dominated by the covalent structure (now rigorously normalized), with small contributions from

VB OUTPUT 2.1

```
*********************
   vb - symbolic
*********************




*****************************************************
      system definitions
      multiplicity              :        1
      number of electrons       :        2
      number of configurations :        3
*****************************************************

configuration      1 :     1   2
configuration      2 :     1   1
configuration      3 :     2   2


***********************************************************
      final results
      number of configurations :        3
      number of structures      :        3
      number of determinants    :        4
      written to the file       : structures
***********************************************************

== list of configurations ==

** configuration      1   **
      1   2
    structure   1  determinants   2
      1 -0.70711   2  0.70711
    ******* determinants *******
      1 : ba    2 : ab

** configuration      2   **
      1   1
    structure   1  determinants   1
      1  1.00000

** configuration      3   **
      2   2
    structure   1  determinants   1
      1  1.00000

======== end of list  =========
```

```
_ _ _ _ _ _ _ _ _ _ _ _ _ _ _ _ _ _ _ _ _ _ _ _ _ _ _ _ _ _ _ _ _ _

final vbscf results after cycle    2 at    0.014 seconds

vb-energy :        -1.13728383494386 hartree

vb-vector :  0.787469  0.133870  0.133870

weights   :  0.784329  0.107836  0.107836

_ _ _ _ _ _ _ _ _ _ _ _ _ _ _ _ _ _ _ _ _ _ _ _ _ _ _ _ _ _ _ _ _ _
_ _ _ _ _ _ _ _ _ _ _ _ _ _ _ _ _ _ _ _ _ _ _ _ _ _ _ _ _ _ _ _ _ _
```

final orbitals :

```
                1            2
            ----------   ----------

    1       1.000000     0.000000
    2       0.000000     1.000000
```

orbital overlap matrix :

```
    1       1.00000000
    2       0.65987316    1.00000000
```

nullity and determinant :   0   0.56457

matrix representation of the hamiltonian :

```
    1       -1.12438723
    2       -0.92376625   -0.75220865
    3       -0.92376625   -0.65716238   -0.75220865
```

corresponding metric :

```
    1       1.00000000
    2       0.77890423    1.00000000
    3       0.77890423    0.43543258    1.00000000
```

```
lowest  3 eigenvectors/values :

          -1.137284   -0.168352    0.483143
          ----------  ----------  ----------

    1      0.787469    0.000000   -2.417515
    2      0.133870    0.941081    1.494602
    3      0.133870   -0.941081    1.494602


natural orbitals :

           1.974668    0.025332
          ----------  ----------

    1      0.548842   -1.212452
    2      0.548842    1.212452

           3           2

           total energy -.113728383494386E+01
       electronic energy -.185238822548440E+01
nuclear repulsion energy 0.715104390540541E+00
        1-electron energy -.248690504350511E+01
          kinetic energy 0.122502929570175E+01
        potential energy -.236231313064561E+01


 virial quotient  -1.9283728


       dipole moments (debye)

          nuclear        electronic     total
    x     0.00000000     0.00000000     0.00000000
    y     0.00000000     0.00000000     0.00000000
    z     0.00000000     0.00000000     0.00000000

    total dipole moment :     0.00000000
```

the two ionic structures. The same information is provided by the values in the line titled "weights", which are the weights of the VB structures, determined from the Chirgwin–Coulson formula (5), as the square of the coefficient plus one-half of the overlap population terms with all the other structures:

$$\omega_i = C_i^2 + \sum_{j \neq i} C_i C_j S_{ij} \tag{2.3}$$

This formula is the VB analogue of the Mulliken population in MO-based calculations; here S_{ij} is the overlap between the VB structures.

The next interesting section is the "matrix representation of the Hamiltonian". Here the data are given in the usual format shown in Equation 2.4:

$$\begin{vmatrix} H_{11} & & \\ H_{21} & H_{22} & \\ H_{31} & H_{32} & H_{33} \end{vmatrix} \tag{2.4}$$

The diagonal terms are the self-energies of the VB structures (in hartree units). It is seen that the H_{11}, the energy of the covalent structure, is much lower than that of the ionic ones, by 235 kcal/mol. Another interesting quantity is the difference between the energy of the most stable structure and the total energy. This difference, which corresponds to 8.3 kcal/mol, is due to the mixing of the ionic structures into the covalent one, and is hence the resonance energy due to covalent–ionic mixing, which is labeled in the book as RE$_{cov-ion}$ (or RE$_{cs}$). In this simple case, this is given by

$$RE_{cov-ion} = |E_{VB-full} - E_{HL}| \tag{2.5}$$

An additional interesting feature of the output is the matrix representation called "corresponding metric", which is nothing else but the overlap matrix of the VB structures. The off-diagonal elements show that the covalent and the ionic structures have a significant overlap of 0.77890423 (which derives from a large AO overlap, but it is not the only source, as will become clear in Chapter 3). These overlap integrals can be coupled with the off-diagonal elements of the Hamiltonian matrix to derive the important matrix element for VB theory, the reduced off-diagonal matrix element, given as

$$H_{ij}^{red} = H_{ij} - E_i S_{ij} \tag{2.6}$$

Here the subscript i refers to the state that is chosen as the origin of energies, (e.g., the HL state). The final features of some interest are the three states that result from diagonalization of the VB Hamiltonian in the structure set and the AO basis set. It is seen that there are three roots; the lowest is the ground state representing the H–H bond (see Eq. 2.2) while the other two represent excited states. Of course, the energies of these excited states are very poor, but nevertheless, the composition of these states in terms of the VB structure set is instructive. It can be seen that the first excited state is made up of the negative combination of the two ionic structures, that is, $\left(\Phi_{ion(1)} - \Phi_{ion(2)}\right)$, while the second excited state is the antibonding combination of the HL and two ionic structures.

2.2   VALENCE BOND MIXING DIAGRAMS

In fact, what we saw in the VB output could have been reasoned qualitatively at the same level and logic of the computational method. This lucidity is part of the beauty inherent in VB theory. This is demonstrated by appeal to Fig. 2.1. Let us start from the VB structures at infinite separation between the atoms. At this asymptotic limit, the ionic structures lie above the covalent one, by an energy quantity given as the difference between the ionization potential of $H^{\bullet}$, I_H, and the corresponding electron affinity, A_H:

$$E(\Phi_{ion}) - E(\Phi_{HL}) = I_H - A_H \qquad r_{HH} = \infty \qquad (2.7)$$

Using experimental values for these quantities, $I_H = 313.6\,kcal/mol$, while $A_H = 17.4\,kcal/mol$; the difference is $\sim 296\,kcal/mol$. As we let the two H fragments approach each other, the HL-structure will be stabilized by the spin-pairing energy, while the ionic structures will go down by virtue of electrostatic interactions. We may assume that the two effects roughly cancel out, so that at the equilibrium distance the gap is still significant, $200\,kcal/mol$ or more.

With such a large energy gap between the VB structures, we can justifiably use perturbation theory to construct the states, and predict the stabilization energy of the ground state by the covalent–ionic resonance energy. This

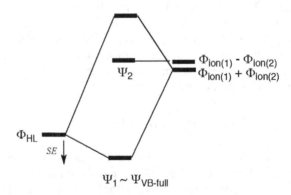

FIGURE 2.1   A VB mixing diagram showing the formation of the states of the H–H bond from the covalent and ionic structures.

discussion can be aided by the VB mixing diagram (6), which is analogous to the orbital interaction diagram used in MO theory. The diagram is shown in Fig. 2.1; it involves the HL structure, and the two symmetry adapted combinations of the ionic structures. The HL structure can mix with the positive combination of the ionic structures and lead to a pair of bonding and antibonding combinations; the bonding structure is the ground state, while the antibonding structure is the second excited state. The first excited state becomes the negative combination of the ionic structures that finds no symmetry match to mix with the HL structure. By using the value of the reduced matrix element (Eq. 2.6, $H_{ij}^{\text{red}} = -30.1\,\text{kcal/mol}$) and the energy gap between the HL structure and the ionic ones, in a perturbation theoretic expression, we can get the mixing coefficients and the stabilization energy (SE) of the ground state as follows:

$$C_{\text{ion}} = H_{ij}^{\text{red}}/[E(\Phi_{\text{HL}}) - E(\Phi_{\text{ion}})] \sim 0.128 \tag{2.8}$$

$$\text{SE} = \left(H_{ij}^{\text{red}}\right)^2 /[E(\Phi_{\text{HL}}) - E(\Phi_{\text{ion}})] = C_{\text{ion}}H_{ij}^{\text{red}} = -3.8\,\text{kcal/mol} \tag{2.9}$$

The $\text{RE}_{\text{cov}-\text{ion}}$ is simply twice the SE term, that is 7.6 kcal/mol.

It is seen that the perturbation treatment and the VB mixing diagram capture the essence of the VB calculations. The major component of the bond energy in H_2 comes from the spin pairing in the HL structure, while the covalent–ionic resonance energy makes a small contribution. The small $\text{RE}_{\text{cov}-\text{ion}}$ arises due to the large energy gap between HL and the ionic structures, as well as the small reduced-matrix element (in our experience, this will be small whenever the AOs overlap is very strong as in H_2). The wave function is dominated by the covalent structure, and as such, the bonding electrons maintain a dominant static correlation, or Coulomb correlation.

2.3   VALENCE BOND OUTPUT FOR THE HF MOLECULE

The second output for HF was calculated using VBSCF/STO-3G, with the program XMVB (7), and is displayed in Output 2.2. The relevant output information begins again with the symbolic representation of the structures, which are written in the same manner as before, "4 5", "4 4", and "5 5", and represent the HL and ionic structures of the H–F bond, as sketched in Scheme 2.1. The number labels of the orbitals are "4" for the bond hybrid of F, and "5" for the AO of H. The rest of the valence electrons are kept in doubly occupied orbitals on fluorine, and the filled subshell is labeled as "1 1 2 2 3 3". This subshell accompanies all the structures. The core electrons are not mentioned specifically, although they are included in the calculations.

Here, the overlap matrix between the structures is called "matrix of overlap", and is followed by "matrix of hamiltonian". It can be seen that the overlap integrals between the VB structures are smaller than in the H_2 example,

VB OUTPUT 2.2

```
************************************************************

   M   M    MM MM    M   M    MMMM        MMM        M
    M M     M M M    M   M    M   M      M   M       M
     M      M M M     M M     MMMM  MMM  M   M       M
    M M     M   M     M M     M   M      M   M       M
   M   M    M   M      M      MMMM        MMM   M   M
************************************************************
```

Cite this work as:

Wei Wu, Lingchun Song, Yirong Mo, and Qianer Zhang, XIAMEN
-- An Ab Initio Spin-free Valence Bond Program, Department
of Chemistry, The State Key Laboratory for Physical
Chemistry of Solid Surfaces, and Institute of Physical
Chemistry, Xiamen University, Xiamen, 361005,
P. R. of China.

```
---------------Input File---------------
FH sto-3G
$ctrl
nbasis=5 norb=5 nelectron=8 nstruct=3 nmul=1
iopt=1 iout=1 itmax=200
$end
$struct
1 1 2 2 3 3 4 5
1 1 2 2 3 3 4 4
1 1 2 2 3 3 5 5
$end
$orb
 2 2 2 2 1
 1 4
 2 3
 3 2
 4 1
 5
$end

---------------End of Input---------------
```

 No Initial Guess Orbital

 FH sto-3G

 OPTIMIZATION METHOD: DFP WITH FORWARD-DIFFERENCE

 Number of Structures: 3

 The following structures are used in calculation:

```
1 *****   1 1 2 2 3 3 4 5
2 *****   1 1 2 2 3 3 4 4
3 *****   1 1 2 2 3 3 5 5
```

 Nuclear Repulsion Energy: -70.652928

```
Total Energy:        -98.596398

First Excited:    -97.6264558334345

The Absulte Value of Gradient:    0.000069

Number of Iteration:        2


                    ****** MATRIX  OF  OVERLAP  ******


             1           2           3
    1    1.000000   -0.556509   -0.556509
    2   -0.556509    1.000000    0.183224
    3   -0.556509    0.183224    1.000000


                   ******  MATRIX  OF  HAMILTONIAN  ******


             1           2           3
    1   -27.827652   15.746156   15.628971
    2    15.746156  -27.497121   -5.289991
    3    15.628971   -5.289991  -27.110184


                  ******  COEFFICIENTS OF STRUCTURES  ******

    1     0.68474  ******   1  1  2  2  3  3  4  5
    2    -0.35152  ******   1  1  2  2  3  3  4  4
    3    -0.13599  ******   1  1  2  2  3  3  5  5


                   ******  WEIGHTS OF STRUCTURES  ******

    1     0.65465  ******   1  1  2  2  3  3  4  5
    2     0.26628  ******   1  1  2  2  3  3  4  4
    3     0.07907  ******   1  1  2  2  3  3  5  5

         Lowdin Weights

    1     0.58657  ******   1  1  2  2  3  3  4  5
    2     0.29322  ******   1  1  2  2  3  3  4  4
    3     0.12021  ******   1  1  2  2  3  3  5  5

         Inverse Weights

    1     0.70215  ******   1  1  2  2  3  3  4  5
    2     0.25907  ******   1  1  2  2  3  3  4  4
    3     0.03877  ******   1  1  2  2  3  3  5  5

Dipole moment (Debye)

  X=    0.000   Y=    0.000   Z=   -1.158   Tol=    1.158

  Kinetic energy:    97.777389
                  Potential energy:  -196.373787

            -V/T =         2.0084
```

reflecting the compact orbitals of fluorine. Furthermore, the diagonal elements of the Hamiltonian matrix show that there are two low energy structures and one much higher than both. The high energy species is structure **3**, which corresponds to the H:$^-$ F$^+$ ionic form. The lowest structure is once again the HL structure, followed by H$^+$ F:$^-$, which lies now 207 kcal/mol higher than the HL structure.

Thus, without much ado, we can immediately see that this bond will be different than H$-$H, in the sense that the wave function will have one ionic structure that is dominant, and a second one that is marginal. Since the HL structure is the lowest, we expect it to have the largest contribution to the wave function. These expectations are indeed born out by the calculations; the table "coefficients of structures" shows that the wave function has the largest coefficient (0.68474) for the HL structure, followed by a significant coefficient (-0.35152) for the dominant ionic structure, H$^+$ F:$^-$, and a smaller one (-0.13599) for the inverse ionic structure, H:$^-$ F$^+$. A seemingly counterintuitive feature is the negative sign of the mixing coefficients. This simply reflects the fact that the bonding AO on F is a 2p$_z$ orbital, which is oriented with its positive lobe along the positive side of the z-axis, while the H atom sits on the negative side of the axis. The same reason explains the negative values in the overlap matrix. Of course, reversing the direction of the 2p$_z$ orbital would yield an equivalent calculation, this time with all positive overlaps and all positive mixing coefficients. The table "weights of structures" further reinforces the picture provided by the coefficients, now showing three sets of weights, one the Coulson$-$Chirgwin weight that is defined above in Equation 2.3, while the others are called "Löwdin weights" and "inverse weights". The latter values are obtained by different formulas (see Chapter 3) than the Chirgwin$-$Coulson formula (Eq. 2.3) (7). All these different weights reveal the same trend; the HL structure has the largest weight followed by H$^+$ F:$^-$ and H:$^-$ F$^+$ that has the smallest weight. Increasing the basis set to 3-21G in Output 2.3 projects the importance of the H$^+$ F:$^-$ ionic structure, which is affected more than the other two structures. Thus, its energy gap relative to the HL structure decreases to 82 kcal/mol, and its weight increases to 0.41$-$0.47 depending on the weight definition.

Finally, the RE$_{cov-ion}$ quantity for H$-$F is $\sim$73/102 kcal/mol in the STO-3G/3-21G basis sets, respectively, much larger than for H$-$H. This is of course expected, at least based on the lower energy of the H$^+$ F:$^-$ structure compared with H$^+$ H:$^-$. But it is clear that this covalent$-$ionic resonance energy is very large and will endow the bond with a special character; it is not merely a polar bond, because its bonding energy is dominated by the covalent$-$ionic resonance energy. Thus, while the VB calculations are very lucid, being in harmony with our chemical intuition, they still reveal to us new features that challenge our intuition. Chapter 10 in the book mentions some more features of this kind of bonding that we recently called charge-shift bonding (8).

VB OUTPUT 2.3

```
*******************************************************************

    M    M     MM MM     M    M     MMMM          MMM        M
     M M      M M M      M    M     M   M        M    M      M
      M       M M M       M  M      MMMM    MMM  M    M      M
     M M      M    M      M  M      M   M        M    M      M
    M    M    M    M       M        MMMM          MMM    M   M
*******************************************************************
---------------Input File---------------
FH 321G
$ctrl
nbasis=10 norb=5 nelectron=8 nstruct=3 nmul=1
iopt=1 iout=1 itmax=200
$end
$struct
1 1 2 2 3 3 4 5
1 1 2 2 3 3 4 4
1 1 2 2 3 3 5 5
$end
$orb
 4 4 4 4 2
 1 4 5 8
 2 3 6 7
 3 2 6 7
 4 1 5 8
 9 10
$end

---------------End of Input---------------

 No Initial Guess Orbital

 FH 321G

 OPTIMIZATION METHOD: DFP WITH FORWARD-DIFFERENCE

 Number of Structures:    3

 The following structures are used in calculation:

   1 *****   1  1  2  2  3  3  4  5
   2 *****   1  1  2  2  3  3  4  4
   3 *****   1  1  2  2  3  3  5  5

 Nuclear Repulsion Energy:    -71.115717

 Max valence electrons:         8

 Total Energy:    -99.480084

 First Excited:   -98.3886839577165

 The Absulte Value of Gradient:   0.000260

 Number of Iteration:    9


       ****** MATRIX  OF  OVERLAP ******
```

	1	2	3
1	1.000000	-0.615789	-0.615789
2	-0.615789	1.000000	0.233956
3	-0.615789	0.233956	1.000000

****** MATRIX OF HAMILTONIAN ******

	1	2	3
1	-28.202649	17.682004	17.392150
2	17.682004	-28.072145	-6.808810
3	17.392150	-6.808810	-27.236124

****** COEFFICIENTS OF STRUCTURES ******

1	0.61766	******	1 1 2 2 3 3 4 5
2	-0.47447	******	1 1 2 2 3 3 4 4
3	-0.03198	******	1 1 2 2 3 3 5 5

****** WEIGHTS OF STRUCTURES ******

1	0.57413	******	1 1 2 2 3 3 4 5
2	0.40914	******	1 1 2 2 3 3 4 4
3	0.01674	******	1 1 2 2 3 3 5 5

Lowdin Weights

1	0.50850	******	1 1 2 2 3 3 4 5
2	0.42249	******	1 1 2 2 3 3 4 4
3	0.06901	******	1 1 2 2 3 3 5 5

Inverse Weights

1	0.52560	******	1 1 2 2 3 3 4 5
2	0.47226	******	1 1 2 2 3 3 4 4
3	0.00215	******	1 1 2 2 3 3 5 5

Dipole moment (Debye)

X= 0.000   Y= 0.000   Z= -2.043   Tol= 2.043
Kinetic energy: 99.292703
Potential energy: -198.772787
-V/T =   2.0019

REFERENCES

1. J. H. van Lenthe, G. G. Balint-Kurti, *J. Chem. Phys.* **78**, 5699 (1983). The Valence-Bond Self-Consistent Field Method (VB–SCF): Theory and Test Calculations.

2. W. Heitler, F. London, *Z. Phys.* **44**, 455 (1927). Wechselwirkung neutraler Atome und homöopolare Bindung nach der Quantenmechanik.

3. For an English translation, see H. Hettema, *Quantum Chemistry Classic Scientific Papers*, World Scientific, Singapore, 2000.

4. J. Verbeek, J. H. Langenberg, C. P. Byrman, F. Dijkstra, J. H. van Lenthe, *TURTLE: An Ab Initio VB/VBSCF Program (1998–2000)*.

5. B. H. Chirgwin, C. A. Coulson, *Proc. R. Soc. Ser. A (London)* **2**, 196 (1950). The Electronic Structure of Conjugated Systems. VI.

6. S. S. Shaik, in *New Theoretical Concepts for Understanding Organic Reactions*, J. Bertran and I. G. Csizmadia, Eds., NATO ASI Series, C267, Kluwer Academic Publishers, 1989, pp. 165–217. A Qualitative Valence Bond Model for Organic Reactions.

7. L. Song, Y. Mo, Q. Zhang, W. Wu, *J. Comput. Chem.* **26**, 514 (2005). XMVB: A Program for Ab Initio Nonorthogonal Valence Bond Computations.

8. S. Shaik, D. Danovich, B. Silvi, D. Lauvergnat, P. C. Hiberty, *Chem. Eur. J.* **11**, 6358 (2005). Charge-Shift Bonding: A Class of Electron-Pair Bonds that Emerges from Valence Bond Theory and Is Supported by the Electron Localization Function Approach.

3 Basic Valence Bond Theory

3.1 WRITING AND REPRESENTING VALENCE BOND WAVE FUNCTIONS

3.1.1 VB Wave Functions with Localized Atomic Orbitals

After looking at the VB outputs for the simple two-center/two-electron (2e/2c) bond, let us get used to the theory by applying it to these bonds to start with. A VB determinant is an antisymmetrized wave function that may or may not also be a proper spin eigenfunction. For example, $|a\bar{b}|$ in Equation 3.1 is a determinant that describes two spin−orbitals a and b, each having one electron; the bar over the b orbital means a β spin, and the absence of a bar means an α spin:

$$|a\bar{b}| = \tfrac{1}{\sqrt{2}}\{a(1)b(2)[\alpha(1)\beta(2)] - a(2)b(1)[\alpha(2)\beta(1)]\} \tag{3.1}$$

The parenthetical numbers 1 and 2 are the electron indexes. By itself, this determinant is not a proper spin-eigenfunction. However, by mixing with $|\bar{a}b|$ there will result two spin-eigenfunctions, one having a singlet coupling and shown in Equation 3.2, the other displaying a triplet situation in Equation 3.3. In both cases, the normalization constants are omitted for the time being.

$$\Phi_{HL} = |a\bar{b}| - |\bar{a}b| \tag{3.2}$$

$$\Psi_{T} = |a\bar{b}| + |\bar{a}b| \tag{3.3}$$

If a and b are the respective AOs of two hydrogen atoms, Φ_{HL} in Equation 3.2 is nothing else but the historical wave function used in 1927 by Heitler and London (1) to treat bonding in the H_2 molecule, hence the subscript descriptor HL. This wave function displays a purely covalent bond in which the two hydrogen atoms remain neutral and exchange their spins (this singlet pairing is represented henceforth by the two dots connected by a line as **1** in Scheme 3.1). Ψ_{T} in Equation 3.3 represents a repulsive triplet interaction (**2**) between two hydrogen atoms having electrons with parallel spins.

A Chemist's Guide to Valence Bond Theory, by Sason Shaik and Philippe C. Hiberty
Copyright © 2008 John Wiley & Sons, Inc.

H•——•H   H↑ ↑H   H⁻ H⁺   H⁺ H⁻

1   2   3   4

Scheme 3.1

The other VB determinants that one can construct in this simple 2e/2c cases are $|a\bar{a}|$ and $|b\bar{b}|$, corresponding to the ionic structures **3** and **4**, respectively. Both ionic structures are spin-eigenfunctions and represent singlet-spin pairing. Note that the rules governing the spin multiplicities and the generation of spin-eigenfunctions from combinations of determinants are the same in VB and MO theories. In a simple 2e case, it is easy to distinguish triplet from singlet eigenfunctions by factorizing the spatial function from the spin function: the singlet spin eigenfunction is antisymmetric with respect to electron exchange, while the triplet is symmetric. Of course, the spatial parts behave in precisely the opposite manner, to keep the complete wave function antisymmetric. For example, the singlet spin function is $\alpha(1)\beta(2) - \beta(1)\alpha(2)$, while the triplet is $\alpha(1)\beta(2) + \beta(1)\alpha(2)$ in Equations 3.2 and 3.3. Note that in Equations 3.2 and 3.3, the complete wave functions are written in an isomorphic manner to the spin wave functions; the singlet HL structure has a negative sign between the determinants as in the corresponding spin function, while the triplet structure has a positive sign as in the corresponding spin function. Keeping in mind this isomorphic relation between the full wave function, in terms of Slater determinants, the corresponding spin function may come in handy for more complicated VB cases.

While the H—H bond in H_2 was considered as purely covalent in Heitler and London's paper (1) (Eq. 3.2 and Structure **1**), as we saw in Chapter **2**, the exact description of H_2 or any homopolar bond ($\Psi_{VB-full}$) involves a small contribution of the ionic structures **3** and **4**, which mix by configuration interaction (CI) in the VB framework. Typically, for homopolar and weakly polar bonds, the weight of the purely covalent structure is 75%, while the ionic structures share the remaining 25%. By symmetry, the wave function maintains an average neutrality of the two bonded atoms (Eq. 3.4).

$$\Psi_{VB-full} = \lambda\left(|a\bar{b}| - |\bar{a}b|\right) + \mu\left(|a\bar{a}| + |b\bar{b}|\right) \quad \lambda > \mu \tag{3.4a}$$

$$H_a - H_b \approx 75\%\,(H_a \bullet\!-\!\bullet H_b) + 25\%\,(H_a^- H_b^+ + H_a^+ H_b^-) \tag{3.4b}$$

For convenience and to avoid confusion, we will symbolize a purely covalent bond between A and B centers as A•——•B, while the notation A—B will be employed for a composite bond wave function like the one displayed in Equation 3.4. In other words, A—B refers to the "real" bond while A•——•B designates its covalent component.

3.1.2 Valence Bond Wave Functions with Semilocalized AOs

One inconvenience of the expression of $\Psi_{VB-full}$ (Eq. 3.4) is its relative complexity compared to the simple Heitler–London function (Eq. 3.2).

Coulson and Fischer (2) proposed an elegant way that combines the simplicity of Φ_{HL} with the "completeness" of $\Psi_{VB\text{-full}}$ in terms of the contributing VB structures. In the Coulson–Fischer wave function, Ψ_{CF}, the 2e bond is described as a formally covalent singlet coupling between two orbitals φ_a and φ_b, which are optimized with freedom to delocalize over the two centers. This is exemplified below for H_2 (once again dropping the normalization factors):

$$\Psi_{CF} = \left|\varphi_a\bar{\varphi}_b\right| - \left|\bar{\varphi}_a\varphi_b\right| \tag{3.5a}$$

$$\varphi_a = a + \varepsilon b \tag{3.5b}$$

$$\varphi_b = b + \varepsilon a \tag{3.5c}$$

Here, a and b are purely localized AOs, while φ_a and φ_b are slightly delocalized or "semilocalized". In fact, experience shows that the Coulson–Fischer orbitals φ_a and φ_b, which result from the optimization of the coefficient ε by energy minimization, are generally not very delocalized ($\varepsilon < 1$), and as such they can be viewed as "distorted" orbitals that remain atomic-like in nature. However, minor as this may look, the slight delocalization renders the Coulson–Fischer wave function equivalent to the VB-full wave function (Eq. 3.4a) with the three classical structures. A straightforward expansion of the Coulson–Fischer wave function leads to the linear combination of the classical structures in Equation 3.6.

$$\Psi_{CF} = (1 + \varepsilon^2)(\left|a\bar{b}\right| - \left|\bar{a}b\right|) + 2\varepsilon(\left|a\bar{a}\right| + \left|b\bar{b}\right|) \tag{3.6}$$

Thus, the Coulson–Fischer representation keeps the simplicity of the covalent picture while treating the covalent–ionic balance by embedding the effect of the ionic terms in a variational way, through the delocalization tails. The Coulson–Fischer idea was subsequently generalized to polyatomic molecules and gave rise to the GVB and SC methods, which were mentioned in Chapter 1 and will be discussed later.

3.1.3 Valence Bond Wave Functions with Fragment Orbitals

Valence bond determinants may involve fragment orbitals (FO) instead of localized or semilocalized AOs. These fragment orbitals may be delocalized, for example, some MOs of the constituent fragments of a molecule. The latter option is an economical way of representing a wave function that is a linear combination of several determinants based on AOs, much as MO determinants are linear combinations of VB determinants (see below). Suppose, for example, that one wanted to treat the recombination of the $CH_3^\bullet$ and $H^\bullet$ radicals in a VB manner, and let $(\varphi_1 - \varphi_5)$ be the MOs of the $CH_3^\bullet$ fragment (φ_5 being singly occupied), and b the AO of the incoming hydrogen. The covalent VB function that describes the active C—H bond in our study just couples the φ_5 and b orbitals in a singlet manner, and is expressed as in Equation 3.7:

$$\Psi(H_3C^\bullet\text{--}^\bullet H) = \left|\varphi_1\bar{\varphi_1}\varphi_2\bar{\varphi_2}\varphi_3\bar{\varphi_3}\varphi_4\bar{\varphi_4}\varphi_5\bar{b}\right| - \left|\varphi_1\bar{\varphi_1}\varphi_2\bar{\varphi_2}\varphi_3\bar{\varphi_3}\varphi_4\bar{\varphi_4}\bar{\varphi_5}b\right| \tag{3.7}$$

in which $\varphi_1 - \varphi_4$ are fully delocalized over the CH_3 fragment. Even the φ_5 orbital is not a pure AO, but may involve some tails on the hydrogens of the CH_3 fragment. It is clear that this option is conceptually simpler than treating all the C—H bonds in a VB way, including those three bonds that are intact during the recombination reaction.

3.1.4   Writing Valence Bond Wave Functions Beyond the 2e/2c Case

Rules for writing VB wave functions in the polyelectronic case are just extensions of the rules for the 2e/2c case above. First, let us consider butadiene **5** in Scheme 3.2, and restrict the description to the π system.

Scheme 3.2

Denoting the π AOs of the C_1-C_4 carbons by a,b,c, and d, respectively, the fully covalent VB wave function for the π system of butadiene displays two singlet couplings: one between a and b, and one between c and d. It follows that the wave function can be expressed in the form of Equation 3.8, as a product of the bond wave functions.

$$\Psi(\mathbf{5}) = \left|(a\bar{b} - \bar{a}b)(c\bar{d} - \bar{c}d)\right| \qquad (3.8)$$

Upon expansion of the product, one gets a sum of four determinants as in Equation 3.9.

$$\Psi(\mathbf{5}) = \left|a\bar{b}\,c\bar{d}\right| - \left|a\bar{b}\,\bar{c}d\right| - \left|\bar{a}b\,c\bar{d}\right| + \left|\bar{a}b\,\bar{c}d\right| \qquad (3.9)$$

The product of bond wave functions in Equation 3.8, involves so-called *perfect pairing*, whereby we take the Lewis structure of the molecule, represent each bond by a HL bond, and finally express the full wave function as a product of all these pair-bond wave functions. As a rule, such a perfectly paired polyelectronic VB wave function having n bond pairs will be described by 2^n determinants, displaying all the possible 2×2 spin permutations between the orbitals that are singlet coupled.

The above rule can readily be extended to other polyelectronic systems, like the π system of benzene (**6**), or to molecules bearing lone pairs as in formamide (**7**). In this latter case, calling n, c, and o, respectively, the π atomic orbitals of nitrogen, carbon, and oxygen, the VB wave function describing the neutral covalent structure is given by Equation 3.10:

$$\Psi_\pi(\mathbf{7}) = \left|n\bar{n}\,c\bar{o}\right| - \left|n\bar{n}\,\bar{c}o\right| \qquad (3.10)$$

In any one of the above cases, improvement of the wave function can be achieved by using Coulson–Fischer orbitals that take into account ionic contributions to the bonds. It should be kept in mind that the number of determinants grows exponentially with the number of covalent bonds (recall, this number is 2^n; n being the number of bonds). Hence, eight determinants are required to describe a Kekulé structure of benzene, and the fully covalent and perfectly paired wave function for methane is made of 16 determinants. This underscores the incentive of using FOs rather than pure AOs, as much as possible, as has been done above (Eq. 3.7). Using FOs to construct VB wave functions is also appropriate when one wants to fully exploit the symmetry properties of the molecule. For example, we can describe all the bonds in methane by constructing group orbitals of the four hydrogens. Subsequently, we can distribute the eight bonding electrons of the molecule into these FOs as well as into the $2s$ and $2p$ AOs of carbon. Then, we can pair up the electrons using orbital symmetry-matched FOs, as shown by the lines connecting these orbital−pairs in Fig. 3.1. The corresponding wave function can be written as the following product of bond pairs:

$$\Psi(CH_4) = \left|(2s\,\overline{\varphi_s} - \overline{\varphi_s}2s)(2p_x\overline{\varphi_x} - \overline{\varphi_x}2p_x)(2p_y\overline{\varphi_y} - \overline{\varphi_y}2p_y)(2p_z\overline{\varphi_z} - \overline{\varphi_z}2p_z)\right|$$

$$(3.11)$$

In this representation, each bond pair is a delocalized covalent two-electron bond, written as a HL-type bond. The VB method that deals with fragment orbitals (FO−VB) is particularly useful in high symmetry cases, for example, ferrocene and other organometallic complexes. Some of its merits are further illustrated at the end of this chapter.

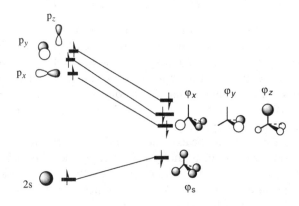

$\Psi_{VB}(CH_4)$

FIGURE 3.1 A VB representation of methane using delocalized FOs. Each line that connects two orbitals represents a bond pair.

3.1.5   Pictorial Representation of Valence Bond Wave Functions by Bond Diagrams

Since we agreed that a bond needs not necessarily involve only two AOs on two centers, we must agree on some pictorial representation of a bond. This bond diagram is in Fig. 3.2, and shows two spin-paired electrons in general orbitals φ_1 and φ_2, by a line connecting these orbitals. This bond diagram represents the wave function in Equation 3.12

$$\Psi_{bond} = |\varphi_1 \overline{\varphi_2}| - |\overline{\varphi_1} \varphi_2| \tag{3.12}$$

where the orbitals can take any shape; the wave function can involve two centers with localized AOs, or two Coulson–Fischer orbitals with delocalization tails, or FOs that span a few centers. In the case of localized orbitals, the bond-diagram represents the full-VB wave function, namely, it implicitly involves the corresponding ionic structures by shifting electrons between the pairs of coupled orbitals.

FIGURE 3.2   A generic bond diagram representation of two spin-paired electrons in orbitals φ_1 and φ_2. The bond pair is indicated by a line connecting the orbitals.

3.2   OVERLAPS BETWEEN DETERMINANTS

A VB calculation is nothing else than a configuration interaction in a space of structures made of AO- or FO-based determinants. As these are in general nonorthogonal to each other, it is essential to derive some basic rules for calculating the overlaps between determinants. The fully general rules have been described in detail elsewhere (3) and will be exemplified here on commonly encountered simple cases. A more systematic presentation can be found in Appendix 3.A.1.

Let us demonstrate the procedure with VB determinants of the type Ω and Ω' in Equation 3.1,

$$\Omega = N|a\,\bar{a}\,b\,\bar{b}|; \quad \Omega' = N'|c\,\bar{c}\,d\,\bar{d}| \tag{3.13}$$

where N and N' are normalization factors. Each determinant is made of a diagonal product of spin orbitals followed by a signed sum of all the permutations of this product, which are obtained by transposing the ordering of the spin–orbitals. Denoting the diagonal products of Ω and Ω' by ψ_d and ψ'_d, respectively, the expression for ψ_d reads:

$$\psi_d = a(1)\,\bar{a}(2)\,b(3)\,\bar{b}(4); \qquad (1, 2, \ldots \text{ are electron indexes}) \qquad (3.14)$$

and an analogous expression can be written for ψ'_d.

The overlap between the (unnormalized) determinants $|a\,\bar{a}\,b\,\bar{b}|$ and $|c\,\bar{c}\,d\,\bar{d}|$ is given by Equation 3.15:

$$\langle |a\,\bar{a}\,b\,\bar{b}|\,||\,c\,\bar{c}\,d\,\bar{d}|\rangle = \langle \psi_d | \Sigma_P (-1)^t P \psi'_d \rangle \qquad (3.15)$$

where the operator P represents a *restricted subset of permutations*: the ones made of pairwise transpositions between spin orbitals of the same spin, and t determines the parity, odd or even, and hence the sign of a given pairwise transposition P will be negative or positive, respectively. Note that the identity permutation is included. In this example, there are four possible permutations in the product ψ'_d

$$\begin{aligned} \Sigma_P(-1)^t P(\psi'_d) = {} & c(1)\,\bar{c}(2)\,d(3)\,\bar{d}(4) - d(1)\,\bar{c}(2)\,c(3)\,\bar{d}(4) \\ & -c(1)\,\bar{d}(2)\,d(3)\,\bar{c}(4) + d(1)\,\bar{d}(2)\,c(3)\,\bar{c}(4) \end{aligned} \qquad (3.16)$$

One then integrates Equation 3.15 electron by electron, leading to Equation 3.17 for the overlap between $|a\,\bar{a}\,b\,\bar{b}|$ and $|c\,\bar{c}\,d\,\bar{d}|$:

$$\langle |a\,\bar{a}\,b\,\bar{b}|\,||\,c\,\bar{c}\,d\,\bar{d}|\rangle = S_{ac}^2 S_{bd}^2 - S_{ad}S_{ac}S_{bc}S_{bd} - S_{ac}S_{ad}S_{bd}S_{bc} + S_{ad}^2 S_{bc}^2 \qquad (3.17)$$

where S_{ac}, for example, is an overlap integral between two orbitals a and c. Generalization to different types of determinants is straightforward (3). As an application, let us obtain the overlap of a VB determinant with itself, and calculate the normalization factor N of the determinant Ω in Equation 3.13:

$$\langle |a\,\bar{a}\,b\,\bar{b}|\,||\,a\,\bar{a}\,b\,\bar{b}|\rangle = 1 - 2S_{ab}^2 + S_{ab}^4 \qquad (3.18)$$

$$\Omega = (1 - 2S_{ab}^2 + S_{ab}^4)^{-1/2}|a\,\bar{a}\,b\,\bar{b}| \qquad (3.19)$$

Generally, normalization factors for determinants are larger than unity, with the exception of those VB determinants that do not have more than one spin–orbital of each spin variety, for example, as is the case of the determinants that compose the HL wave function. For these latter determinants the normalizing factor is unity, that is, $N = 1$.

3.3 VALENCE BOND FORMALISM USING THE EXACT HAMILTONIAN

Let us turn now to the calculation of energetic quantities using exact VB theory by considering the simple case of the H_2 molecule. The exact electronic

Hamiltonian is of course the same as in MO theory, and is composed in this case of two core terms and a bielectronic repulsion:

$$H = h(1) + h(2) + 1/r_{12} \qquad (3.20)$$

where the h operator represents the kinetic energy and the attraction between one electron and the nuclei, and r_{12} is the interelectronic distance. The molecular Hamiltonian is derived from the electronic one (Eq. 3.20) by adding the nuclear repulsion, here $1/R$, that is, the inverse of the distance between the nuclei.

3.3.1 Purely Covalent Singlet and Triplet Repulsive States

In the VB framework, some particular notations are traditionally employed to designate the various energies and matrix elements:

$$Q = \langle |a\,\bar{b}||H||a\,\bar{b}\rangle = \langle a|h|a\rangle + \langle b|h|b\rangle + \langle ab|1/r_{12}|ab\rangle \qquad (3.21)$$

$$K = \langle |a\,\bar{b}||H||\bar{b}a|\rangle = \langle ab|1/r_{12}|ba\rangle + 2S_{ab}\langle a|h|b\rangle \qquad (3.22)$$

$$\langle |a\,\bar{b}|||b\,\bar{a}|\rangle = S_{ab}^2 \qquad (3.23)$$

Here, Q is the electronic energy of a single determinant $|a\bar{b}|$, K is the spin exchange term that will be dealt with later, and S_{ab} is the overlap integral between the two AOs a and b.

The quantity $Q + 1/R$ has an interesting property: This quantity is quasiconstant along the interatomic distance, from infinite distance to the equilibrium bonding distance R_{eq} of H_2. It corresponds to the energy of two hydrogen atoms when brought together without exchanging their spins. Such a pseudo-state (which is not a spin-eigenfunction) is called the "quasiclassical state" of H_2 (Ψ_{QC} in Fig. 3.3), because all the terms of its energy have an analogue in classical (not quantum) physics. Turning now to real states, that is, spin-eigenfunctions, the total energy of the ground state of H_2, in the fully covalent approximation of Heitler–London, is readily obtained:

$$E(\Psi_{HL}) = \frac{\langle(|a\,\bar{b}| - |\bar{a}\,b|)|H|(|a\,\bar{b}| - |\bar{a}\,b|)\rangle}{\langle(|a\,\bar{b}| - |\bar{a}\,b|)|(|a\,\bar{b}| - |\bar{a}\,b|)\rangle} + \frac{1}{R} = \frac{Q + K}{1 + S_{ab}^2} + \frac{1}{R} \qquad (3.24)$$

Plotting the $E(\Psi_{HL})$ curve as a function of the distance now gives a qualitatively correct Morse curve behavior (Fig. 3.3), with a reasonable bonding energy, even if a deeper potential well can be obtained by allowing further mixing with the ionic terms (Ψ_{exact} in Fig. 3.3). This figure shows that, in the covalent approximation, all the bonding comes from the K terms. Thus, *the physical phenomenon responsible for the bond is the resonance between the two spin arrangement patterns $a\bar{b}$ and $\bar{a}b$.*

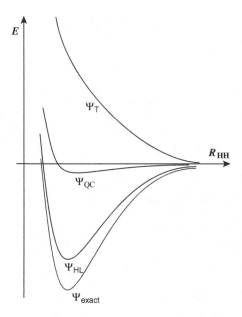

FIGURE 3.3 Energy curves for H_2 as a function of internuclear distance. The curves displayed, from top to bottom, correspond to the triplet state, Ψ_T, the quasiclassical state, Ψ_{QC}, the HL state, Ψ_{HL}, and the exact full (full CI) curve, Ψ_{exact}.

The term K in Equation 3.22 has been called exchange (1), but it needs not be confused with the exchange integral in MO theory. The VB term K is composed of two contributions: One is a repulsive exchange integral (which is akin to the exchange integral of MO theory). This term is positive, but necessarily small (unlike Coulomb two-electron integrals). The second is a negative term, given by the product of the overlap S_{ab} and an integral that is called the "resonance integral", itself nearly proportional to S_{ab}. Replacing Ψ_{HL} by Ψ_T in Equation 3.24 leads to the energy of the triplet state, Equation 3.25.

$$E(\Psi_T) = \frac{\langle(|a\,\bar{b}| + |\bar{a}\,b|)|H|(|a\,\bar{b}| + |\bar{a}\,b|)\rangle}{\langle(|a\,\bar{b}| + |\bar{a}\,b|)|(|a\,\bar{b}| + |\bar{a}\,b|)\rangle} + \frac{1}{R} = \frac{Q - K}{1 - S_{ab}^2} + \frac{1}{R} \qquad (3.25)$$

Recalling that $Q + 1/R$ is a quasiconstant from R_{eq} to infinite distance, this quantity remains nearly equal to the energy of the separated fragments and can serve, at any distance, as a reference for the bond energy, itself having a zero-bonding energy. It follows from Equations 3.24 and 3.25 that, if we neglect overlap in the denominator, the triplet state (Ψ_T in Figure 3.3) is repulsive by the same quantity ($-K$) as the singlet is bonding ($+K$). Thus, at any distance $>R_{eq}$, the bonding energy is about one-half of the singlet–triplet energy gap. This property will be used later in applications to reactivity problems.

3.3.2   Configuration Interaction Involving Ionic Terms

By using the expression of the exact Hamiltonian in Equation 3.20, the self-energy of the ionic terms and the off-diagonal Hamiltonian matrix elements are readily obtained

$$\langle|a\,\bar{a}||H||a\,\bar{a}|\rangle = 2\langle a|h|a\rangle + J_{aa} \tag{3.26}$$

$$\langle|a\,\bar{a}||H||a\,\bar{b}|\rangle = \langle a|h|b\rangle + S_{ab}\langle a|h|a\rangle + \langle aa|1/r_{12}|ab\rangle \tag{3.27}$$

$$\langle|a\,\bar{a}||H||b\,\bar{b}|\rangle = 2S_{ab}\langle a|h|b\rangle + \langle aa|1/r_{12}|ab\rangle = K \tag{3.28}$$

By using these matrix elements and the calculated overlaps between determinants, the accurate VB wave function of H_2 (Eq. 3.4) can be variationally determined by 3×3 nonorthogonal CI.

3.4   VALENCE BOND FORMALISM USING AN EFFECTIVE HAMILTONIAN

The use of the exact Hamiltonian for calculating matrix elements between VB determinants leads, in the general case, to complicated expressions involving numerous bielectronic integrals, owing to the $1/r_{ij}$ terms. Thus, for practical qualitative or semiquantitative applications, one uses an effective molecular Hamiltonian in which the nuclear repulsion and the $1/r_{ij}$ terms are only implicitly taken into account, in an averaged manner. Then, one defines a Hamiltonian made of a sum of independent monoelectronic Hamiltonians, much as in simple MO theory:

$$H^{\text{eff}} = \Sigma_i\, h(i) \tag{3.29}$$

where the summation runs over the total number of electrons. Here, the operator h has a meaning different from Equation 3.20 since it is now an effective monoelectronic operator that incorporates part of the electron–electron and nuclei–nuclei repulsions (3). Going back to the 4e example above (Section 3.2), the determinants Ω and Ω' are coupled by the following effective Hamiltonian matrix element:

$$\langle\Omega|H^{\text{eff}}|\Omega'\rangle = \langle\Omega|h(1) + h(2) + h(3) + h(4)|\Omega'\rangle \tag{3.30}$$

It is apparent that the above matrix element is made of a sum of four terms, which are calculated independently (also consult Appendix 3.A.1). The calculation of each of these terms, for example, the first one ($h(1)$), is quite analogous to the calculation of the overlap in Equation 3.17, except that the first monoelectronic overlap S in each product is replaced by a monoelectronic Hamiltonian term:

$$\langle|a\,\bar{a}\,b\,\bar{b}||h(1)||c\,\bar{c}\,d\,\bar{d}|\rangle = h_{ac}S_{ac}S_{bd}^2 - h_{ad}S_{ac}S_{bc}S_{bd} - h_{ac}S_{ad}S_{bd}S_{bc} + h_{ad}S_{ad}S_{bc}^2 \tag{3.31a}$$

$$h_{ac} = \langle a|h|c\rangle, \text{ and so on} \tag{3.31b}$$

The same type of calculation is repeated for $h(2)$, $h(3)$, and $h(4)$. The obtained integrals are then summed up to get the H^{eff} matrix element of Equation 3.30.[†] In Equation 3.31b, the monoelectronic integral accounts for the interaction that takes place between two overlapping orbitals. A diagonal term of the type h_{aa} is interpreted as the energy of the orbital a, and will be noted ε_a in the following equations. By using Equations 3.30 and 3.31, it is easy to calculate the energy of the determinant $|a\,\bar{a}\,b\,\bar{b}|$:

$$E(|a\,\bar{a}\,b\bar{b}|) = N^2(2\varepsilon_a + 2\varepsilon_b - 2\varepsilon_a S_{ab}^2 - 2\varepsilon_b S_{ab}^2 - 4h_{ab}S_{ab} + 4h_{ab}S_{ab}^3) \qquad (3.32)$$

where N is the normalization factor of the determinant, shown in Equation 3.19.

An application of the above rules is the calculation of the energy of a spin-alternant determinant like **8** in Scheme 3.3 for butadiene. Such a determinant, in which the spins are arranged so that two neighboring orbitals always display opposite spins, is referred to as a quasiclassical (QC) state and is a generalization of the QC state that we already encountered above for H_2. The rigorous formulation for its energy involves some terms that arise from permutations between orbitals of the same spins, which are necessarily nonneighbors. Neglecting interactions between nonnearest neighbors, the energy of the QC state is given by the simple expression below:

$$E(|a\,\bar{b}\,c\,\bar{d}|) = \varepsilon_a + \varepsilon_b + \varepsilon_c + \varepsilon_d \qquad (3.33)$$

8

Scheme 3.3

Generalizing: The energy of a spin-alternant determinant is always the sum of the energies of its constituting orbitals. In the QC state, the interaction between overlapping orbitals is therefore neither stabilizing nor repulsive. *This is a nonbonding state*, which can be used for defining a reference state, with zero energy, in the framework of VB calculations of bonding energies or repulsive interactions.

Note that the rules and formulas that are expressed above in the framework of qualitative VB theory are independent of the type of orbitals

---

[†]In all rigor, the calculation of the $h(1)$ matrix element would necessitate the permutations of the orbital products to be generated for *both* determinants, leading to the generation of 16 nonidentical terms. However, these terms would be redundant with those arising from the calculation of the other $h(i)$ matrix elements, and this is why Equation 3.31 has as few terms as it does. This detail, however, does not matter as the calculation of the $h(1)$ matrix element alone is only an intermediate step in the calculations of the whole H^{eff} matrix element, involving all $h(i)$ terms, in which case it is sufficient to consider the permutations in the right-hand determinant alone, as done in Equation 3.31a and in Appendix 3.A.1.

that are used in the VB determinants: purely localized AOs, FOs, or Coulson–Fischer semilocalized orbitals. Depending on which kind of orbitals are chosen, the h and S integrals take different values, but the formulas remain the same.

3.5 SOME SIMPLE FORMULAS FOR ELEMENTARY INTERACTIONS

In qualitative VB theory, it is customary to take the average value of the orbital energies as the origin for various quantities. With this convention, and using some simple algebra (3), one can define a reduced monoelectronic Hamiltonian matrix element between two orbitals, just as done in previous chapters with off-diagonal matrix elements between VB structures (Eq. 2.6). This reduced matrix element, β_{ab}, is nothing else but the so-called and familiar "reduced resonance integral":

$$\beta_{ab} = h_{ab} - 0.5(h_{aa} + h_{bb})S_{ab} \qquad (3.34)$$

It is important to note that these β integrals, which we use in the VB framework, are the same as those used in simple MO models, such as extended Hückel theory.

Based on the new energy scale, the sum of orbital energies is set to zero, that is:

$$\Sigma_i \varepsilon_i = 0 \qquad (3.35)$$

In addition, since the energy of the QC determinant is given by the sum of orbital energies, its energy becomes then zero:

$$E(|a\,\bar{b}\,c\,\bar{d}|) = 0 \qquad (3.36)$$

3.5.1 The Two-Electron Bond

By application of the qualitative VB theory, Equation 3.37 expresses the HL bond energy of two electrons in AOs a and b, which belong to the atomic centers A and B. The binding energy is defined relative to the quasiclassical state $|a\bar{b}|$ or to the energy of the separate atoms, which is one and the same thing within the approximation scheme. In terms of β_{ab} and S_{ab}, noted β and S for short from now on, the two-electron bonding energy is expressed as Equation 3.37:

$$D_e(A - B) = 2\beta S/(1 + S^2) \qquad (3.37)$$

Note that if instead of using purely localized AOs for a and b, we use semilocalized Coulson–Fischer orbitals, Equation 3.37 will no more be the simple HL bond energy, but would represent *the bonding energy of the real A—B bond* that includes its optimized covalent and ionic components. In this case, the origin of the energy would still correspond to the QC determinant with the localized orbitals. Unless otherwise specified, in what follows we always use qualitative VB theory in this latter convention.

$$A{\uparrow}\ {\uparrow}B \qquad\qquad A: \ :B \qquad\qquad A{\uparrow}\ :B$$

$$\textbf{9} \qquad\qquad\qquad \textbf{10} \qquad\qquad\qquad \textbf{11}$$

Scheme 3.4

3.5.2   Repulsive Interactions in Valence Bond Theory

By using the above definitions, one gets the following expression for the repulsion energy of the triplet state (**9**, in Scheme 3.4):

$$\Delta E_T(A{\uparrow}{\uparrow}B) = -2\beta S/(1 - S^2) \tag{3.38}$$

Thus, the triplet repulsion arises due to the Pauli exclusion principle and is often referred to as Pauli repulsion.

For a situation where we have four electrons on the two centers (**10**), VB theory predicts a doubling of the Pauli repulsion, and the following expression is obtained by complete analogy to qualitative MO theory:

$$\Delta E(A{::}B) = -4\beta S/(1 - S^2) \tag{3.39}$$

One can in fact very simply generalize the rules for Pauli repulsion. Thus, the electronic repulsion in an AO-based determinant is equal to the quantity,

$$\Delta E_{\mathrm{rep}} = -2n\beta S/(1 - S^2) \tag{3.40}$$

n being the number of electron pairs with identical spins.

Consider, for example, VB structures with three electrons on two centers, $(A: {}^{\bullet}B)$ and $(A^{\bullet} :B)$, each being described as a single AO-based determinant (see Exercise 3.3). The interaction energy that takes place between A and B in each one of these structures by itself (e.g., **11**) is repulsive and following Equation 3.40 will be given by the Pauli repulsion term in Equation 3.41:

$$\Delta E((A: {}^{\bullet}B) \quad \text{and} \quad (A^{\bullet} :B)) = -2\beta S/(1 - S^2) \tag{3.41}$$

For an interacting system that is described by a VB structure involving more than one determinant (see Exercise 3.4), Equation 3.40 can still be applied in an approximate form if squared overlaps are neglected (i.e., $S^2 = 0$):

$$\Delta E_{\mathrm{rep}} \approx -2n\beta S \tag{3.42}$$

Equation 3.42 will be used below in Section 3.5.4.

3.5.3 Mixing of Degenerate Valence Bond Structures

Whenever a wave function is written as a normalized resonance hybrid between two VB structures of equivalent energies, for example, as in Equation 3.43, the energy of the hybrid is given by the normalized self-energies of the constituent resonance structures and the interaction matrix element, H_{12}, between the structures in Equation 3.44.

$$\Psi = N[\Phi_1 + \Phi_2]; \qquad N = 1/[2(1 + S_{12})]^{1/2} \qquad (3.43)$$

$$E(\Psi) = 2N^2 E_{ind} + 2N^2 H_{12}; \qquad H_{12} = \langle \Phi_1 | H | \Phi_2 \rangle, \qquad E_{ind} = E(\Phi_1) = E(\Phi_2) \qquad (3.44)$$

where Φ_1 and Φ_2 are the normalized wave functions for the individual VB structures. Such mixing causes stabilization relative to the energy of each individual (E_{ind}) VB structure, by a quantity called "resonance energy" (RE):

$$RE = [H_{12} - E_{ind}S_{12}]/(1 + S_{12}); \qquad S_{12} = \langle \Phi_1 | \Phi_2 \rangle \qquad (3.45)$$

The resonance energy is nothing else but the difference between the energy of the resonance hybrid and that of a reference state. This definition is general, and the reference state is taken as any one of the two individual VB structures if they are degenerate, and as the lowest of the two if they have different energies. Equation 3.45 expresses the RE for the case where the two limiting structures Φ_1 and Φ_2 have equal or nearly equal energies, which is the most favorable situation for maximum stabilization. However, if the energies E_1 and E_2 are significantly different, then according to the usual rules of perturbation theory, the stabilization will still be finite, albeit smaller than in the degenerate case (see, e.g., the VB interaction diagram in Fig. 2.1).

A typical situation, where the VB wave function is written as a resonance hybrid, is odd-electron bonding (1e or 3e bonds). For example, a 1e bond A•B is a situation where only one electron is shared by two centers A and B (Eq. 3.46), while three electrons are distributed over the two centers in a 3e bond A∴B (Eq. 3.47):

$$A{\bullet}B = A^{+{\bullet}}B \leftrightarrow A^{{\bullet}+}B; \qquad \Psi(A{\bullet}B) = N(|\cdots a| + |\cdots b|) \qquad (3.46)$$

$$A\therefore B = A^{\bullet}{:}B \leftrightarrow A{:}\,{}^{\bullet}B \qquad \Psi(A\therefore B) = N'(|\cdots a\bar{a}b| + |\cdots a\bar{b}b|) \qquad (3.47)$$

Simple algebra (see Exercise 3.3) shows that in both cases, the overlap between the two interacting VB structures is equal to S (the $\langle a|b \rangle$ orbital overlap)[‡] and that resonance energy follows Equation 3.48:

$$RE = \beta/(1 + S) = D_e(A^{+{\bullet}}B \leftrightarrow A^{{\bullet}+}B) \qquad (3.48)$$

[‡]Writing Φ_1 and Φ_2 so that their positive combination is the resonance-stabilized one. For the 3e case, this implies that the two determinants are written in such a way that they exhibit maximum orbital and spin correspondence, as in Equation 3.47. See also, Appendix 3.4.2.

Equation 3.48 also gives the bonding energy of a 1e bond. Combining Equations. 3.41 and 3.48, we get the bonding energy of the 3e bond, Equation 3.49:

$$D_e(A^\bullet :B \leftrightarrow A: {}^\bullet B) = -2\beta S/(1 - S^2) + \beta/(1 + S) = \beta(1 - 3S)/(1 - S^2) \quad (3.49)$$

These equations for odd-electron bonding energies are good for cases where the forms are degenerate or nearly so. In cases where the two structures are not identical in energy, one should use the perturbation theoretic expression (3).

For more complex situations, general guidelines for derivation of matrix elements between polyelectronic determinants are given in Appendices 3.A.1 and 3.A.2. Alternatively, one could follow the protocol given in the original literature (3,4).

3.5.4 Nonbonding Interactions in Valence Bond Theory

Some situations are encountered where one orbital bears an unpaired electron in the vicinity of a bond, like **12** in Scheme 3.5.

A↑ B•——•C   A•——•B   C•——•D

12   **13**

Scheme 3.5

Since the $A^\bullet$ B•——•C structure displays a singlet coupling between orbitals b and c, Equation 3.50 gives its wave function:

$$A^\bullet \text{ B•——•C} = N(|ab\bar{c}| - |a\bar{b}c|) \quad (3.50)$$

in which it is apparent that the first determinant involves a triplet repulsion (between the electrons in a and b) while the second one is a spin-alternant determinant. The energy of this state, relative to a situation where A and BC are separated, can be estimated by means of Equation 3.42, leading to Equation 3.51:

$$E(A^\bullet \text{ B•——•C}) - [E(A^\bullet) + E(\text{B•——•C})] \approx -\beta S \quad (3.51)$$

which means that bringing an unpaired electron in the vicinity of a covalent bond results in one-half of the full triplet repulsion (for the calculation of the exact nonbonding interaction energy between $A^\bullet$ and B•——•C, see Exercise 3.4). This property will be used below when we discuss VB correlation diagrams for radical reactions. The repulsion is the same if we bring two covalent bonds, A•——•B and C•——•D, close to each other, as in **13**:

$$E(\text{A•——•B}\dots\text{C•——•D}) - [E(\text{A•——•B}) + E(\text{C•——•D})] \approx -\beta S \quad (3.52)$$

Equation 3.52 can be used to calculate the total π energy of one canonical structure of a polyene, for example, **14** of butadiene (Scheme 3.6).

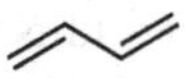

14

Scheme 3.6

Since there are two covalent bonds (each accounting for $2\beta S$) and one nonbonded repulsive interaction $(-\beta S)$ in this VB structure, its energy simply expresses a balance between the two corresponding energy quantities, namely:

$$E(\mathbf{14}) \approx 4\beta S - \beta S = 3\beta S \qquad (3.53)$$

As an application, let us compare the energies of two isomers of hexatrienes. The linear *s-trans* conformation can be described as a resonance between the canonical structure **15** and "long-bond" structures **16–18** (Scheme 3.7), where one short bond is replaced by a long one. On the other hand, the branched isomer is made of only structures **19–21**, since it lacks an analogous structure to **18**.

It is apparent that the canonical structures **15** and **19** have the same energies (three bonds, two nonbonded repulsions in both cases), and that structures **16–18**, **20**, and **21** are also degenerate (two bonds, three nonbonded repulsions). Furthermore, if one omits structure **18**, the matrix elements between the remaining long-bond structures and the canonical ones are all the same (see Appendix 3.A.2). Thus, elimination of structure **18** will make the two isomers isoenergetic. If, however, we take structure **18** into account, it will mix and increase, however slightly, the RE of the linear polyene that becomes thermodynamically more stable than the branched one. This subtle prediction, which is in agreement with experiment, can also be demonstrated in the framework of the Heisenberg Hamiltonian (see later).

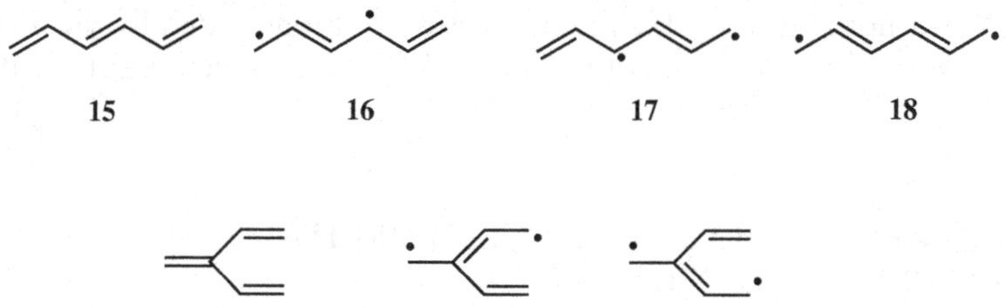

15   **16**   **17**   **18**

19   **20**   **21**

Scheme 3.7

3.6   STRUCTURAL COEFFICIENTS AND WEIGHTS
OF VALENCE BOND WAVE FUNCTIONS

Once a wave function Ψ is available and is written as a linear combination of VB structures Φ_K, as in Equation 3.54, the major VB structures can be distinguished from the minor ones by consideration of the magnitudes of their respective coefficients.

$$\Psi = \sum_K C_K \Phi_K \qquad (3.54)$$

More generally, one has to consider the "weights" of VB structures, which are quantitatively related to physical properties like electron densities, net charges, and so on. According to the popular Chirgwin–Coulson formula (5), the weight of a given structure, Φ_K, is defined as the square of the coefficient plus one-half of the overlap population terms with all the other structures:

$$W_K = \sum_L C_K C_L \langle \Phi_K | \Phi_L \rangle \qquad (3.55)$$

This formula is the VB analogue of the Mulliken population in MO-based calculations. The VB weights sum to unity if the wave function Ψ, in Equation 3.54, is normalized. However, Equation 3.55 can be used even if the Φ_K VB structures are not normalized, or even if Φ_K represents an AO-based determinant rather than a VB structure. In such a case, it is useful to note that with the definition of the weights as in the Chirgwin–Coulson formula, the weight of a VB structure is equal to the sum of the weights of its constituting VB determinants.

Other definitions have also been proposed, such as the Löwdin weights (6), Equation 3.56, and the inverse weights (7), Equation 3.57.

$$W_K^{\text{Lowdin}} = \sum_{I,J} S_{KI}^{1/2} C_I S_{KJ}^{1/2} C_J \qquad (3.56)$$

$$W_K^{\text{INV}} = N C_K^2 / (S^{-1})_{KK}; \qquad N = 1 / \sum_K C_K^2 / (S^{-1})_{KK} \qquad (3.57)$$

where N is a normalization factor. All these definitions generally yield results that are consistent with one another, as can be seen in some outputs displayed in Chapter 2.

3.7   BRIDGES BETWEEN MOLECULAR ORBITAL
AND VALENCE BOND THEORIES

After reviewing basic elements of VB theory, we would like to create bridges between the popular and widely used MO theory and the less familiar VB theory. The goal here is not to demonstrate that one theory is "better" than the other, but actually to show that by borrowing insights from MO theory, VB theory itself becomes easier to handle, more predictive and more widely applicable to chemical problems.

3.7.1   Comparison of Qualitative Valence Bond and Molecular Orbital Theories

Some (not all) of the elementary interaction energies that are discussed above have also qualitative MO expressions, which in some cases match the VB expressions. In qualitative MO theory, the interaction between two overlapping AOs leads to a pair of bonding and antibonding MOs, the former being stabilized by a quantity $\beta/(1 + S)$ and the latter destabilized by $-\beta/(1 - S)$ relative to the nonbonding level. The stabilization–destabilization of the interacting system relative to the separate fragments are then calculated by summing up the occupancy-weighted energies of the MOs. A comparison of the qualitative VB and MO approaches is given in Table 3.1, where the energetics of the elementary interactions are expressed with both methods. It is apparent that both qualitative theories give identical expressions for the odd-electron bonds, the 4e repulsion, and the triplet repulsion. This is not surprising if one notes that the MO and VB wave functions for these four types of interaction are identical (see Exercise 3.5 and Section 3.7.3 about the relationship between MO and VB wave functions). On the other hand, the expressions for the MO and VB 2e bonding energies are different; the difference is related to the fact that MO and VB wave functions are themselves different in this case (see next Section 3.7.2). There follows a rule that may be useful if one is more familiar with MO theory than with VB.

Whenever the VB and MO wave functions of an electronic state are equivalent, the VB energy can be estimated using qualitative MO theory.

TABLE 3.1   Elementary Interaction Energies in the Qualitative MO and VB Models

Type of Interaction	Stabilization or Destabilization (MO Model)	Stabilization or Destabilization (VB Model)
1-electron	$\beta/(1 + S)$	$\beta/(1 + S)$
2-electron	$2\beta/(1 + S)$	$2\beta S/(1 + S^2)$
3-electron	$\beta\,(1 - 3S)/(1 - S^2)$	$\beta(1 - 3S)/(1 - S^2)$
4-electron	$-4\beta S/(1 - S^2)$	$-4\beta S/(1 - S^2)$
triplet repulsion	$-2\beta S/(1 - S^2)$	$-2\beta S/(1 - S^2)$
3-electron repulsion		$-2\beta S/(1 - S^2)$

3.7.2   The Relationship between Molecular Orbital and Valence Bond Wave Functions

What is the difference between the MO and VB descriptions of an electronic system, at the simplest level of both theories? As we will see, in the cases of 1e, 3e, and 4e interactions between two centers, there is no difference between the two theories, except for the representations that look different. On the other hand, the two theories differ in their description of the 2e bond. Once again let us take the example of H_2, with its two AOs a and b, and examine the VB description first, dropping normalization factors for simplicity.

As has been said already, at the equilibrium distance the bonding is not 100% covalent, and it requires some ionic component to be described accurately. On the other hand, at long distances the HL wave function is the correct state, as the ionic components necessarily drop to zero and each hydrogen atom carries one electron away through homolytic bond breaking. The HL wave function *dissociates correctly*, but is quantitatively inaccurate at bonding distances. Therefore, the way to improve the HL description is straightforward: by simply mixing Φ_{HL} with the ionic determinants and optimizing the coefficients variationally, by CI. One then gets the wave function $\Psi_{VB\text{-full}}$, in Equation 3.4, which contains a major covalent component and a minor ionic one.

Now let us turn to the MO description. Bringing together two hydrogen atoms leads to the formation of two MOs, σ and σ^*, respectively bonding and antibonding (Eq. 3.58, dropping normalization constants).

$$\sigma = a + b; \qquad \sigma^* = a - b \qquad (3.58)$$

At the simple MO level, the ground state of H_2 is described by Ψ_{MO}, in which the bonding σ MO is doubly occupied. Expansion (see Chapter 4 for a general method in the polyelectronic case) of this MO determinant into its AO determinant constituents leads to Equation 3.59 (again dropping normalization constants):

$$\Psi_{MO} = |\sigma\,\bar{\sigma}| = (|a\,\bar{b}| - |\bar{a}\,b|) + (|a\,\bar{a}| + |b\,\bar{b}|) \qquad (3.59)$$

From Eq. 3.59, it is apparent that the first one-half of the expansion is nothing else but the Heitler–London function Φ_{HL} (Eq. 3.2), while the remaining part is ionic. It follows that the MO description of the homonuclear 2e bond will always be half-covalent and half-ionic, irrespective of the bonding distance. Qualitatively, it is already clear that in the MO wave function, the ionic weight is excessive at bonding distances, and becomes absurd at long distances, where the weight of the ionic structures should drop to zero to accord with the homolytic cleavage. The simple MO description *does not dissociate correctly*. This is the reason why it is inappropriate for the description of stretched bonds, as, for example, those found in transition states. The remedy for this poor description is CI, specifically the mixing of the ground configuration, σ^2, with the diexcited one, σ^{*2}. The reason why this mixing re-sizes the covalent *vs*. ionic weights is the following: If one expands the diexcited configuration, Ψ_D, into its VB constituents, one finds the same covalent and ionic components as in Equation 3.59, but coupled with a negative sign as in Equation 3.60:

$$\Psi_D = |\sigma^*\overline{\sigma^*}| = -\left(|a\,\bar b| - |\bar a\,b|\right) + \left(|a\,\bar a| + |b\,\bar b|\right) \tag{3.60}$$

It follows that mixing the two configurations Ψ_{MO} and Ψ_D with different coefficients, as in Equation 3.61, will lead to a wave function Ψ_{MO-CI} in which the covalent and ionic components

$$\Psi_{MO-CI} = c_1|\sigma\,\bar\sigma| - c_2|\sigma^*\overline{\sigma^*}| \qquad c_1, c_2 > 0 \tag{3.61}$$

have unequal weights, as shown by an expansion of Ψ_{MO-CI} into AO determinants in Equation 3.62:

$$\Psi_{MO-CI} = (c_1 + c_2)\left(|a\,\bar b| - |\bar a\,b|\right) + (c_1 - c_2)\left(|a\,\bar a| + |b\,\bar b|\right) \tag{3.62a}$$

$$c_1 + c_2 = \lambda; \qquad c_1 - c_2 = \mu \tag{3.62b}$$

Since c_1 and c_2 are variationally optimized, expansion of Ψ_{MO-CI} should lead to exactly the same VB function as $\Psi_{VB\text{-full}}$ in Equation 3.4, leading to the equalities expressed in Equation 3.62 and to the equivalence of Ψ_{MO-CI} and $\Psi_{VB\text{-full}}$ (see Exercise 3.1) The equivalence also includes the Coulson–Fischer wave function Ψ_{CF} (Eq. 3.5) which, as we have seen, is equivalent to the VB-full description (see Exercise 3.2).

$$\Psi_{MO} \neq \Psi_{VB}; \qquad \Psi_{MO-CI} \equiv \Psi_{VB-full} \equiv \Psi_{CF} \tag{3.63}$$

To summarize, the simple MO level describes the bond as being too ionic, while the simple VB level (Heitler–London) defines it as being purely covalent. Both theories converge to the right description when CI is introduced.

The accurate description of 2e bonding is half-way in between the simple MO and simple HLVB levels; elaborated MO and VB levels become equivalent and converge to the right description, in which the bond is mostly covalent, but has a substantial contribution from ionic structures.

This equivalence clearly projects that the MO–VB rivalry, discussed in Chapter 1, is unfortunate and senseless. Both VB and MO theories are not so diametrically different that they exclude each other, but rather two representations of reality, which are mathematically equivalent. The best approach is to use these two representations jointly and benefit from their complementary insight. In fact, from the above discussion of how to write a VB wave function, it is apparent that there is a spectrum of orbital representations that stretches between the fully local VB representations through semilocalized CF orbitals, to the use of delocalized fragment orbitals VB (FO–VB), and all the way to the fully delocalized MO representation (in the MO−CI language). Based on the problem at hand, the choice representation from this spectrum of possibilities should be the one that gives the clearest and most portable insight into the problem.

Up to this point, we restricted ourselves to the simple case of determinants involving no more than two orbitals. However, the MO−VB correspondence is general, and in fact, any MO or MO−CI wave function can be exactly transformed into a VB wave function, provided it is a spin-eigenfunction (i.e., not a spin-unrestricted wave function). While this is a trivial matter for small determinants, larger ones require a bit of algebra and a systematic method is discussed in Chapter 4 for the interested or advanced reader.

3.7.3   Localized Bond Orbitals: A Pictorial Bridge between Molecular Orbital and Valence Bond Wave Functions

The standard MO wave function involves canonical MOs (CMOs), which are permitted to delocalize over the entire molecule. However, it is well known (8,9) that an MO wave function based on CMOs can be transformed to another MO wave function that is based on localized MOs (LMOs), known also as localized bond orbitals (LBOs) (10). This transformation is called unitary transformation, and as such, it changes the representation of the orbitals without affecting the total energy or the total MO wave function. This equivalence is expressed in Equation 3.64:

$$\left| \ldots \varphi_i^{cmo} \ldots \varphi_j^{cmo} \ldots \right| = \left| \ldots \varphi_i^{lbo} \ldots \varphi_j^{lbo} \ldots \right| \tag{3.64}$$

where φ_i^{cmo} corresponds to a CMO while φ_i^{lbo} is an LBO.

A unitary transformation involves simple subtractions and additions of orbitals within the complete set of the occupied CMOs. To illustrate such a transformation, we choose a simple molecule, BeH_2, for which the procedure may be done in a pictorial manner without resort to equations. Figure 3.4 shows the valence occupied CMOs of BeH_2, the lowest of the two is made from the bonding combination of the 2s AO of Be and the positive combination of the 1s AOs of the two hydrogen atoms, while the higher one is the bonding orbital between the $2p_z$(Be) orbital and the negative combination of the 1s(H) AOs. We can now make two linear combinations of these orbitals, one negative and one positive, as in Equation 3.65, dropping normalization constants:

$$\sigma_R = \varphi_1^{cmo} + \varphi_2^{cmo}; \qquad \sigma_L = \varphi_1^{cmo} - \varphi_2^{cmo} \tag{3.65}$$

These linear combinations, shown on the right-hand side of Figure 3.4, are seen to generate two LBOs made from sp hybrids on the Be and the 1s AOs of the hydrogens. One of these LBOs, σ_R, is a two-center bonding orbital localized on the right-hand side of the molecule, while the other, σ_L, is equivalent to the former, but localized on the left-hand side. Of course, since the coefficients of the hydrogens in φ_1^{cmo} and φ_2^{cmo} are not exactly equal in absolute value, the localization is not perfect, and each LBO contains a small component out of the bonding region, called "delocalization tail", which is, however, very small. The wave function based on these localized orbitals possesses two doubly occupied LBOs and is completely equivalent to the starting wave function based on CMOs, as expressed in Equation 3.66:

$$\Psi(\text{BeH}_2) = \left|\varphi_1^{cmo}\overline{\varphi_1^{cmo}}\,\varphi_2^{cmo}\overline{\varphi_2^{cmo}}\right| = \left|\sigma_R\overline{\sigma_R}\,\sigma_L\overline{\sigma_L}\right| \qquad (3.66)$$

This LBO-based wave function is not a VB wave function. Nevertheless, it represents a Lewis structure, and hence also a pictorial analogue of a perfect-pairing VB wave function. The difference between the LBO and VB wave functions is that the latter involves electron correlation while the former does not. As such, in a perfectly paired VB wave function, based on CF orbitals, each localized Be—H bond would involve an optimized covalent–ionic combination as we demonstrated above for H_2 and generalized for other 2e bonds. In contrast, the LBOs in Equation 3.65 possess some constrained combination of these components, with exaggeration of the bond ionicity.

Of course, the LBO wave function in Equation 3.65 can be upgraded to a proper VB wave function quite easily, by first localizing the vacant orbitals of BeH_2, in much the same way as we just did for the occupied ones, and as illustrated in the upper right-hand side of Figure 3.4. By using these vacant

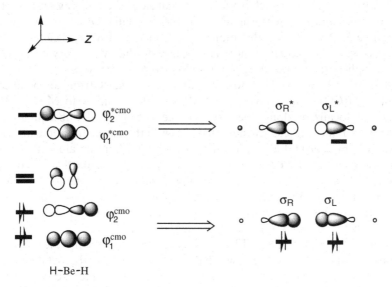

FIGURE 3.4 Transformation of the valence orbitals of BeH_2, from canonical MOs (left-hand side) to localized bond orbitals (right-hand side). This transformation leaves the polyelectronic Hartree–Fock function unchanged.

σ^*-LBOs, we can improve the LBO wave function in Equation 3.66 by CI, in the same manner as discussed above for H_2. Now the resulting wave function will be equivalent to a VB wave function involving two localized bond pairs with CF orbitals, and would correspond to a perfect pairing GVB wave function for the molecule. Thus, the LBO wave function can be considered qualitatively as a crude VB wave function, just one step before the improvement of the covalent–ionic components of each bond. As such, we will occasionally be using LBO-based wave function in our various applications of VB theory to chemical reactivity and as an entry to bonding in organometallic complexes. Some relevant exercises are given in the end of the chapter.

For molecules involving many bonds, the localizing unitary transformations are more complicated than in the BeH_2 case, and are usually done by means of a computer program. This program is available in all current *ab initio* codes. As there are an infinite number of unitary transformations of orbitals that leave the Slater determinant unchanged, the localizing transformations are determined so as to best satisfy some specific criteria, for example, by requiring that the total spread of the localized orbitals be minimal, as in the Foster–Boys method (9). On the other hand, it is impossible to find a set of well-localized orbitals for molecules whose electronic system is intrinsically delocalized, like benzene or, to a much lesser extent, butadiene (see Exercise 3.8).

As an example of using the LMO and VB–FO concepts to gain insight into bonding in a complex molecule, we selected the organometallic compound, $Fe(CO)_4[\eta^2-C_2H_4]$ in which we intend to consider the bonding between iron and ethylene, and the stereochemistry of the molecule. Of course, the use of qualitative MO theory for this molecule would have been sufficiently simple and successful. The intention here is to illustrate that VB theory can become widely applicable by importing key insights from MO theory.

Elian and Hoffman (11) showed that one can start from an octahedral complex, $M(CO)_6$, and convert the CMOs of the complex to M-CO LBOs, which are localized along the axes of the octahedron. Subsequently, by successively removing CO ligands, they show that each ligand removal leaves behind a hybrid orbital localized on the metal and pointing along the axis of the missing site of the octahedron. The two hybrids are part of the d-block orbitals, which now has three low lying orbitals from the t_{2g} set in the octahedron, and two hybrid orbitals that replace the e_g set of the octahedron. Subsequently, this has formed the basis for the now well known "isolobal analogy" between organometallic and organic fragments (12).

Following the Hoffmann–Elian strategy, the $Fe(CO)_4$ fragment has two hybrids (h_1 and h_2) pointing toward the missing axes of the octahedron, as shown in Fig. 3.5a. Since Fe in oxidation state zero has eight valence electrons, the d-block orbitals will have a filled "t_{2g}" set, and singly occupied hybrids (h_1 and h_2) pointing toward the missing sites of the octahedron. Now we can bring ethylene and try to bind it with $Fe(CO)_4$. One way to do that is to use the two localized hybrids on $Fe(CO)_4$, to uncouple the electrons of the π-bond in ethylene, and form two new bonds using covalent and ionic structures, as we did in the chapter.

Since we are interested in building bridges to MO theory, we are going to use FOs, and exploit their symmetry in order to generate the VB wave function. This is done in Fig. 3.5b, where the "t_{2g}" set is omitted for clarity; first we form two linear combinations from the two localized hybrids (exactly the opposite procedure of the localization in the preceding exercises), one symmetric and one antisymmetric with respect to the plane of symmetry that includes the (CO)—Fe—(CO) axis. As amply discussed (11,12), the antisymmetric combination is dominated by the 3d orbital of iron, while the symmetric combination has a large component of 4s and 4p, Therefore, the latter orbital is higher in energy than the former. Each of these new orbitals has a single electron, capable of making two bonds with the π-electrons on ethylene. As shown in the scheme, a perfect pairing bond diagram between the two fragments requires uncoupling of the π-electrons of ethylene. Thus, in order to form the maximum number of bonds, the ethylene molecule must be promoted to a triplet $\pi\pi^*$ state, so that the electrons in the symmetry-matched orbitals can form bond pairs between the two fragments. The resulting bonding scheme describes a metallacyclopropane–iron tetracarbonyl complex. It is further seen that in order to maintain two bond pairs, the ethylene must occupy the equatorial plane of the molecule. Rotation of the ethene to the axial plane will break one bond pair (the one between π^* and $h_1 - h_2$) and will encounter a significant rotational barrier (the experimental value (13) is $18 - 25$ kcal/mol). Instant recognition of stereochemistry is one of the advantages of using FO–VB representation over AO–VB. Later we will see other advantages of the FO–VB representation.

As discussed above, the bond diagram represents the HL-type coupling between the FOs, as well as the ionic structures that can be generated from them. This is done by simply shifting electrons between the orbitals that form the bond pair in the fundamental perfect-pairing diagram (see Fig. 3.5b). Some of the so generated structures are shown in Fig. 3.5c. One can see two ionic structures (Φ_{ion}) that are generated by transferring one electron either from the ethylene to the $Fe(CO)_4$ fragment or vice versa. The third structure in Fig. 3.5c is generated from the fundamental one by transferring two electrons, but this generates a no-bond wave function (Φ_{nb}), which by itself is nonbonded.

The wave function of the complex will be a linear combination of the four structures in Figs. 3.5b and c. In a series of olefins, we may expect to see a spectrum of cases. For example, in a series of olefins where the singlet-to-triplet $\pi\pi^*$ excitation is gradually lowered we may see an increasing metallacyclic character up to complexes, where the C—C distance is that of a single bond. With olefins that are good electron donors, we may see a wave function dominated by a mixture of $\Phi_{ion(1)}$ and Φ_{nb}, while for powerful electron acceptors, we may expect a wave function dominated by $\Phi_{ion(2)}$ and Φ_{nb}.

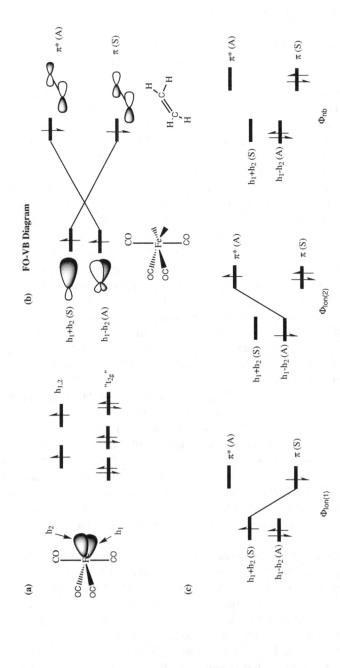

FIGURE 3.5 The FO−VB representation of the bonding between $Fe(CO)_4$ and ethylene: (a) The localized hybrids of $Fe(CO)_4$, (b) the FO−VB bond diagram that describes perfect pairing between two fragments, (c) different VB−FO contributions to the bonding due to charge transfer between the two fragments.

APPENDIX

3.A.1   NORMALIZATION CONSTANTS, ENERGIES, OVERLAPS, AND MATRIX ELEMENTS OF VALENCE BOND WAVE FUNCTIONS

This appendix describes a scheme for enumerating VB terms in a didactic manner (3,4). The Hamiltonian that is used for calculating energies and matrix elements is the effective polyelectronic Hamiltonian H, which is expressed as a sum of monoelectronic Hamiltonians $h(i)$, one per electron:

$$H = h(1) + h(2) + h(3) + h(4) + \cdots \qquad (3.A.1)$$

This scheme involves no approximations and can be used to obtain all the contributing terms, if one so wishes. Let us exemplify the procedure with the following VB function, which involves a unique determinant, preceded by a normalization constant N.

$$\Omega = N|a\,\bar{a}b\bar{b}| \qquad (3.A.2)$$

The corresponding diagonal spin–orbital product is,

$$\psi_\mathrm{d} = a(1)\bar{a}(2)b(3)\bar{b}(4); \qquad (1, 2, \cdots \text{ are electron indexes}) \qquad (3.A.3)$$

To begin with, there are a total of 24 permutations on this diagonal element. These can be minimized by eliminating all the permutations that transpose spin–orbitals of a different spin variety (α and β), because these permutations have zero contributions to all the title quantities. We can therefore group the spin–orbitals of the determinant into two subsets and define elementary permutations that act exclusively within the subsets. From there, we can build up more complex permutations, until all the contributing permutations are included.

The elementary permutations are the identity that leaves unchanged the ψ_d of any determinant, and the permutations that cause a single pairwise transposition in the order of the spin–orbitals. The sign of the elementary permutations is given by $(-1)^t$, where t is the number of pairwise transpositions. For the diagonal element in Equation 3.A.3 the elementary permutations are,

$$(-1)^t P_\mathrm{elem} = P_0, -P_{ab}, -P_{\bar{a}\bar{b}}; \qquad (P_0 = \text{identity}) \qquad (3.A.4)$$

where the subscript of the permutation defines its applied transposition.

Successive applications of the elementary permutations are used to construct more complex permutations. In this manner, we generate composite permutations that perform two, three, or more, pairwise transpositions within each spin–orbital subset, or composite permutations which perform the transpositions simultaneously on the two spin–orbital subsets. For our example, the only composite permutation is,

$$(-1)^t P_{\text{comp}} = (-P_{ab}), (-P_{\bar{a}\bar{b}}) = P_{ab}P_{\bar{a}\bar{b}} \qquad (3.A.5)$$

The total number of permutations that are selected by this process are then,

$$(-1)^t P_i = P_0, -P_{ab}, -P_{\bar{a}\bar{b}}, P_{ab}P_{\bar{a}\bar{b}} \qquad (3.A.6)$$

Thus we have minimized the number of permutations in our example from 24 to 4. For a 6×6 determinant, this selection process leaves 36 of the 720 possible permutations.

3.A.1.1   Energy and Self-Overlap of an Atomic Orbital-Based Determinants

Having the permutations, we now set a table including them (Table 3.A.1). The first column of the table lists all the permutations with their signs. The title line in the second column of the table is the diagonal spin–orbital product of Equation 3.A.3, and lined below it are all the permuted products that result after the permutations that are indicated in the first column, in each line. In the third column we list the contributions of each permutation to the energy of the determinant.

Table 3.A.1

$(-1)^t P_i$	$a\ \bar{a}\ b\ \bar{b}$	Energy Terms
(1) P_0	$a\ \bar{a}\ b\ \bar{b}$	$2\varepsilon_a + 2\varepsilon_b$
(2) $-P_{ab}$	$-b\ \bar{a}\ a\ \bar{b}$	$-\varepsilon_a S_{ab}^2 - \varepsilon_b S_{ab}^2 - 2h_{ab}S_{ab}$
(3) $-P_{\bar{a}\bar{b}}$	$-a\ \bar{b}\ b\ \bar{a}$	$-\varepsilon_a S_{ab}^2 - \varepsilon_b S_{ab}^2 - 2h_{ab}S_{ab}$
(4) $P_{ab}P_{\bar{a}\bar{b}}$	$b\ \bar{b}\ a\ \bar{a}$	$+4h_{ab}S_{ab}^3$

The third column of the table includes energy terms, which correspond to the following integral,

$$\langle \psi_{\text{d}} | h(1) + h(2) + h(3) + h(4) | (-1)^t P_i \psi_{\text{d}} \rangle \qquad (3.A.7)$$

Accordingly, the total energy contribution, in each of the table lines, must include a total of four terms. The terms can be deduced by a digit-to-digit inspection of any of the spin–orbital products against the title product of the table.

Consider, for example, the permuted orbital product of the second line of the table, $-\bar{b}\bar{a}a\bar{b}$, and let us integrate the monoelectronic Hamiltonian $h(1)$. This Hamiltonian applies to the first orbital of the orbital products: a for the diagonal product, and b for the permuted product. This yields the term h_{ab}. This term is multiplied by a product of overlaps between the remaining orbitals:

$$\langle a\bar{a}b\bar{b} | h(1) | -\bar{b}\bar{a}a\bar{b} \rangle = -\langle a|h|b\rangle \langle \bar{a}|\bar{a}\rangle \langle b|a\rangle \langle \bar{b}|\bar{b}\rangle = -h_{ab}S_{ab} \qquad (3.A.8)$$

Now, let us integrate the monoelectronic Hamiltonian $h(2)$, which applies to the second orbital of the products. This time, the orbital is the same, a, for the diagonal and permuted products. This yields a matrix element of diagonal type, h_{aa}, which is interpreted as the energy ε_a of the spin–orbital a. Once again, the other orbitals contribute to overlap terms:

$$\langle a\bar{a}b\bar{b}|h(2)| - b\bar{a}a\bar{b}\rangle = -\langle a|b\rangle\langle \bar{a}|h|\bar{a}\rangle\langle b|a\rangle\langle \bar{b}|\bar{b}\rangle = -\varepsilon_a S_{ab}^2 \qquad (3.A.9)$$

Then, $h(3)$ and $h(4)$ are integrated in the same way. This yields the final energy term in the second line of Table 3.A.1:

$$\langle a\bar{a}b\bar{b}|h(1) + h(2) + h(3) + h(4)| - b\bar{a}a\bar{b}\rangle = -2h_{ab}S_{ab} - (\varepsilon_a + \varepsilon_b)S_{ab}^2 \quad (3.A.10)$$

The same calculations can be repeated for the third line of the table and yield identical results. The energy terms corresponding to the two remaining permutations are calculated the same way. In the first row of the table, for example, the permutation is identity, so that all the digits are identical to those of the title diagonal product. Therefore, all the energy terms are of the ε type, and all the orbital overlaps are unity, yielding an energy term that is nothing else but a sum of the spin–orbital monoelectronic energies:

$$\langle a\bar{a}b\bar{b}|h(1) + h(2) + h(3) + h(4)|a\bar{a}b\bar{b}\rangle = 2\varepsilon_a + 2\varepsilon_b \qquad (3.A.11)$$

The energy term in the fourth row is determined in the same manner, but now all the orbitals of the permuted product are different from those of the diagonal product. As a consequence, all energy terms are of the h_{ab} type, and all overlaps are different from unity.

$$\langle a\bar{a}b\bar{b}|h(1) + h(2) + h(3) + h(4)|b\bar{b}a\bar{a}\rangle = 4h_{ab}S_{ab}^3 \qquad (3.A.12)$$

The energy terms in the third row are then summed up yielding the energy of the AO-based determinant:

$$E(|a\bar{a}b\bar{b}|) = 2\varepsilon_a + 2\varepsilon_b - 4h_{ab}S_{ab} - 2(\varepsilon_a + \varepsilon_b)S_{ab}^2 + 4h_{ab}S_{ab}^3 \qquad (3.A.13)$$

Of course, this determinant is not normalized, as its self-overlap is different from unity. This self-overlap is calculated from the same formulas as for integrating a monoelectronic Hamiltonian, for example, $h(1)$, by replacing the h_{ab} terms by orbital overlaps S_{ab}, and the ε terms by unity. Equivalently, one may take the formula that gives the energy of the determinant (Eq. 3.A.13), replace h_{ab} by S_{ab} and ε by 1, and divide by the number of electrons. This yields

$$\langle |a\,\bar{a}\,b\,\bar{b}|\,||a\,\bar{a}\,b\,\bar{b}|\rangle = (4 - 4S_{ab}^2 - 4S_{ab}^2 + 4S_{ab}^4)/4 = 1 - 2S_{ab}^2 + S_{ab}^4 \qquad (3.A.14)$$

From this self-overlap, the square of the normalization factor of Ω (Eq. 3.A.2) is readily calculated

$$N^2 = 1/[1 - 2S_{ab}^2 + S_{ab}^4] \tag{3.A.15}$$

and the energy of Ω is the energy of the determinant $|a\,\bar{a}\,b\,\bar{b}|$ multiplied by the N^2 term:

$$E(\Omega) = N^2 E(|a\,\bar{a}\,b\,\bar{b}|) \tag{3.A.16}$$

It must be emphasized that the energy terms due to the bielectronic part of the exact Hamiltonian can be enumerated by use of the same table.

3.A.1.2 Hamiltonian Matrix Elements and Overlaps between Atomic Orbital-Based Determinants

Matrix elements between two different determinants, for example, $|a\,\bar{a}\,b\,\bar{b}|$ and $|c\,\bar{c}\,d\,\bar{d}|$, follow from the equation,

$$\langle|a\,\bar{a}bb||H||c\,\bar{c}d\bar{d}|\rangle = \langle\psi_d|H|(-1)^t P_i\psi'_d\rangle \tag{3.A.17}$$

where ψ'_d is the diagonal product of the determinant $|c\,\bar{c}\,d\,\bar{d}|$. The rules for calculating an off-diagonal matrix element are the same as those for calculating the self-energy of a determinant. The energy terms are collected in Table 3.A.2, for the general case when all the orbitals of the second determinant are different from those of the first one.

Table 3.A.2

$(-1)^t P_i$	a	$\bar{a}$	b	$\bar{b}$	Energy Terms
(1) P_0	c	$\bar{c}$	d	$\bar{d}$	$2h_{ac}S_{ac}S_{bd}^2 + 2h_{bd}S_{bd}S_{ac}^2$
(2) $-P_{cd}$	$-d$	$\bar{c}$	c	$\bar{d}$	$-h_{ad}S_{ac}S_{bc}S_{bd} - h_{ac}S_{ad}S_{bc}S_{bd} - h_{bc}S_{ad}S_{ac}S_{bd} - h_{bd}S_{ad}S_{ac}S_{bc}$
(3) $-P_{\bar{c}\bar{d}}$	$-c$	$\bar{d}$	d	$\bar{c}$	$-h_{ac}S_{ad}S_{bd}S_{bc} - h_{ad}S_{ac}S_{bd}S_{bc} - h_{bd}S_{ac}S_{ad}S_{bc} - h_{bc}S_{ac}S_{ad}S_{bd}$
(4) $P_{cd}P_{\bar{c}\bar{d}}$	d	$\bar{d}$	c	$\bar{c}$	$2h_{ad}S_{ad}S_{bc}^2 + 2h_{bc}S_{bc}S_{ad}^2$

The Hamiltonian matrix element between the two determinants is then calculated by summing up the energy terms of the third column of Table 3.A.2.

As before, the overlap between the two determinants can be calculated from their Hamiltonian matrix element, by replacing the h terms by S terms and dividing the result by the number of electrons:

$$\langle|a\,\bar{a}\,b\,\bar{b}||c\,\bar{c}\,d\,\bar{d}|\rangle = S_{ac}^2 S_{bd}^2 - 2S_{ad}S_{ac}S_{bc}S_{bd} + S_{ad}^2 S_{bc}^2 \tag{3.A.18}$$

3.A.2 SIMPLE GUIDELINES FOR VALENCE BOND MIXING

Derivation of matrix elements between polyelectronic VB determinants follows from the discussion in the text and the preceding appendix. This can be done by enumerating all the permutations of the respective diagonal terms, as in Equation 3.A.19. Subsequently, one must define the reduced matrix element in Equation 3.A.20.

$$\langle \Omega | H^{\text{eff}} | \Omega' \rangle = \langle \psi_d | \Sigma h(i) | \Sigma(-1)^t P(\psi'_d) \rangle \tag{3.A.19}$$

$$\langle \Omega | H^{\text{eff}} | \Omega' \rangle_{\text{reduced}} = \langle \Omega | H^{\text{eff}} | \Omega' \rangle - 0.5(E(\Omega) + E(\Omega'))\langle \Omega | \Omega' \rangle \tag{3.A.20}$$

As just seen, the retention of overlap leads to many energy and overlap terms that need to be collected and organized, making this procedure quite tedious. A practice that we found useful is to focus on the leading term of the matrix element and use reduced matrix elements, labeled hereafter as β. In this respect, we show a few qualitative guidelines that were derived in detail in the original paper (3) and discussed elsewhere (4). Initially, one has to arrange the two VB determinants with maximum correspondence of their spin–orbitals. Then, one must find out the number of spin–orbitals that are different in the two determinants, and apply the following rules:

1. The first and foremost rule is that the entire matrix element between two VB determinants is signed as the corresponding determinant overlap and has the same power in AO overlap. For example, the overlap between the two determinants of a HL bond, $|a\bar{b}|$ and $|\bar{a}b|$ is $-S_{ab}^2$. Hence, the matrix element is negatively signed and given as $-2\beta_{ab}S_{ab}$; since β_{ab} is proportional to S_{ab}, both the matrix element and the determinant-overlap involve AO overlap to the power of 2. For the one-electron bond case (Eq. 3.46), the overlap between the determinants is $+S_{ab}$ and the matrix element $+\beta_{ab}$.

2. When the VB determinants differ by the occupancy of one spin–orbital, say orbital a in one determinant is replaced by b in the other (keeping the ordering of the other orbitals unchanged), the leading term of the matrix element will be proportional to β_{ab}. Both the 1e and 3e bonds are cases that differ by a single electron occupancy and the corresponding matrix elements are indeed $\pm\beta$, with a sign as the corresponding overlap between the determinants. In the 3e case, the overlap between the determinants exhibiting maximum spin–orbital correspondence, $|a\bar{a}b|$ and $|a\bar{b}b|$, is S_{ab} and the matrix element is $+\beta_{ab}$. If one prefers to consider the determinants written as $|a\bar{a}b|$ and $|b\bar{b}a|$, then the overlap is $-S_{ab}$ and the matrix element is likewise $-\beta_{ab}$. Note that the sign is not important, but the relative signs for two cases are important. It is therefore always advised to use determinants with maximum correspondence, when one wants to deduce trends that depend on the sign of the matrix element (see later in Chapter 5 about aromaticity–antiaromaticity).

3. When the VB determinants differ by the occupancy of two spin orbitals, the leading term of the matrix element will be the sum of the

corresponding $\beta_{ij}S_{ij}$ terms, with the appropriate sign. An example is the matrix element $-2\beta_{ab}S_{ab}$ between the $|a\bar{b}|$ and $|\bar{a}b|$ determinants, which differ by the occupancy of two spin orbitals, a and b.

4. The above considerations are the same whether the spin orbitals are AOs, CF orbitals, or FOs.

REFERENCES

1. W. Heitler, F. London, *Z. Phys.* **44**, 455 (1927). Wechselwirkung neutraler Atome und homöopolare Bindung nach der Quantenmechanik.

2. C. A. Coulson, I. Fischer, *Philos. Mag.* **40**, 386 (1949). Notes on the Molecular Orbital Treatment of the Hydrogen Molecule.

3. S. S. Shaik, in *New Theoretical Concepts for Understanding Organic Reactions*, J. Bertrán, I. G. Csizmadia, Eds., NATO ASI Series, C267, Kluwer Academic Publishers, 1989, pp. 165–217. A Qualitative Valence Bond Model for Organic Reactions.

4. S. S. Shaik, E. Duzy, A. Bartuv, *J. Phys. Chem.* **94**, 6574 (1990). The Quantum Mechanical Resonance Energy of Transition States: An Indicator of Transition State Geometry and Electronic Structure.

5. B. H. Chirgwin, C. A. Coulson, *Proc. R. Soc. Ser. A. (London)* **2**, 196 (1950). The Electronic Structure of Conjugated Systems. VI.

6. P.-O. Löwdin, *Ark. Mat. Astr. Fysik* **A35**, 1 (1947). A Quantum Mechanical Calculation of the Cohesive Energy, the Interionic Distance, and the Elastic Constants of Some Ionic Crystals.

7. G. A. Gallup, J. M. Norbeck, *Chem. Phys. Lett.* **21**, 495 (1973). Population Analyses of Valence-Bond Wave Functions and BeH_2.

8. C. Edmiston, K. Ruedenberg, *Rev. Mod. Phys.* **35**, 457 (1963). Localized Atomic and Molecular Orbitals.

9. S. F. Boys, in *Quantum Theory of Atoms, Molecules, and the Solid State*, P.-O. Löwdin, Ed., Academic Press, New York, 1968, p. 253.

10. E. Honegger, E. Heilbronner, in *Theoretical Models of Chemical Bonding*, Vol. 3, Z. B. Maksic, Ed., Springer Verlag, Berlin- Heidelberg, 1991, pp. 100–151. The Equivalent Bond Orbital Model and the Interpretation of PE Spectra.

11. M. Elian, R. Hoffmann, *Inorg. Chem.* **14**, 1058 (1975). Bonding Capabilities of Transition Metal Carbonyl Fragments.

12. R. Hoffmann, *Angew. Chem. Int. Ed. Engl.* **21**, 711 (1982). Building Bridges Between Inorganic and Organic Chemistry (Nobel Lecture).

13. T. A. Albright, R. Hoffmann, J. C. Thibeault, D. L. Thorn, *J. Am. Chem. Soc.* **101**, 3801 (1979). Ethylene Complexes. Bonding, Rotational Barriers, and Conformational Preferences.

EXERCISES

3.1. The coefficients of the σ and σ^* MOs of H_2, in STO-3G basis set, are given below as functions of the atomic orbitals a and b.

	σ	σ^*
a	0.54884	1.21245
b	0.54884	-1.21245

a. Based on these coefficients, express the normalized expression of the Hartree−Fock configuration $|\sigma\bar{\sigma}|$ in terms of AO-determinants. Do the same for the diexcited configuration $|\sigma^*\bar{\sigma^*}|$.

b. After 2×2 CI in the space of the MO configurations, the wave function $\Psi_{\text{MO−CI}}$ reads

$$\Psi_{\text{MO−CI}} = 0.99365|\sigma\bar{\sigma}| - 0.11254|\sigma^*\bar{\sigma^*}| \qquad (3.\text{Ex}.1)$$

Express $\Psi_{\text{MO−CI}}$ in terms of AO determinants. Show that CI reduces the coefficients of the ionic structures. How do these coefficients compare with those resulting from the VB calculations in Equation 2.2?

c. Assuming that the expression of $\Psi_{\text{MO−CI}}$ in terms of AO determinants is equivalent to $\Psi_{\text{VB-full}}$ in Equation 2.2, calculate the normalization constant N of the HL wave function below:

$$\Phi_{\text{HL}} = N(|a\bar{b}| - |\bar{a}b|) \qquad (3.\text{Ex}.2)$$

3.2. The wave function of H_2 is expressed below as a formally covalent VB structure Ψ_{CF} using Coulson−Fischer (CF) orbitals φ_a and φ_b:

$$\Psi_{\text{CF}} = N(|\varphi_a\bar{\varphi_b}| - |\bar{\varphi_a}\varphi_b|) \qquad (3.\text{Ex}.3)$$

where N is a normalization constant. The coefficients of the CF orbitals as functions of the atomic orbitals a and b are given in the following table:

	φ_a	φ_b
a	0.90690	0.13344
b	0.13344	0.90690

a. Knowing that the overlap between the orbitals φ_a and φ_b is $S \sim 0.7963$, calculate the overlap between the two CF determinants and the normalization constant of the wave function.

b. Express Ψ_{CF} in terms of pure AO determinants, and show that it is equivalent to $\Psi_{\text{MO−CI}}$ in Exercise 3.1.

3.3. Consider two bonded atoms A and B, with atomic orbitals a and b, respectively.

a. Use VB theory with an effective Hamiltonian (Eq. 3.29 and Appendix 3.A.1), and express the energy of the unnormalized determinant $|a\bar{a}b|$

as a function of the orbital energies ε_a and ε_b, the off-diagonal monoelectronic Hamiltonian matrix element h_{ab}, and the overlap S_{ab} between orbitals a and b. Calculate the matrix element $\langle|a\bar{a}b||H||a\bar{b}b|\rangle$. Calculate the self-overlap of $|a\bar{a}b|$ and the overlap between $|a\bar{a}b|$ and $|a\bar{b}b|$. Express the normalization constant N_1 of the normalized wave function for the VB structure A: •B

$$\Psi(A: \bullet B) = N_1|a\bar{a}b|$$

Calculate the energy of $\Psi(A: \bullet B)$.
Calculate the overlap between the normalized wave functions for the VB structures A: •B and A• :B

b. A and B are now two identical atoms. We take ε_a and ε_b as the origin for the orbital energies, that is,

$$\varepsilon_a = \varepsilon_b = 0$$

Knowing that with this convention, h_{ab} is replaced by a reduced resonance integral β_{ab} in the expression of the energy terms, express the energies of $\Psi(A: \bullet B)$ and the matrix element $\langle|a\,\bar{a}b||H||a\,\bar{b}b|\rangle$ in terms of β_{ab} and S_{ab} (β and S for short)

c. Express the energy of $\Psi(A\therefore B)$, the normalized wave function for the 3e-bonded state (A∴B = A: •B ↔ A• :B)

$$\Psi(A\therefore B) = N_2(|a\bar{a}b| + |a\bar{b}b|)$$

Compare the expressions for the energies of A: •B and A∴B (relative to the separate fragments) to Equations 3.41 and 3.49.

3.4. One wishes to calculate exactly the energy of A• B•—•C (Eq. 3.50) relative to a situation where A• and B•—•C are separated, in the effective VB Hamiltonian framework, as in the preceding exercise. Rewrite Equation 3.50 so that the two determinants exhibit maximum orbital and spin correspondence. Calculate the energies of the unnormalized determinants $|a\bar{b}c|$ and $|a\bar{c}b|$, and the Hamiltonian matrix element $\langle|a\bar{b}c||H||a\bar{c}b|\rangle$. The following simplifications will be used

$$h_{ab} = h_{bc} = h, \qquad S_{ab} = S_{bc} = S \qquad h_{ac} = S_{ac} = 0$$

Calculate the overlap between $|a\bar{b}c|$ and $|a\bar{c}b|$ and their self-overlap. By setting all orbital energies to zero and replacing h_{ab} by β, calculate the energy of A• B•—•C, and the difference $E(A\bullet B\bullet—\bullet C) - E(A\bullet) - E(B\bullet—\bullet C)$. Compare the result with Equation 3.51.

3.5. The atoms A and B are two bonded and identical atoms, with atomic orbitals a and b, respectively. In the MO framework, the A—B interaction forms two MOs, a bonding combination σ and an antibonding combination σ^*, expressed below (dropping normalization factors):

$$\sigma = a + b$$
$$\sigma^* = a - b$$

Ψ_{MO} and Ψ_{VB} are the wave functions that represent a 3e interaction between A and B, respectively, in the MO and VB framework.

$$\Psi_{MO}(A\therefore B) = |\sigma\bar{\sigma}\sigma^*|$$

$$\Psi_{VB}(A\therefore B) = |a\bar{a}b| + |a\bar{b}b| \text{ (unnormalized)}$$

By expanding Ψ_{MO} into AO determinants, prove that the two wave functions are identical.

Show the same MO–VB identities for the 1e interaction A•B, the triplet 2e repulsive interaction A ↑↑ B and the 4e repulsive interaction A::B.

3.6. Let φ_1^{lmo} and φ_2^{lmo} be two LBOs obtained from the canonical orbitals φ_1^{cmo} and φ_2^{cmo} by the following unitary transformation.

$$\varphi_1^{lmo} = (\cos\theta)\varphi_1^{cmo} + (\sin\theta)\varphi_2^{cmo} \quad (3.Ex.4a)$$

$$\varphi_2^{lmo} = -(\sin\theta)\varphi_1^{cmo} + (\cos\theta)\varphi_2^{cmo} \quad (3.Ex.4b)$$

Prove that the transformation leaves the 2e Slater determinant unchanged, as expressed in Equation 3.Ex.5, irrespective of the value of θ.

$$\left|\varphi_1^{lmo}\varphi_2^{lmo}\right| = \left|\varphi_1^{cmo}\varphi_2^{cmo}\right| \quad (3.Ex.5)$$

3.7. The occupied valence CMOs of water, $\varphi_1^{cmo} - \varphi_4^{cmo}$, are represented in Scheme 3.Ex.1.

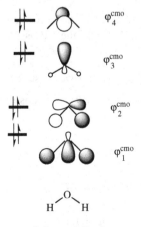

Scheme 3.Ex.1

Show pictorially how a unitary transformation converts these CMOs to the LBO picture that is taught in many freshmen textbooks.

Hint: Divide the CMOs into two sets: one set involving the bonding orbitals, the other involving the nonbonding ones. Do the localization separately in each set.

3.8. Given below are the occupied Hückel π-CMOs for butadiene.

$$\varphi_1 = a\chi_1 + b\chi_2 + b\chi_3 + a\chi_4 \qquad (3.Ex.6a)$$

$$\varphi_2 = b\chi_1 + a\chi_2 - a\chi_3 - b\chi_4 \qquad (3.Ex.6b)$$

$a = 0.37$, $b = 0.60$, χ_{1-4} are the p AOs of butadiene, perpendicular to the plane.

We will use a unitary transformation that attempts to localize these orbitals and produce two π-LBOs Π_1 and Π_2:

$$\Pi_1 = c_{11}c_1 + c_{21}\chi_2 + c_{31}\chi_3 + c_{41}\chi_4 \qquad (3.Ex.7a)$$

$$\Pi_2 = c_{12}\chi_1 + c_{22}\chi_2 + c_{32}\chi_3 + c_{42}\chi_4 \qquad (3.Ex.7b)$$

As a criterion for localization, we will require that in each LBO the product of the coefficients of the contributing AOs to a given LBO would be maximized on the two carbons that are linked by a formal π-bond in the Kekulé structure of butadiene:

$$\text{For } \Pi_1 : (c_{11}) \times (c_{21}) = \text{max}$$

$$\text{For } \Pi_2 : (c_{32}) \times (c_{42}) = \text{max}$$

Give the resulting expression of Π_1 and Π_2 in terms of the AOs. Are these orbitals perfectly localized (i.e., with negligible tails)? And if not, why?

Hint: The unitary transformation will be expressed as a rotation θ in the space generated by the CMOs. As such, the LBOs π_1 and π_2 will be expressed as follows:

$$\Pi_1 = (\cos \theta)\varphi_1 + (\sin \theta)\varphi_2 \qquad (3.Ex.8a)$$

$$\Pi_2 = -(\sin \theta)\varphi_1 + (\cos \theta)\varphi_2 \qquad (3.Ex.8b)$$

Answers

Exercise 3.1

a. The determinants $|\sigma\bar{\sigma}|$ and $|\sigma^*\bar{\sigma^*}|$ are both normalized. Inserting the LCAO expression of the MOs into these Slater determinants and multiplying out the diagonal terms, converts the MO-based determinants to AO determinants, which after normalization lead to the following expressions:

$$|\sigma\bar{\sigma}| = 0.30123(|a\bar{b}| - |\bar{a}b|) + 0.30123(|a\bar{a}| + |b\bar{b}|)$$

$$|\sigma^*\bar{\sigma^*}| = -1.47004(|a\bar{b}| - |\bar{a}b|) + 1.47004(|a\bar{a}| + |b\bar{b}|)$$

b. $$\Psi_{MO-CI} = 0.46476(|a\bar{b}| - |\bar{a}b|) + 0.13388(|a\bar{a}| + |b\bar{b}|)$$

Thus, the ionic component has decreased while the covalent component has increased in Ψ_{MO-CI}, relative to the Hartree–Fock configuration $|\sigma\bar{\sigma}|$.

c. Recalling the expression of $\Psi_{VB\text{-full}}$ in Equation 2.2,

$$\Psi_{VB-full} = 0.787469\Phi_{HL} + 0.133870(\Phi_{ion(1)} + \Phi_{ion(2)})$$

it appears that the coefficients of the ionic structures are the same in Ψ_{MO-CI} and $\Psi_{VB\text{-full}}$. Moreover, by equating the HL components of both equations, one finds

$$0.787469\,\Phi_{HL} = 0.46476(|a\bar{b}| - |\bar{a}b|)$$

$$\Phi_{HL} = 0.59019(|a\bar{b}| - |\bar{a}b|)$$

Exercise 3.2

a. According to Section 3.2, the overlap between the two determinants is

$$\langle|\varphi_a\overline{\varphi_b}|\,\|\,\overline{\varphi_a}\varphi_b|\rangle = -\langle|\varphi_a\overline{\varphi_b}|\,\|\,\varphi_b\overline{\varphi_a}|\rangle = -S^2 = -0.63409$$

from which we deduce the normalization constant of the Ψ_{CF} wave function:

$$N = \frac{1}{\sqrt{2 + 2S^2}} = 0.5531$$

b. We can then expand Ψ_{CF} in AO determinants (by inserting the expressions of the CF orbitals in terms of the pure AOs and then multiplying the terms):

$$\Psi_{CF} = 0.5531[(0.90690^2 - 0.13344^2)(|a\bar{b}| - |\bar{a}b|) + 2 \times 0.90690 \times 0.13344(|a\bar{a}| + |b\bar{b}|)]$$

$$\Psi_{CF} = 0.46476(|a\bar{b}| - |\bar{a}b|) + 0.13387(|a\bar{a}| + |b\bar{b}|)$$

which is equivalent to Ψ_{MO-CI} in Exercise 3.1.

Exercise 3.3

a. The energy terms of the unnormalized determinant $|a\,\bar{a}b|$ are displayed in Table 3.Ans.1:

Table 3.Ans.1

$(-1)^tP_i$		a	$\bar{a}$	b		Energy Terms
(1) P_0		a	$\bar{a}$	b		$2\varepsilon_a + \varepsilon_b$
(2) $-P_{ab}$		$-b$	$\bar{a}$	a		$-2h_{ab}S_{ab} - \varepsilon_a S_{ab}^2$

$$E(|a\bar{a}b|) = 2\varepsilon_a + \varepsilon_b - 2h_{ab}S_{ab} - \varepsilon_a S_{ab}^2$$

$\langle|a\bar{a}b||H||a\bar{b}b|\rangle$ is calculated from Table 3.Ans.2:

Table 3.Ans.2

$(-1)^tP_i$		a	$\bar{a}$	b		Energy Terms
(1) P_0		a	$\bar{b}$	b		$\varepsilon_a S_{ab} + \varepsilon_b S_{ab} + h_{ab}$
(2) $-P_{ab}$		$-b$	$\bar{b}$	a		$-3h_{ab}S_{ab}^2$

$$\langle|a\bar{a}b||H||a\bar{b}b|\rangle = \varepsilon_a S_{ab} + \varepsilon_b S_{ab} + h_{ab} - 3h_{ab}S_{ab}^2$$

The self-overlap of $|a\bar{a}b|$ is deduced from $E(|a\bar{a}b|)$ by replacing ε by 1, h_{ab} by S_{ab}, and dividing by 3:

$$\langle|a\bar{a}b||a\bar{a}b|\rangle = 1 - S_{ab}^2$$

By the same method, one can deduce $\langle|a\,\bar{a}\,b||a\,\bar{b}b|\rangle$ from $\langle|a\,\bar{a}b||H||a\,\bar{b}b|\rangle$:

$$\langle|a\,\bar{a}b||a\,\bar{b}b|\rangle = S_{ab}(1 - S_{ab}^2)$$

The normalized wave function for the VB structure A: •B reads

$$\Psi(\text{A: •B}) = (1 - S_{ab}^2)^{-1/2}|a\,\bar{a}\,b|$$

$$E(\text{A: •B}) = (2\varepsilon_a + \varepsilon_b - 2h_{ab}S_{ab} - \varepsilon_a S_{ab}^2)/(1 - S_{ab}^2)$$
$$= 2\varepsilon_a + \varepsilon_b + [(\varepsilon_a + \varepsilon_b)S_{ab}^2 - 2h_{ab}S_{ab}]/(1 - S_{ab}^2)$$

Using the definition $\beta_{ab} = h_{ab} - S_{ab}(\varepsilon_a + \varepsilon_b)/2$
This becomes

$$E(\text{A: •B}) = 2\varepsilon_a + \varepsilon_b - 2\beta_{ab}S_{ab}/(1 - S_{ab}^2)$$

The normalized VB structure for A• :B reads

$$\Psi(\text{A• :B}) = (1 - S_{ab}^2)^{-1/2}|a\,\bar{b}\,b|$$

The overlap between the normalized VB structures for A: •B and A• :B follows:

$$\langle \Psi(A: {}^{\bullet}B)|\Psi(A^{\bullet} :B)\rangle = \langle |a\,\bar{a}\,b|\,||a\,\bar{b}\,b|\rangle/(1 - S_{ab}^2) = S_{ab}$$

b.
$$E(|a\,\bar{a}\,b|) = -2\beta S$$

Energy of $\Psi(A: {}^{\bullet}B)$:

$$E(A: {}^{\bullet}B) = -2\beta S/(1 - S^2)$$

Matrix element $\langle |a\,\bar{a}b|\,||H||a\,\bar{b}b|\rangle = \beta(1 - 3S^2)$

c. Energy of $\Psi(A\therefore B)$:

$$E(A\therefore B) = E(|a\,\bar{a}b| + |a\,\bar{b}b|)/\langle(|a\,\bar{a}b| + |a\,\bar{b}b|)|(|a\,\bar{a}b| + |a\,\bar{b}b|)\rangle$$
$$= \frac{-4\beta S + 2\beta(1 - 3S^2)}{2(1 + S)(1 - S^2)} = \beta(1 - 3S)/(1 - S^2)$$

d. With our conventions, $E(A:) = E({}^{\bullet}B) = 0$. Thus, the energies of $\Psi(A: {}^{\bullet}B)$ and $\Psi(A\therefore B)$ are the energies of the A: •B and A∴B structures relative to the separate fragments. These formulas match Equations 3.41 and 3.49.

Exercise 3.4 $\qquad A^{\bullet}\,B{\bullet}\!-\!\!{\bullet}\,C = N(|a\,\bar{b}c| + |a\,\bar{c}b|)$

The energies of $|a\,\bar{b}c|$ and $|a\,\bar{c}b|$ and their off-diagonal Hamiltonian matrix element are calculated as usual with the tables of permutations.

Table 3.Ans.3:
Self-Energies

$(-1)^t P_i$	a	$\bar{b}$	c	Energy Terms
(1) P_0	a	$\bar{b}$	c	$\varepsilon_a + \varepsilon_b + \varepsilon_c$
(2) $-P_{ac}$	$-c$	$\bar{b}$	a	0

$(-1)^t P_i$	a	$\bar{c}$	b	Energy Terms
(1) P_0	a	$\bar{c}$	b	$\varepsilon_a + \varepsilon_c + \varepsilon_b$
(2) $-P_{ab}$	$-b$	$\bar{c}$	a	$-2h_{ab}S_{ab} - \varepsilon_c S_{ab}^2$

Table 3.Ans.4:
Off-diagonal Hamiltonian matrix element

$(-1)t P_i$	a	$\bar{b}$	c	Energy Terms
(1) P_0	a	$\bar{c}$	b	$\varepsilon_a S_{bc}^2 + 2h_{bc}S_{bc}$
(2) $-P_{ab}$	$-b$	$\bar{c}$	a	0

It follows:

$$E(|a\,\bar{b}\,c|) = \varepsilon_a + \varepsilon_b + \varepsilon_c$$

$$E(|a\,\bar{c}\,b|) = \varepsilon_a + \varepsilon_b + \varepsilon_c - 2hS - \varepsilon_c S^2$$

$$\langle|a\,\bar{b}\,c|\,|H|\,|a\,\bar{c}\,b|\rangle = \varepsilon_a S^2 + 2hS$$

The overlaps are deduced from the energy terms by replacing ε by 1, and h by S, and dividing the result by 3:

$$\langle|a\,\bar{b}\,c|\,|\,|a\,\bar{b}\,c|\rangle = 1$$

$$\langle|a\,\bar{c}\,b|\,|\,|a\,\bar{c}\,b|\rangle = 1 - S^2$$

$$\langle|a\,\bar{b}\,c|\,|\,|a\,\bar{c}\,b|\rangle = S^2$$

Setting orbital energies to zero and replacing h by β lead to

$$E(|a\,\bar{b}\,c|) = 0$$

$$E(|a\,\bar{c}\,b|) = -2\beta S$$

$$\langle|a\,\bar{b}\,c|\,|H|\,|a\,\bar{c}\,b|\rangle = +2\beta S$$

$$E(\text{A}^\bullet\text{ B}\bullet\!\!-\!\!\bullet\text{C}) = \frac{E(|a\,\bar{b}\,c|) + E(|a\,\bar{c}\,b|) + 2(|a\,\bar{b}\,c|\,|H|\,|a\,\bar{c}\,b|)}{\langle(|a\,\bar{b}\,c| + |a\,\bar{c}\,b|)|(|a\,\bar{b}\,c| + |a\,\bar{c}\,b|)\rangle}$$

$$= \frac{-2\beta S + 4\beta S}{1 + 1 - S^2 + 2S^2} = \frac{2\beta S}{2 + S^2}$$

$$E(\text{A}^\bullet\text{ B}\bullet\!\!-\!\!\bullet\text{C}) - E(\text{A}^\bullet) - E(\text{B}\bullet\!\!-\!\!\bullet\text{C}) = \frac{2\beta S}{2 + S^2} - \frac{2\beta S}{1 + S^2} \approx -\beta S$$

Exercise 3.5 Normalization constants are dropped everywhere.
Three-electron bond

$$\Psi_{\text{MO}}(\text{A}\,\therefore\,\text{B}) = |(a + b)\,\overline{(a + b)}(a - b)|$$
$$= -|a\,\bar{a}\,b| - |a\,\bar{b}\,b| + |b\,\bar{a}\,a| + |b\,\bar{b}\,a|$$

$$-\Psi_{\text{MO}}(\text{A}\,\therefore\,\text{B}) = |a\,\bar{a}\,b| + |a\,\bar{b}\,b| = \Psi_{\text{VB}}(\text{A}\,\therefore\,\text{B})$$

Note that if we write the two determinants as $|a\,\bar{a}\,b|$ and $|b\,\bar{b}\,a|$ the sign of their combination will be negative (see comments in Appendix 3.A.2).
One-electron bond

$$\Psi_{\text{MO}}(\text{A}\bullet\text{B}) = |\ldots(a + b)|$$
$$= |\ldots a| + |\ldots b| = \Psi_{\text{VB}}(\text{A}\bullet\text{B})$$

Four-electron repulsion

$$\Psi_{MO}(A::B) = |(a+b)\,\overline{(a+b)}(a-b)\,\overline{(a-b)}|$$
$$= |a\,\bar{a}\,b\,\bar{b}| - |a\,\bar{b}\,b\,\bar{a}| - |b\,\bar{a}\,a\,\bar{b}| + |b\,\bar{b}\,a\,\bar{a}|$$
$$= |a\,\bar{a}\,b\,\bar{b}| = \Psi_{VB}(A::B)$$

By removing the β spin-orbitals in the preceding equations, one would demonstrate in the same way the MO-VB identity for the triplet repulsive interaction.

Exercise 3.6

$$|\varphi_1^{lmo}\varphi_2^{lmo}| = |(\cos\theta \times \varphi_1^{cmo} + \sin\theta \times \varphi_2^{cmo})(-\sin\theta \times \varphi_1^{cmo} + \cos\theta \times \varphi_2^{cmo})|$$
$$= -2\cos\theta\sin\theta|\varphi_1^{cmo}\varphi_1^{cmo}| + 2\cos\theta\sin\theta|\varphi_2^{cmo}\varphi_2^{cmo}| + (\cos^2\theta + \sin^2\theta)|\varphi_1^{cmo}\varphi_2^{cmo}|$$

As the two first determinants each have two identical columns in the latter equation, they cancel out. There remains

$$|\varphi_1^{lmo}\varphi_2^{lmo}| = |\varphi_1^{cmo}\varphi_2^{cmo}|$$

The identity holds whatever is the value of θ.

Exercise 3.7   The set of the bonding CMOs involves φ_1^{cmo} and φ_2^{cmo}, which represent the O-H bonds. The nonbonding CMOs are φ_3^{cmo} and φ_4^{cmo}. Then, within each set, we add and substract the orbitals, as in Scheme 3.Ans.1.

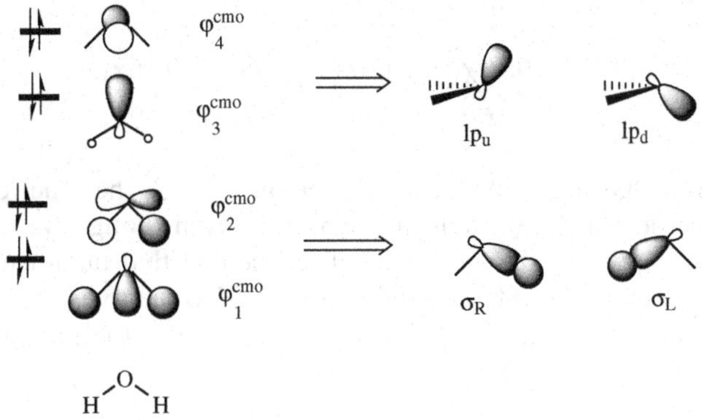

Scheme 3.Ans.1

With omission of the small tails we get from the first set, the LBOs σ_R and σ_L, which describe the two localized O-H bond orbitals. The addition and subtraction of the second set mix two CMOs, one in the molecular plane and the other perpendicular to it. This mixing amounts to a rotation of the orbital to a position in between the two plans, giving rise to lp_u and lp_d that describe the lone-pairs pointing up and down, respectively, relative to the plane of the molecule.

Exercise 3.8   Applying the rotation (3.Ex.8) to φ_1 and φ_2 in Equation 3.Ex.6 yields the following coefficients for Π_1 and Π_2, following the notations of Equation 3.Ex.7:

$$c_{11} = -c_{32} = a(\cos\theta) + b(\sin\theta)$$
$$c_{21} = -c_{42} = b(\cos\theta) + a(\sin\theta)$$
$$c_{31} = c_{12} = b(\cos\theta) - a(\sin\theta)$$
$$c_{41} = c_{22} = a(\cos\theta) - b(\sin\theta)$$

The localization criterion expresses as:

$$F = [a(\cos\theta) + b(\sin\theta)] \times [b(\cos\theta) + a(\sin\theta)] = \max$$
$$\frac{\partial F}{\partial \theta} = [a(\cos\theta) + b(\sin\theta)] \times [-b(\sin\theta) + a(\cos\theta)] + [-a(\sin\theta) + b(\cos\theta)]$$
$$\times [b(\cos\theta) + a(\sin\theta)]$$
$$= a^2\cos^2\theta - b^2\sin^2\theta + b^2\cos^2\theta - a^2\sin^2\theta$$
$$= (a^2 + b^2)(\cos^2\theta - \sin^2\theta) = 0$$

This equation yields a value of $\pi/4$ for θ. Maximizing the product $c_{32} \times c_{42}$ in Π_2 would lead to the same equations and the same value of θ. The final expressions for Π_1 and Π_2 read:

$$\Pi_1 = 0.69\chi_1 + 0.69\chi_2 - 0.16\chi_3 + 0.16\chi_4$$
$$\Pi_2 = 0.16\chi_1 - 0.16\chi_2 + 0.69\chi_3 + 0.69\chi_4$$

The resulting LBOs are fairly localized, one on C_1—C_2 the other on C_3—C_4. However, the delocalization tails are significant even though we used Hückel orbitals. These large localization tails reflect the fact that butadiene has some conjugation between the π-bonds and in terms of VB theory is describable by a linear combination of the major Kekulé structure and the minor long bond structure.

4 Mapping Molecular Orbital–Configuration Interaction to Valence Bond Wave Functions

As seen in Chapter 3, there are quite a few bridges between MO and VB theories, and one of these bridges was the transformation of an MO-based wave function into a VB form. Any MO or MO–CI wave function can be exactly transformed into a VB wave function provided it is a spin-eigenfunction (i.e., not a spin-unrestricted wave function). While this is a trivial matter for the two-electron-two-orbital case (see Chapter 3), the polyelectronic case requires a more elaborate method that is presented in this chapter. The general procedure consists of projecting the MO–CI wave function onto a basis of VB structures, and involves the following steps: (a) determine a complete and linearly independent basis set of VB structures for the electronic system at hand; (b) expand the MO-determinants of the MO–CI wave function as linear combinations of AO determinants; (c) organize the expression obtained in step (b) in terms of the basis set of VB structures generated in step (a). Before proceeding any further, we remind the reader that this chapter is a bit more technical than the others; it is meant for those who intend to delve a bit more into the subject, and acquire another tool for analyzing computational data. At the same time, the general reader may choose to skip this chapter with no serious consequences on the comprehension of the rest of the book or on the ability to utilize VB theory qualitatively or quantitatively.

4.1 GENERATING A SET OF VALENCE BOND STRUCTURES

In the general case, a convenient set of VB structures, on which to project an MO–CI wave function, cannot include all the VB structures that one could possibly imagine. Such a set would be over-complete (with redundancies), and the result of the projection would not be unique. For the projection to be exact and nonarbitrary, it is necessary to generate a complete and linearly

A Chemist's Guide to Valence Bond Theory, by Sason Shaik and Philippe C. Hiberty
Copyright © 2008 John Wiley & Sons, Inc.

independent set of VB structures. This can be done following a convenient and chemically appealing method proposed by Rumer (1). Let us start with a neutral system, with n electrons and n orbitals $(\chi_1 \cdots \chi_n)$, for which we want to generate a Rumer basis of neutral structures. The orbitals are graphically displayed on a circle, which is merely a device with no implications of the shape of the molecule. The orbitals on this circle are then connected in all possible ways, *so that the connecting lines do not cross*. In this manner, one generates a number of drawings, each of which represents a mode of spin coupling of the set of orbitals into pairs, and the ensemble of such coupling diagrams represents the full basis of linearly independent VB structures.

Let us take the π-electronic system of benzene as an example, with six AOs $(\chi_1 - \chi_6)$ and six electrons. The five possible Rumer diagrams that can be drawn are shown in Scheme 4.1.

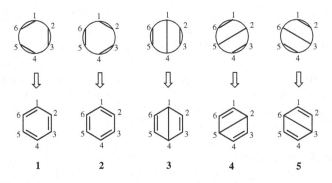

Scheme 4.1

It is clear that the first diagrams, **1** and **2**, correspond to the familiar Kekulé structures, and the three remaining ones, **3–5**, to the Dewar structures.

For the π-system of butadiene, shown in Scheme 4.2, the four AOs are still disposed on a circle, although the molecule is linear. Here the first structure, **6**, corresponds to the unique Kekulé structure of butadiene. The second one, **7**, links together χ_1 and χ_4, and therefore represents a "long bond" structure, which is nothing else but a singlet diradical. Finally, structure **8** is eliminated, as it does not satisfy Rumer's rule, namely, that the connecting lines should not cross. Mathematically, this means that **8** is a combination of **6** and **7** and is therefore a redundant structure. Note that the four AOs could have been set in a different order on the Rumer circle, for example, χ_1-χ_3-χ_2-χ_4 or χ_1-χ_4-χ_2-χ_3. The basis set would still be mathematically correct, but it would be made of structures (**7**, **8**) or (**6**, **8**), which would not be very natural from a chemical point of view. Therefore, the choice of structures **6** and **7** as the linearly independent basis is the only chemically appealing one.

Of course, Schemes 4.1 and 4.2 focus only on the covalent structures. However, Rumer diagrams for ionic VB structures can be generated in the same manner. Negative and positive charges are first assigned to some specific atoms,

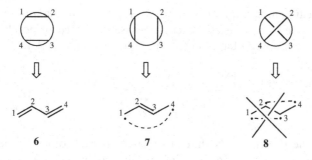

Scheme 4.2

and the remaining system of orbitals and electrons are paired using the graphical Rumer method. Then another distribution of charges is chosen, and the Rumer method is applied to the rest of the system, and so on, until all possibilities are covered (see Exercise 4.1). If the number of orbitals differs from the number of electrons, the Rumer diagrams are generated as for an ionic system, even if the whole molecule itself is neutral (e.g., the π system of ozone, Exercise 4.2).

4.2 MAPPING A MOLECULAR ORBITAL–CONFIGURATION INTERACTION WAVE FUNCTION INTO A VALENCE BOND WAVE FUNCTION

Once the VB structure set is defined, one can proceed to the next step, the expansion of the MO–CI wave function into VB determinants.

4.2.1 Expansion of Molecular Orbital Determinants in Terms of Atomic Orbital Determinants

Let Ω_{MO} be a single determinant involving molecular spin-orbitals φ_i and φ_j, which can be of α or β spins:

$$\Omega_{\mathrm{MO}} = \left| \ldots \varphi_i \ldots \varphi_j \ldots \right| \tag{4.1}$$

$$\varphi_i = \sum_\mu C_{\mu i} \chi_\mu \tag{4.2}$$

$$\varphi_j = \sum_\nu C_{\nu j} \chi_\nu \tag{4.3}$$

Replacing φ_i and φ_j in Equation 4.1 by their expansions in terms of AOs in Equations 4.2–4.3, Ω_{MO} can be expanded into a linear combination of AO-based determinants. The procedure is carried out in the same manner as expansion of the diagonal elements of the MO determinant, that is, by multiplying out the diagonal terms into a sum of orbital products. One proceeds along the following steps: (a) Replace the MOs by their expression based on linear combination of AOs. (b) Multiply out the diagonal MO term to obtain AO based products, (c) each AO-based term corresponds then to a diagonal term of an AO based determinant as in Equation 4.4.

$$\left| \ldots \varphi_i \ldots \varphi_j \ldots \right| = \left| \ldots \left(\sum_{\mu} c_{\mu i} \chi_{\mu} \right) \ldots \left(\sum_{\nu} c_{\nu j} \chi_{\nu} \right) \ldots \right| \qquad (4.4)$$

$$= \sum_{\mu} c_{\mu i} \ldots \sum_{\nu} c_{\nu j} \ldots \left| \ldots \cdot \chi_{\mu} \ldots \chi_{\nu} \ldots \right|$$

(d) Gather the determinants that are related to each other by permutation (e.g., $\left| \ldots a\bar{b} \right| = -\left| \ldots \bar{b}a \right|$). In so doing, recall that interchanging the order of AOs or MOs in the diagonal term would correspond to a change of the sign of the corresponding determinant. Thus, many AO determinants in Equation 4.4 can be regrouped after permuting their orbitals and changing their signs.

While this multiplication and regrouping task is a trivial matter for a small number of electrons, the procedure becomes more complicated for larger systems, owing to the number of AO-based determinants that can be generated. To keep the method tractable by hand, and/or to render it efficient in terms of memory storage by a computer program, it is convenient to use a mathematical intermediary that we call "half-determinant" (2).

Let us consider the Slater determinant Ω_{MO}, in Equation 4.5, which involves α spin-orbitals ($\varphi_i \ldots \varphi_j$) and β spin-orbitals ($\overline{\varphi_k} \ldots \overline{\varphi_l}$), as one determinant that is composed of two "half-determinants", h_{MO}^{α} and h_{MO}^{β}, one regrouping the spin-orbitals of α spins and the other those of β spins.

$$\Omega_{MO} = \left| \ldots \varphi_i, \ldots, \varphi_j, \ldots \overline{\varphi_k}, \ldots \overline{\varphi_l}, \ldots \right| \qquad (4.5)$$

The two half-determinants, h_{MO}^{α} and h_{MO}^{β}, which involve the α and β spin-orbitals, respectively, can then be defined as actual determinants of a particular spin variety, in Equations 4.6 and 4.7:

$$h_{MO}^{\alpha} = \left| \ldots \varphi_i \ldots \varphi_j \ldots \right| \qquad (4.6)$$

$$h_{MO}^{\beta} = \left| \ldots \overline{\varphi_k} \ldots \overline{\varphi_l} \ldots \right| \qquad (4.7)$$

Each half-determinant is a determinant made of a diagonal product of spin-orbitals followed by a signed sum of all the permutations of this product, with the restriction that a half-determinant must involve spin-orbitals that are all of the same spin. It will be convenient for the problem at hand to consider a full Slater determinant as the union of its two constituent half-determinants:

$$\Omega_{MO} = \left| h_{MO}^{\alpha}, h_{MO}^{\beta} \right| \qquad (4.8)$$

Each of these MO-based half-determinants can be expanded into AO-based half-determinants, following the prescription leading to Equation 4.4, by replacing the MOs by their expressions in terms of AOs. After orbital permutations the AO-based half-determinants that are equivalent are regrouped, leading thereby to a collection of AO-based half-determinants h_r^α, each having a unique set of AOs:

$$h_{MO}^\alpha = |\ldots \varphi_i \ldots \varphi_j \ldots| = \sum_r C_r^\alpha h_r^\alpha \qquad (4.9)$$

$$h_r^\alpha = |\ldots \chi_\mu \ldots \chi_\nu \ldots| \qquad (4.10)$$

where the label r designates a given set of AOs.

The coefficients C_r^α of each of these AO half-determinants in the expansion (4.9) are given by Equation 4.11:

$$C_r^\alpha = \sum_P (-1)^t P(\ldots \times C_{\mu i} \times \ldots \times C_{\nu j} \ldots) \qquad (4.11)$$

where P is a permutation between indices μ and ν and t is the parity of the permutation. As such, Equation 4.11 can be rewritten in a more convenient form, in terms of an $n \times n$ determinant, in Equation 4.12:

$$C_r^\alpha = \begin{bmatrix} C_{11} \ldots \ldots \ldots C_{1n} \\ \ldots C_{\mu i} \ldots C_{\nu i} \ldots \\ \ldots C_{\mu j} \ldots C_{\nu j} \ldots \\ C_{n1} \ldots \ldots \ldots C_{nn} \end{bmatrix} \qquad (4.12)$$

n being the number of orbitals in the half-determinants. This determinant can be constructed as follows: One first sets a table displaying all the MOs in terms of the AOs, for example, Table 4.1 below in Section 4.2.3. Then, to find the coefficient of the AO-based half-determinant h_r^α in h_{MO}^α, one extracts a subtable restricted to the MOs of h_{MO}^α and the AOs of h_r^α. This subtable is isomorphous to the determinant in Equation 4.12, which determines the coefficient C_r^α.

The advantage of expanding half-determinants rather than full determinants is that the former are much less numerous than the latter. Then, by regrouping two AO-based half-determinants h_r^α and h_s^β, one gets the full AO-based determinant (h_r^α, h_s^β) for which the coefficient in the expansion of Ω_{MO} is just the product of the coefficients of its two half-determinants:

$$\Omega_{MO} = \sum_{r,s} C_r^\alpha C_s^\beta |h_r^\alpha, h_s^\beta| \qquad (4.13)$$

4.2.2 Projecting the Molecular Orbital–Configuration Interaction Wave Function Onto the Rumer Basis of Valence Bond Structures

Having the coefficients of the AO-determinants in the MO wave function (Eq. 4.13), and the expressions of the VB structures in terms of AO determinants (see Section 3.1.4), one can now project the MO wave function onto the Rumer basis of linearly independent VB structures. First, let us express an MO-determinant Ω_{MO} in terms of VB structures:

$$\Omega_{MO} = \sum_K C_K^0 \Phi_K^0 + \sum_L C_L^1 \Phi_L^1 + \sum_M C_M^2 \Phi_M^2 + \cdots \qquad (4.14)$$

where the superscript designates the order of ionicity of the VB structure: 0 if the VB structures involve no pair of charges $(+/-)$, 1 if it involves one pair of charges, and so on. Then, by expressing each VB structure (Φ_K^0, Φ_L^1, etc.) in Equation 4.14 in terms of AO-determinants, and equating the resulting expression of Ω_{MO} with that of Equation 4.13, one gets the coefficients C_K^0, C_L^1, and so on. Note that the projection can be truncated at any desired level, for example, neutral structures, neutral plus monoionic structures.

In the case of an MO–CI wave function (Ψ_{MO-CI}) made of several determinants, the projection procedure is repeated for each determinant (Eq. 4.14), and the results of these projections are combined to yield the final expression of Ψ_{MO-CI} in terms of VB structures (see Exercise 4.1).

4.2.3   An Example: The Hartree–Fock Wave Function of Butadiene

The MOs of butadiene are listed in Table 4.1. At the Hartree–Fock level, only the two first MOs are occupied, and the wave function reads:

$$\Omega_{MO} = |\ldots \pi_1 \bar{\pi}_1 \pi_2 \bar{\pi}_2| \qquad (4.15)$$

where the σ orbitals are not specifically indicated.

TABLE 4.1   Coefficients of the MOs of 1,3-Butadiene[a]

	π_1	π_2	π_3	π_4
χ_1	0.367	0.527	0.629	0.502
χ_2	0.479	0.402	−0.467	−0.699
χ_3	0.479	−0.402	−0.467	0.699
χ_4	0.367	−0.527	0.629	−0.502

[a]Optimized at the B3LYP/cc-pVTZ level, with C−C bond lengths of 1.3340 and 1.4528 Å, respectively.

As shown above in Scheme 4.2, the Rumer basis for butadiene is made of the VB structures **6** and **7**. From Section 3.1.4, the VB function corresponding to a given bonding scheme is the one that involves singlet coupling between the AOs that are paired in this scheme. This, however, can be done in two ways. In the first way, the AOs are kept in the same order in the various determinants, and the determinants display all the possible 2×2 spin permutations between the orbitals that are singlet coupled. This is the convention used in Equation 4.16, which is similar to Equation 3.9. In this case, the determinant has a positive or negative sign, depending on the parity of the permutation (see Section 3.1.4).

$$\Psi(6) = N_1(|\chi_1\bar{\chi}_2\chi_3\bar{\chi}_4| - |\bar{\chi}_1\chi_2\chi_3\bar{\chi}_4| - |\chi_1\bar{\chi}_2\bar{\chi}_3\chi_4| + |\bar{\chi}_1\chi_2\bar{\chi}_3\chi_4|) \qquad (4.16)$$

In the second way, one can decide that all determinants will have their spins arranged in alternated order, that is, α, β, α, etc., in which case the spin-coupling is characterized by permutations of the AOs. In such a case, the coefficients of the AO-determinants are all positive for a singlet state. Both methods are strictly equivalent, and the first one is generally employed in this book. However, the method of half-determinants is easier to apply with the second convention, which will be used in this section. Accordingly, the VB functions for structures **6** and **7** can be rewritten as in Equation 4.17 and 4.18:

$$\Psi(6) = N_1(|\chi_1\bar{\chi}_2\chi_3\bar{\chi}_4| + |\chi_2\bar{\chi}_1\chi_3\bar{\chi}_4| + |\chi_1\bar{\chi}_2\chi_4\bar{\chi}_3| + |\chi_2\bar{\chi}_1\chi_4\bar{\chi}_3|) \qquad (4.17)$$

$$\Psi(7) = N_2(|\chi_1\bar{\chi}_2\chi_3\bar{\chi}_4| + |\chi_1\bar{\chi}_3\chi_2\bar{\chi}_4| + |\chi_4\bar{\chi}_2\chi_3\bar{\chi}_1| + |\chi_4\bar{\chi}_3\chi_2\bar{\chi}_1|) \qquad (4.18)$$

In Equations 4.16–4.18, N_1 and N_2 are normalization constants that can be calculated exactly by using Equation 3.15 in Section 3.2, knowing the overlaps between AOs. We may, however, skip this calculation for the sake of simplicity. Indeed, given that the AO-determinants in Equations 4.17 and 4.18 all differ from one another by replacements of at least two orbitals, their mutual overlap is expected to be weak (see Appendix 3.4.2), and can be neglected, leading to the approximate value 0.5 for both N_1 and N_2 (exact values are 0.473 and 0.491, respectively).

The VB functions $\Psi(6)$ and $\Psi(7)$ involve six different half-determinants, of α or β spin: $|\chi_1, \chi_2|$, $|\chi_1, \chi_3|$, $|\chi_1, \chi_4|$, $|\chi_2, \chi_3|$, $|\chi_2, \chi_4|$, and $|\chi_3, \chi_4|$. Knowing the expressions of π_1 and π_2 in Table 4.1, and using Equation 4.12, it becomes a straightforward matter to find the coefficients of the six half-determinants. For example, the coefficient of $|\chi_1, \chi_2|$ in $|\pi_1, \pi_2|$ is given by the determinants of the AO coefficients of the corresponding AOs (χ_1,χ_2) in the MOs π_1 and π_2, that is, $(0.367 \times 0.402 - 0.479 \times 0.527 = -0.105)$. The other coefficients are listed in Table 4.2.

TABLE 4.2   Coefficients of the AO-Half-Determinants in the Expansion of the Hartree–Fock Wave Function of Butadiene

coefficient	$|\chi_1, \chi_2|$	$|\chi_1, \chi_3|$	$|\chi_1, \chi_4|$	$|\chi_2, \chi_3|$	$|\chi_2, \chi_4|$	$|\chi_3, \chi_4|$
	−0.105	−0.400	−0.387	−0.385	−0.400	−0.105

From these values, the coefficients of the AO-determinants are calculated by means of Equation 4.13. For example, the first determinant in Equation 4.17 is made of the half-determinants $|\chi_1, \chi_3|$ and $|\chi_2, \chi_4|$. Its coefficient is therefore: $(-0.400) \times (-0.400)$. Note that the order of the orbitals in a half-determinant is important: two half-determinants that are related to each other by a permutation of a pair of orbitals have opposite coefficients. Following this rule, the coefficient of $|\chi_4, \chi_3|$ is $+0.105$, according to Table 4.2. Once this transformation is complete, the Hartree–Fock wave function can be rewritten in terms of the AO-based determinants:

$$\Omega_{MO} = 0.160|\chi_1\bar{\chi}_2\chi_3\bar{\chi}_4| + 0.149|\chi_2\bar{\chi}_1\chi_3\bar{\chi}_4| + 0.149|\chi_1\bar{\chi}_2\chi_4\bar{\chi}_3| + 0.160|\chi_2\bar{\chi}_1\chi_4\bar{\chi}_3|$$
$$+ 0.011|\chi_1\bar{\chi}_3\chi_2\bar{\chi}_4| + 0.011|\chi_4\bar{\chi}_2\chi_3\bar{\chi}_1| \tag{4.19}$$

The projection of Ω_{MO} on the Rumer basis consists of combining Equation 4.19 with the final expression of Ω_{MO} in terms of VB structures, Equation 4.20:

$$\Omega_{MO} = C_1\Psi(\mathbf{6}) + C_2\Psi(\mathbf{7}) \tag{4.20}$$

By using the expressions of $\Psi(\mathbf{6})$ and $\Psi(\mathbf{7})$ in Equations 4.17 and 4.18, one finds

$$C_1 = 0.149/N_1 \sim 0.298; \qquad C_2 = 0.011/N_2 \sim 0.022 \tag{4.21}$$

Thus, structure **6** is the dominant covalent structure, while **7** is much less important for the ground state of butadiene.

4.3   USING HALF-DETERMINANTS TO CALCULATE OVERLAPS BETWEEN VALENCE BOND STRUCTURES

Since a half-determinant is a signed sum of orbital products, it is easy to define an overlap between two half-determinants of the same spin. Let us consider, for example, two half-determinants of α-spin-orbitals, h_r^α (Eq. 4.10) and h_u^α in Equation 4.22.

$$h_u^\alpha = |\ldots\chi_\kappa\cdots\chi_\lambda\ldots| \tag{4.22}$$

The overlap between these two half-determinants is given by Equation 4.23:

$$\langle h_r^\alpha|h_u^\alpha\rangle = \langle(\ldots\chi_\mu\cdots\chi_\nu\ldots)|(-1)^t P(\ldots\chi_\kappa\cdots\chi_\lambda\ldots)\rangle \tag{4.23}$$

where P is a permutation of the orbitals in the diagonal orbital product, and t is the parity of this permutation. Of course, the final quantity of interest is the overlap between two AO-based determinants, Ω_{rs} and $\Omega_{\mu\nu}$. This overlap can be obtained as the product of the overlaps of the constituent half-determinants, as in Equations 4.24–4.26:

$$\Omega_{rs} = |h_r^\alpha, h_s^\beta| \tag{4.24}$$

$$\Omega_{uv} = |h_u^\alpha, h_v^\beta| \tag{4.25}$$

$$\langle\Omega_{rs}|\Omega_{uv}\rangle = \langle h_r^\alpha|h_u^\alpha\rangle \times \langle h_s^\beta|h_v^\beta\rangle \tag{4.26}$$

Once again, the advantage of dealing with half-determinants is their small number (the square root of the number of AO-based determinants), and their small dimension, which facilitate the calculation of overlaps. As an example, let us consider the determinants $|a\bar{a}b\bar{b}|$ and $|c\bar{c}d\bar{d}|$, whose mutual overlap has already been expressed in Equation 3.17, and let us recalculate this overlap by using half-determinants. The two determinants are first split into their constituent half-determinants:

$$|a\bar{a}b\bar{b}| = |h_1^\alpha, h_1^\beta|; \qquad h_1^\alpha = |ab|; \qquad h_1^\beta = |\bar{a}\bar{b}| \qquad (4.27)$$

$$|c\bar{c}d\bar{d}| = |h_2^\alpha, h_2^\beta|; \qquad h_2^\alpha = |cd|; \qquad h_2^\beta = |\bar{c}\bar{d}| \qquad (4.28)$$

The overlap between half-determinants is then calculated according to Equation 4.23. As each half-determinant contains only two orbitals, only one permutation is possible, in addition to identity, from the diagonal orbital product:

$$\langle h_1^\alpha | h_2^\alpha \rangle = \langle a(1)b(2)|c(1)d(2) - d(1)c(2)\rangle = S_{ac}S_{bd} - S_{ad}S_{bc} \qquad (4.29)$$

$$\langle h_1^\beta | h_2^\beta \rangle = \langle \bar{a}(1)\bar{b}(2)|\bar{c}(1)\bar{d}(2) - \bar{d}(1)\bar{c}(2)\rangle = S_{ac}S_{bd} - S_{ad}S_{bc} \qquad (4.30)$$

where the S terms are orbital overlaps. Finally, the overlap between determinants is calculated according to Equation 4.26:

$$\langle |a\bar{a}b\bar{b}| \, || \, c\bar{c}d\bar{d}| \rangle = \langle h_1^\alpha | h_2^\alpha \rangle \times \langle h_1^\beta | h_2^\beta \rangle = (S_{ac}S_{bd} - S_{ad}S_{bc})^2 \qquad (4.31)$$

where it is seen that the result is identical to that of Equation 3.17.

The overlap of $|a\bar{a}b\bar{b}|$ with itself would be calculated in a similar way, which is in agreement with Equation 3.18:

$$\langle |a\bar{a}b\bar{b}| \, || \, a\bar{a}b\bar{b}| \rangle = \langle h_1^\alpha | h_1^\alpha \rangle \times \langle h_1^\beta | h_1^\beta \rangle = (1 - S_{ab}^2)^2 \qquad (4.33)$$

REFERENCES

1. G. Rumer, *Got. Nachr.* 337 (1932). Zum Theorie der Spinvalenz.
2. P. C. Hiberty, C. Leforestier, *J. Am. Chem. Soc.* **100**, 2012 (1978). Expansion of Molecular Orbital Wave Functions into Valence Bond Wave Functions. A Simplified Procedure.

EXERCISES

4.1 Generate the 12 monoionic VB structures of the π-electronic system of butadiene.

4.2 Find the 6 VB structures of the π-electronic system of ozone and write their wave functions. The overlaps between AO-based determinants can be neglected.

4.3 (a) Express the Hartree–Fock wave function of ozone in terms of the VB structures of the preceding exercise, given the list of the π MOs below:

	π_1	π_2	π_3
χ_1	0.368	0.710	0.614
χ_2	0.764	0.0	−0.671
χ_3	0.368	−0.710	0.614

(b) The 2×2 CI wave function of ozone reads

$$\Psi_{2\times2} = 0.908\,|\pi_1\bar\pi_1\,\pi_2\bar\pi_2| - 0.418\,|\pi_1\bar\pi_1\,\pi_3\bar\pi_3| \qquad (4.Ex.1)$$

Express this wave function in terms of VB structures, and show that CI has the effect of increasing the weight of the diradical structure, and lowering those of the ionic structures, especially the 1,3-dipolar ones.

(c) Given the overlaps between the π AOs of ozone below, calculate the weight of the diradical structure in the 2×2 CI wave function $\Psi_{2\times2}$, by means of the Chirgwin–Coulson formula (Eq. 3.55), taking all overlaps into account. We recall that, according to Equation 3.55, the weight of a VB structure can be calculated as the sum of the weights of its constituting AO-based determinants.

	χ_1	χ_2	χ_3
χ_1	1.	0.12738	0.00813
χ_2		1.	0.12738
χ_3			1.

Answers

Exercise 4.1   The 12 zwitterionic structures are displayed in Scheme 4.3.

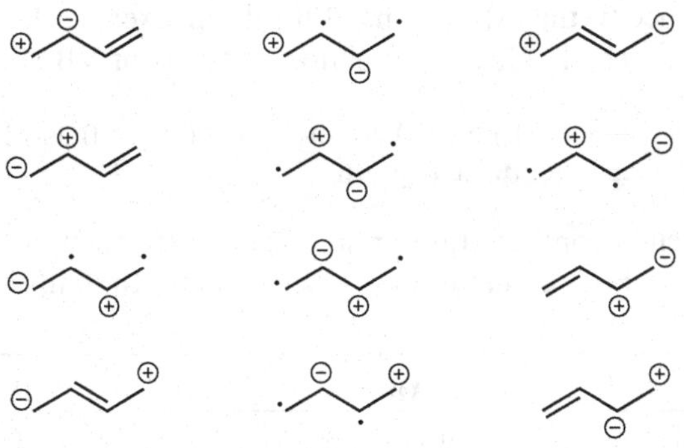

Scheme 4. Ans.1

Exercise 4.2 The π-electronic system of ozone is a 4-electron/3-orbital system. The six VB structures are Φ_1-Φ_6 below:

$\Phi_1 = 0.707(|\chi_2\bar{\chi}_2\chi_1\bar{\chi}_3| + |\chi_2\bar{\chi}_2\chi_3\bar{\chi}_1|)$ (diradical structure $\dot{O} - \ddot{O} - \dot{O}$)

$\Phi_2 = 0.707(|\chi_1\bar{\chi}_2\chi_3\bar{\chi}_3| + |\chi_2\bar{\chi}_1\chi_3\bar{\chi}_3|)$ (ionic structure $O{=}O^+ - O^-$)

$\Phi_3 = 0.707(|\chi_1\bar{\chi}_1\chi_2\bar{\chi}_3| + |\chi_1\bar{\chi}_1\chi_3\bar{\chi}_2|)$ (ionic structure $O - O^+{=}O^-$)

$\Phi_4 = |\chi_1\bar{\chi}_1\chi_2\bar{\chi}_2|$ (dipolar ionic structure $O^- - \ddot{O} - O^+$)

$\Phi_5 = |\chi_2\bar{\chi}_2\chi_3\bar{\chi}_3|$ (dipolar ionic structure $O^+ - \ddot{O} - O^-$)

$\Phi_6 = |\chi_1\bar{\chi}_2\chi_3\bar{\chi}_3|$ (di-ionic structure $O^- - O^{2+} - O^-$)

The overlaps between AO-determinants have been neglected in the calculations of the normalization factors in Φ_1-Φ_3.

Exercise 4.3

(a) There are three possible half-determinants. Their coefficients in the Hartree–Fock half-determinant are listed in the following table.

	$|\chi_1, \chi_2|$	$|\chi_1, \chi_3|$	$|\chi_2, \chi_3|$
Coefficient	−0.542	−0.523	−0.542

The coefficients of the AO-determinants are as follows:

	$\begin{array}{c}|\chi_2\bar{\chi}_2\chi_1\bar{\chi}_3|\\|\chi_2\bar{\chi}_2\chi_3\bar{\chi}_1|\end{array}$	$\begin{array}{c}|\chi_1\bar{\chi}_1\chi_2\bar{\chi}_3|\\|\chi_1\bar{\chi}_1\chi_3\bar{\chi}_2|\end{array}$	$\begin{array}{c}|\chi_1\bar{\chi}_2\chi_3\bar{\chi}_3|\\|\chi_2\bar{\chi}_1\chi_3\bar{\chi}_3|\end{array}$	$|\chi_1\bar{\chi}_1\chi_2\bar{\chi}_2|$	$|\chi_2\bar{\chi}_2\chi_3\bar{\chi}_3|$	$|\chi_1\bar{\chi}_1\chi_3\bar{\chi}_3|$
Coefficient	−0.294	0.283	0.283	0.294	0.294	0.274

By reference to the expressions of Φ_1-Φ_6 in Exercise 4.2, the expression of the Hartree−Fock wave function in terms of VB structures is

$$|\pi_1\bar{\pi}_1\pi_2\bar{\pi}_2| = -0.416\Phi_1 + 0.400\Phi_2 + 0.400\Phi_3 + 0.294\Phi_4$$
$$+ 0.294\Phi_5 + 0.274\Phi_6 \qquad (4.\text{Ans}.1)$$

(b) The diexcited configuration $|\pi_1\bar{\pi}_1\pi_3\bar{\pi}_3|$ is expanded in the same way as the Hartree-Fock configuration. Coefficients of the half-determinants are:

	$|\chi_1,\chi_2|$	$|\chi_1,\chi_3|$	$|\chi_2,\chi_3|$
Coefficient	−0.716	0.0	0.716

$$|\pi_1\bar{\pi}_1\pi_3\bar{\pi}_3| = 0.726\Phi_1 + 0.0\Phi_2 + 0.0\Phi_3 + 0.513\Phi_4 + 0.513\Phi_5 + 0.0\,\Phi_6$$
$$(4.\text{Ans}.2)$$

Combining Equations 4.Ans.1 and 4.Ans.2 with the expression of $\Psi_{2\times2}$ in terms of MO configurations (Eq. 4.Ex.1) leads to

$$\Psi_{2\times2} = -0.681\,\Phi_1 + 0.363\,\Phi_2 + 0.363\,\Phi_3 + 0.053\,\Phi_4 + 0.053\,\Phi_5 + 0.249\,\Phi_6$$
$$(4.\text{Ans}.3)$$

(c) First let us establish a table of overlaps between half-determinants:

	$|\chi_1,\chi_2|$	$|\chi_1,\chi_3|$	$|\chi_2,\chi_3|$
$|\chi_1,\chi_2|$	0.98377	0.12634	0.00810
$|\chi_1,\chi_3|$		0.99993	0.012634
$|\chi_2,\chi_3|$			0.98377

By means of Equation 4.26, these six values suffice to calculate the 45 overlaps between the nine possible AO-based determinants. To save space, we will use a simplified notation for the latter: for example, the determinant $|\chi_2\bar{\chi}_2\chi_1\bar{\chi}_3|$, which is made of the half-determinants $|\chi_2\chi_1|$ and $|\bar{\chi}_2\bar{\chi}_3|$, will be noted $21\bar{2}\bar{3}$. One then expresses $\Psi_{2\times2}$ as a function of the AO-based determinants:

$$\Psi_{2\times2} = -0.481(21\bar{2}\bar{3} + 23\bar{2}\bar{1}) + 0.256(13\bar{2}\bar{3} + 23\bar{1}\bar{3}) + 0.256(12\bar{1}\bar{3} + 13\bar{1}\bar{2})$$
$$+ 0.053(12\bar{1}\bar{2}) + 0.053(23\bar{2}\bar{3}) + 0.249(13\bar{1}\bar{3})$$

One then calculates the overlap of the first determinant with itself and with all the others:

$21\overline{23}$	$23\overline{21}$	$13\overline{23}$	$23\overline{13}$	$12\overline{13}$	$13\overline{12}$	$12\overline{12}$	$23\overline{23}$	$13\overline{13}$
$21\overline{23}$ 0.96780	0.00007	−0.12429	−0.00102	−0.12429	−0.00102	−0.00797	−0.00797	−0.01596

Then, the weight of $21\overline{23}$ is calculated by means of the Chirgwin–Coulson formula Eq. 3.55:

$$W(21\overline{23}) = 0.257$$

The weight of $23\overline{21}$ is equal to that of $21\overline{23}$ and need not be recalculated, since these two determinants correspond to each other by spin inversions. The weight of Φ_1 is the sum of the weights of these two determinants:

$$W(\Phi_1) = W(21\overline{23}) + W(23\overline{21}) = 0.514.$$

Note that this is the result of 2×2 CI. A complete CI in the π space would further increase the weight of the diradical structure, to 0.60 (see Ref. 2).

5 Are The "Failures" of Valence Bond Theory Real?

5.1 INTRODUCTION

As mentioned in the introductory part, VB theory has been stamped with a few so-called "failures" that are occasionally used to dismiss the theory, and have caused it unwarranted disrepute. It is clear that a good VB theory is like MO−CI since the two theories ultimately converge when electron correlation is fully taken into account. As such, it should come as no surprise that *ab initio* VB calculations are free of these "failures" much as good level MO−CI calculations are. However, it is interesting to realize that even qualitative semiempirical VB theory, such as the one described in Chapter 3, is free of these "failures". It is important to demonstrate this point before turning to practical applications of the VB model to actual problems of chemical reactivity or structure. Therefore, this chapter examines the various alleged VB "failures" and shows that the simple qualitative guidelines presented in the previous chapters make perfectly good predictions.

5.2 THE TRIPLET GROUND STATE OF DIOXYGEN

One of the most highlighted "failures" that has been associated with VB theory concerns the ground state of the dioxygen molecule, O_2. It is true that a naive application of hybridization followed by perfect pairing (simple Lewis pairing) would predict a $^1\Delta_g$ ground state, that is, the diamagnetic doubly bonded molecule O=O. This is likely to be the origin of the notion that VB theory makes a flawed prediction that contradicts experiment. However, this attitude is not particularly scientific, since as early as the 1970's Goddard et al. (1) performed GVB calculations and demonstrated that VB theory leads to a triplet $^3\Sigma_g^-$ ground state. This was followed by the same outcome in papers by McWeeny (2) and Harcourt (3). In fact, any VB calculation, at any imagined level, will lead to the same result, so that the myth of "failure" is definitely baseless.

A Chemist's Guide to Valence Bond Theory, by Sason Shaik and Philippe C. Hiberty
Copyright © 2008 John Wiley & Sons, Inc.

Goddard et al. (1) and subsequently the present authors (4) also provided a simple VB explanation for the choice of the ground state. Let us reiterate this explanation based on our qualitative VB theory, outlined in Chapter 3.

Apart from one σ bond and one σ lone pair on each oxygen atom, the dioxygen molecule has six π electrons to be distributed in the two π planes, say π_x and π_y (see Fig. 5.1). There are two possible modes of distribution, shown in Fig. 5.1 along with their corresponding VB functions. The question is what is the most favorable one? Is it **1**, where three electrons are placed in each π plane, or maybe it is **2**, where two electrons are allocated to one plane and four to the other? Obviously, **1** is a diradical structure displaying one 3e bond in each of the π planes, whereas **2** exhibits a singlet π bond, in one plane, and a 4e repulsion, in the other. A naive application that neglects the repulsive 3e and 4e interactions would predict that structure **2** is preferred, leading to the above-mentioned mythical failure of VB theory, namely, that the theory predicts the ground state of O_2 to be the singlet closed-shell structure, O=O. However, a mere inspection of the repulsive interactions shows that they are of the same order of magnitude or even larger than the bonding interactions; that is, the neglect of these repulsions is unjustified. The right answer immediately pops out, if we account for the VB energy correctly, including the repulsion and bonding interactions for structures **1** and **2**.

Knowing the VB wave function for a 3e bond (Eq. 3.47), we can construct the wave function for structure **1** by considering this electronic distribution as made of two independent three-electron/two-orbital systems orthogonal to

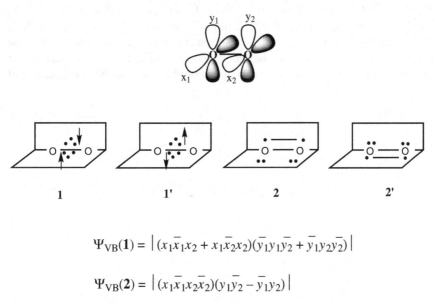

$$\Psi_{VB}(\mathbf{1}) = \left| (x_1\bar{x}_1x_2 + x_1\bar{x}_2x_2)(\bar{y}_1y_1\bar{y}_2 + \bar{y}_1y_2\bar{y}_2) \right|$$

$$\Psi_{VB}(\mathbf{2}) = \left| (x_1\bar{x}_1x_2\bar{x}_2)(\bar{y}_1y_2 - \bar{y}_1y_2) \right|$$

FIGURE 5.1   The four possible distributions of six electrons in four atomic p_π orbitals of dioxygen, leading to a total spin component $S_z = 0$.

each other, one in the π_x plane, the other in the π_y plane. This leads to $\Psi_{VB}(\mathbf{1})$ in Fig. 5.1 (see Exercise 5.1), where the total S_z spin component has been set to zero so as to make this function suitable for generating both singlet and triplet states. Accordingly, the energy of structure **1** is twice that of a 3e interaction that, we recall, has the same expression in VB and MO theories (see Eq. 3.49 and Table 3.1), Equation 5.1:

$$E(\mathbf{1}) = 2\beta(1 - 3S)/(1 - S^2) \tag{5.1}$$

On the other hand, the energy of structure **2** is given by Equation 5.2, in the VB framework, where the two terms represent, respectively, the 2e bonding energy and the 4e repulsion:

$$E(\mathbf{2}) = 2\beta S/(1 + S^2) - 4\beta S/(1 - S^2) \tag{5.2}$$

Comparison of Equations 5.1 and 5.2 clearly demonstrates that the diradical structure **1** is more stable than the doubly bonded Lewis structure **2** (or **2'**).

$$E(\mathbf{2}) - E(\mathbf{1}) = -2\beta(1 - S)^2/(1 - S^4) > 0 \tag{5.3}$$

So far, we have not considered the effect of mixing **1** and **2** with **1'** and **2'**, respectively; the two sets are symmetry equivalent, related by inversion of the π_x and π_y planes and vice versa. The interactions between the two sets of determinants yield two pairs of resonant–antiresonant combinations that constitute the final low lying states of dioxygen (represented in Fig. 5.2). Of course, our effective VB theory was chosen to disregard the bielectronic terms, and therefore the theory, as such, will not tell us what is the lowest spin state in the O_2 diradical. This, however, is a simple matter, because allowing for the bielectronic exchange terms in the theory would lead us to the right answer, or even more simply stated, we can appeal to Hund's rule, which is precisely what

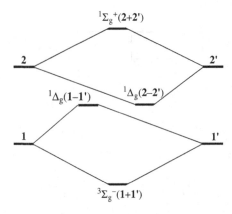

FIGURE 5.2 A VB mixing diagram for the formation of the symmetry-adapted states of O_2 from the biradical (**1**, **1'**) and perfectly paired (**2**, **2'**) structures.

qualitative MO theory has to do in order to predict the triplet nature of the O_2 ground state. Accordingly, the in- and out-of-phase combinations of the diradical determinants **1** and **1'** lead to a triplet $^3\Sigma_g^-$ and a singlet $^1\Delta_g$ states, the former being the lowest lying state by virtue of favorable bielectronic exchange energy.

Similarly, **2** and **2'** yield a resonant $^1\Delta_g$ combination and an antiresonant $^1\Sigma_g^+$ one. Thus, it is seen that simple qualitative VB considerations not only predict the ground state of O_2 to be a triplet, but in addition they yield a correct energy ordering for the remaining low lying excited states.

5.3   AROMATICITY–ANTIAROMATICITY IN IONIC RINGS $C_nH_n^{+/-}$

As discussed in Chapter 1, simple resonance theory completely fails to predict the fundamental differences between, for example, $C_5H_5^+$ and $C_5H_5^-$, $C_3H_3^+$ and $C_3H_3^-$, $C_7H_7^+$ and $C_7H_7^-$. Hence, a decisive blow was dealt to resonance theory when, during the 1950s–1960s, organic chemists were finally able to synthesize these transient molecules and establish their stability patterns, which followed the Hückel rules, with no guide or insight coming from resonance theory. We will now demonstrate, what has been known for quite a while, (5–7), namely, that the simple VB theory outlined above is capable of deriving the celebrated $4n + 2/4n$ dichotomy for ions, and even goes beyond this prediction.

As an example, we compare the singlet and triplet states of the cyclopropenium molecular ions, $C_3H_3^+$ and $C_3H_3^-$, in Figs. 5.3 and 5.4. The VB configurations needed to generate the singlet and triplet states of the equilateral triangle $C_3H_3^+$ are shown in Fig. 5.3. It is seen that all the structures can be generated from one another by shifting single electrons from a singly occupied p_π orbital to a vacant one. By using the guidelines for VB matrix elements (see Appendix 3.A.2), we deduce that the leading matrix element between any pair of structures with singlet spins is $+\beta$, while for any pair with triplet spin the matrix element is $-\beta$ (see Exercise 5.2). The corresponding configurations of $C_3H_3^-$ are shown in Fig. 5.4. In this case, the signs of the matrix elements are inverted compared with the case of the cyclopropenium cation, and the matrix elements are $-\beta$ for any pair of singlet VB structures, while being $+\beta$ for any triplet pair of structures.

If we symbolize the VB configurations by heavy dots we can present these resonance interactions graphically, as shown in the mid-parts of Figs. 5.3 and 5.4. These interaction graphs are all triangles and have the topology of corresponding Hückel and Möbius AO interactions (8). Of course, one could go ahead to diagonalize the corresponding Hückel–Möbius matrices and obtain energy levels and wave functions. There is, however, a shortcut based on the well-known mnemonic of Frost and Musulin (9). Thus, the triangle is inscribed within a circle having a radius $2|\beta|$, and the energy levels are obtained from the points where the vertices of the triangle touch the circle. The use of this mnemonic for VB mixing shows that the ground state of $C_3H_3^+$ is a singlet

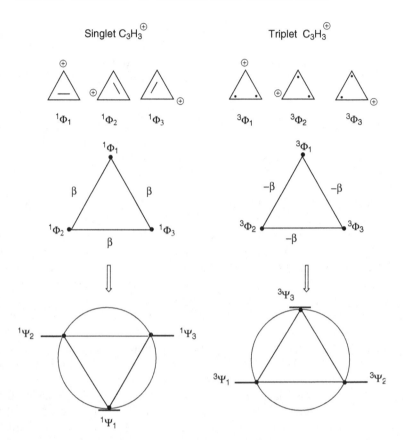

FIGURE 5.3   The VB structures for singlet and triplet states of $C_3H_3^+$, along with the graphical representation of their interaction matrix elements. The spread of the states is easily predicted from the circle mnemonic used in simple Hückel theory. The expressions for the VB structures (dropping normalization) are deduced from each other by circular permutations: $^1\Phi_1 = |a\bar{b}| - |\bar{a}b|$, $^1\Phi_2 = |b\bar{c}| - |\bar{b}c|$, $^1\Phi_3 = |c\bar{a}| - |\bar{c}a|$, $^3\Phi_1 = |ab|$, $^3\Phi_2 = |bc|$, $^3\Phi_3 = |ca|$.

state, while the triplet state is higher lying and doubly degenerate. In contrast, the ground state of $C_3H_3^-$ is a triplet state, while the singlet state is higher lying, doubly degenerate, and hence Jahn-Teller unstable. Thus, $C_3H_3^+$ is aromatic, while $C_3H_3^-$ is antiaromatic (5). In a similar manner, the VB states for $C_5H_5^+$ and $C_5H_5^-$ can be constructed. Restricting the treatment to the lowest energy structures, there remain five structures for each spin state, and the sign of the matrix elements will be inverted compared to the $C_3H_3^{+,-}$ cases. Now the cation will have $-\beta$ matrix elements for the singlet configurations and $+\beta$ for the triplets, while the anion will have precisely the opposite signs. The VB mnemonics will show instantly that $C_5H_5^-$ possesses a singlet ground state, and in contrast, $C_5H_5^+$ has a triplet ground state, whereas its singlet state is higher in energy and Jahn-Teller unstable. Now, in the cyclopentadienyl ions,

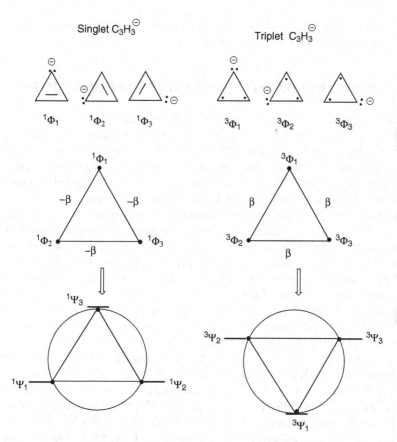

FIGURE 5.4 The VB structures for singlet and triplet states of $C_3H_3^-$, along with the graphical representation of their interaction matrix elements. The spread of the states is easily predicted from the circle mnemonic used in simple Hückel theory. The expressions for the VB structures (dropping normalization) are deduced from each other by circular permutations: $^1\Phi_1 = |ab\bar{c}c| - |\bar{a}bcc|$, $^1\Phi_2 = |\bar{b}caa| - |bc\bar{a}a|$, $^1\Phi_3 = |c\bar{a}bb| - |\bar{c}abb|$, $^3\Phi_1 = |abc\bar{c}|$, $^3\Phi_2 = |bca\bar{a}|$, $^3\Phi_3 = |cabb|$.

the cation is antiaromatic while the anion is aromatic. Moving next to the $C_7H_7^{+,-}$ species, the sign patterns of the matrix element will be inverted again and be the same as those in the corresponding $C_3H_3^{+,-}$ cases. As such, the VB mnemonic will lead to similar conclusions, that the cation is aromatic, while the anion is antiaromatic with a triplet ground state. Thus, the sign patterns of the matrix element, and hence the ground state's stability of molecular ions, obey the $4n/4n + 2$ dichotomy.

Clearly, a rather simple VB theory is required to reproduce the rules of aromaticity and antiaromaticity of the molecular ions, and to provide the correct relative energy levels of the corresponding singlet and triplet states. This VB treatment is virtually as simple as HMO theory itself, with the exception of the need to know the sign of the VB matrix element. But with some practice this can be learnt easily.

5.4 AROMATICITY/ANTIAROMATICITY IN NEUTRAL RINGS

Another mythical failure that has become associated with VB theory is the inability of simple resonance theory to distinguish between molecules such as benzene, on the one hand, and CBD and COT on the other hand. Recall that resonance theory, which simply enumerates resonance structures without any consideration of the size or sign of their matrix elements, considers that since benzene, CBD, and COT can all be expressed as resonance hybrids of their respective Kekulé structures, they should have similar properties, which they obviously do not. This "failure" to predict the right answer stuck to VB theory even though it has been known for quite some time that *ab initio* VB theory, even with modest basis sets, correctly predicts the geometric properties and stability patterns of aromatic and antiaromatic neutral rings. Indeed, at the *ab initio* level, Voter and Goddard (10) demonstrated that GVB calculations correctly predict the properties of CBD. Subsequently, Gerratt and co-workers (11,12) showed that spin-coupled VB theory correctly predicts the geometries and ground states of CBD and COT. In 2001, the authors of this book and their co-workers used VB theory to demonstrate (13) that the vertical resonance energy (RE) of benzene is larger than that of CBD and COT, and hence, the standard Dewar resonance energy (DRE) of benzene is substantial (e.g., 21 kcal/mol in VBSCF), while that for CBD is negative, in perfect accord with experimental estimates. Thus, properly done *ab initio* VB theory indeed predicts the antiaromatic nature of CBD and COT as opposed to the aromatic one for benzene. It does not fail.

The question is whether or not these reliable predictions of quantitative VB theory may also arise from a qualitative VB theory. Early semiempirical HLVB calculations by Wheland (14,15) and for that matter any VB calculations with only HL structures, incorrectly predict that CBD has resonance energy larger than that of benzene. Wheland, who analyzed the CBD problem, concluded that ionic structures play an important role, and that their inclusion would probably correct the VB predictions. Indeed the above mentioned successful *ab initio* VB calculations implicitly include ionic structures due to the use of CF orbitals in the VB descriptions of benzene, CBD, and COT. As will be immediately seen, ionic structures are indeed essential for understanding the difference between aromatic and antiaromatic species, such as benzene, CBD, and COT. Furthermore, the inclusion of ionic structures bring in some novel insight into other features of these molecules, such as ring currents, and so on (see Exercise 5.4).

First, let us see how VB theory explains the symmetry properties of $4n$ versus $4n + 2$ electronic systems in neutral rings. Consider a ring involving $2N$ atomic p_π orbitals, labeled from 1 to $2N$, (**3** in Scheme 5.1) and having one

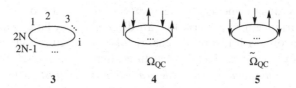

Scheme 5.1

electron per center. Each covalent Kekulé structure K_i^{cov} can be written as a product of the HL wave functions of all the π-bonded pairs in the structure. Thus, the pairs are $1-2, 3-4, \cdots, (2N-1)-(2N)$ for the first Kekulé structure, K_1^{cov}, as expressed in Equation 5.4, while for the second Kekulé structure, the pairs are $2-3, 4-5, \cdots, (2N)-1$, as in Equation 5.5 (normalization constants are dropped).

$$K_1^{cov} = |(1\bar{2} - \bar{1}2)(3\bar{4} - \bar{3}4) \cdots ((2N-1)\overline{2N} - \overline{(2N-1)}2N)| \qquad (5.4)$$

$$K_2^{cov} = |(2\bar{3} - \bar{2}3) \cdots ((2N-2)\overline{(2N-1)} - \overline{(2N-2)}(2N-1))((2N)\bar{1} - 1\overline{(2N)})| \qquad (5.5)$$

Expansion of Equation 5.4 or 5.5 results in a linear combination of determinants, which describe the different patterns of arranging N electrons with spin-up and N electrons with spin-down in a ring with $2N$ AOs. Each covalent Kekulé structure contains 2^N such determinants with equal coefficients except for their signs as shown in Equations 5.6 and 5.7:

$$K_1^{cov} = (\Omega_{QC} + (-1)^N \tilde{\Omega}_{QC}) + \sum_{i=1}^{2N-1} \pm \Omega'_i \qquad (5.6)$$

$$K_2^{cov} = (-1)^{N-1} \left[(\Omega_{QC} + (-1)^N \tilde{\Omega}_{QC}) + \sum_{i=1}^{2N-1} \pm \Omega''_i \right] \qquad (5.7)$$

Two of the determinants, labeled Ω_{QC} and $\tilde{\Omega}_{QC}$ are unique, and as depicted in **4** and **5** in Scheme 5.1, the electrons are arranged in these determinants in an alternant manner, spin-up–spin-down, and so on. In contrast, in the other determinants for example, Ω''_i, there are pairs of identical spins on adjacent carbon atoms. The spin-alternant determinants, Ω_{QC} and $\tilde{\Omega}_{QC}$, were earlier called the antiferromagnetic determinants, the Neel state (16), or the (QC) state (17). The corresponding wave functions in Equations 5.8 and 5.9 use the QC shorthand notation for these determinants:

$$\Omega_{QC} = |1\bar{2}3\bar{4}\ldots(2N-1)\overline{2N}| \qquad (5.8)$$
$$\tilde{\Omega}_{QC} = |\bar{1}2\bar{3}4\ldots\overline{(2N-1)}2N| \qquad (5.9)$$

These spin-alternant determinants are the only ones that are common to the two Kekulé structures in Equations 5.6 and 5.7. Moreover, they represent the lowest energy arrangements of the spin-system. All the other determinants possess spin arrangement patterns that are destabilized by Pauli repulsion among adjacent electrons with identical spins. It follows therefore that, as a rule, the lowest combination of the Kekulé structures, that is, *the ground state, will be the one that retains the spin-alternant determinants in contrast to the excited state combination that annihilates them.* Consequently, according to Equations 5.6 and 5.7, the lowest energy combination of covalent Kekulé structures must be signed by $(-1)^{N-1}$. Thus, the ground state for aromatics $(2N = 4n + 2)$ is always the positive combination, which also transforms as the totally symmetric representation in the D_{mh} point group (18). On the other hand, the ground state for antiaromatic species $(2N = 4n)$ involves the negative combination of their Kekulé structures and transforms as the B_{1g} representation (10,18–24).

Now, let us use this symmetry information to discuss the covalent–ionic mixing. The 60 monoionic structures of benzene fall into groups, which are distinguished by the distance between the ionic centers as shown in Fig. 5.5. The ortho-ionic structures are labeled as $\Phi_{ion}(1,2)$, the meta-ionic as $\Phi_{ion}(1,3)$, and the para-ionic as $\Phi_{ion}(1,4)$. For uniformity with other species, the latter will also be called the diagonal-ionic structures, $\Phi_{ion}(diagonal)$. Symmetry classification of these structures shows that each type of ionic structure has an A_{1g} combination, and this is also the case for structures with higher ionicity (di-ionic, etc.). In total, the entire set of 170 ionic structures of benzene

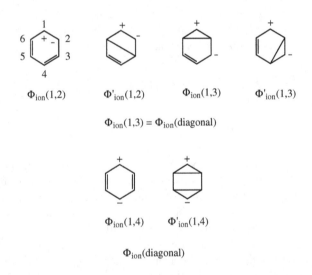

$$\Gamma_{ion}(1,2) = \Gamma_{ion}(1,3) = A_{1g} + A_{2g} + 2E_{2g} + B_{1u} + B_{2u} + 2E_{1u}$$

$$\Gamma_{ion}(1,4) = A_{1g} + E_{2g} + B_{1u} + E_{1u}$$

FIGURE 5.5 Monoionic structure types for benzene, and their reducible symmetry representations. (Adapted with permission from Ref. 24.)

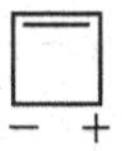

$\Phi_{ion}(1,2)$ 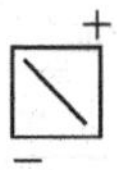 $\Phi_{ion}(1,3)$ = Φ_{ion}(diagonal)

$$\Gamma_{ion}\ (1,2) = B_{1g} + A_{1g} + A_{2g} + B_{2g} + 2E_u$$

$$\Gamma_{ion}\ (1,3) = A_{1g} + B_{2g} + E_u$$

FIGURE 5.6 Monoionic structure types for cyclobutadiene, and their reducible symmetry representations (Adapted with permission from Ref. 24.)

contains 20 ionic combinations with A_{1g} symmetry, and spanning all ranks and types of ionicity. These ionic combinations are able in turn, to mix into the covalent-state combination of the Kekulé structures that contains the spin-alternant determinants and has A_{1g} symmetry. Explicit VB calculations (24) showed that indeed *all* of these ionic types mix into the ground state of benzene.

Performing the same symmetry analysis for CBD shows that the diagonal-ionic structures, shown in Fig. 5.6, do not contain any B_{1g} combination to mix into the covalent ground state. Thus, the diagonal-ionics are excluded by symmetry from mixing into the ground state of cyclobutadiene in the D_{4h} uniform geometry. A similar exclusion appears in the di-ionic structures (24). Therefore, it follows that *although both benzene and square CBD are bond-delocalized, the stabilization associated with electronic delocalization, in their uniform geometries, is not the same when one considers the patterns of covalent-ionic mixing.* A similar analysis shows that 1,5 diagonal ionic structures are excluded from the mixing with the covalent Kekulé structures in COT, and so on (24).

It follows from the above symmetry analysis that aromatic versus antiaromatic species cannot be distinguished at the HLVB level, which uses only covalent structures defined with pure AOs. On the other hand, if the Kekulé structures are defined with CF orbitals, thus implicitly taking the ionic structures into account, then the RE arising from the mixing of CF-based Kekulé structures fits the following correct order for $4n$ *versus* $4n + 2$ electronic systems:

$$RE(4n) < RE(4n + 2) \tag{5.10}$$

Thus, the "failure" of VB theory to account for the instability of CBD arises from the use of an oversimplified version of VB theory, the HLVB method, *while modern VB theory successfully predicts a 4n/4n + 2 rule, already at the qualitative level.* As a numerical confirmation, we show in Table 5.1 the energies required to distort some C_mH_m molecules from D_{mh} to $D_{(1/2)mh}$, as calculated at different VB levels (24). It is apparent that at the HLVB level,

TABLE 5.1 Distortion Energies $\Delta E_{dis}{}^{a,b}$

Entry[c]	Species	ΔE_{dis}(cov)	ΔE_{dis} (CF)
1a	C_4H_4	2.2	−2.6
1b		1.2	−2.6
2a	C_6H_6	3.3	3.0
2b		3.6	4.0
3a	C_8H_8	2.3	−5.8

[a]Calculated (from Ref. 24) as the difference between the energy at the bond-alternated geometry and the uniform geometry, $\Delta E_{dis} = E(D_{(1/2)mh}) - E(D_{mh})$. A negative value signifies that the species is stabilized by distortion from the regular geometry (equal C−C bond lengths) to the bond alternated one. A positive value signifies the opposite.
[b]In kcal/mol.
[c]The entries labeled as "a" stand for calculations with STO-3G basis set, whereas "b" stands for calculations with the 6-31G basis set.

using covalent VB structures with localized AOs, the distortion energy is always positive for benzene, CBD and COT, showing a seemingly uniform aromatic behavior. On the other hand, if the same calculations are done at a modern VB level, using CF orbitals for the formally covalent structures, the distortion energy remains positive for benzene, but becomes negative for CBD and COT, in agreement with the familiar behavior of the aromatic and antiaromatic compounds with respect to bond alternation.

The role of ionic structures is crucial, and it can be appreciated based on a simple symmetry analysis. Indeed, as shown recently (24), the mixing of all ionic structure types into the spin-alternant determinant in benzene mediates the circular electronic motion (between the two determinants) and induce the diamagnetic ring current in a magnetic field. By contrast, the exclusion of the diagonal ionics in CBD excludes this motion. Exercise 5.4 demonstrates qualitatively this dichotomy of ring currents.

5.5 THE VALENCE IONIZATION SPECTRUM OF CH$_4$

As discussed in Chapter 1, the development of PES showed that the spectra could be simply interpreted if one assumed that electrons occupy delocalized molecular orbitals (25,26). In contrast, VB theory, which uses localized bond orbitals (LBOs), seems completely useless for interpretation of PES. Additionally, since VB theory describes equivalent electron pairs that occupy LBOs, the experimental PES results seem to be in discord with this theory. An iconic example of this "failure" of VB theory is the PES of methane that displays two different ionization peaks. These peaks correspond to the a_1 and t_2 MOs, but not to the four equivalent C−H LBOs in Pauling's hybridization theory.

Now, let us examine the problem carefully, in terms of LBOs, to demonstrate that VB gives the right result for the right reason. A physically correct representation of the $CH_4{}^+$ cation would be a linear combination of

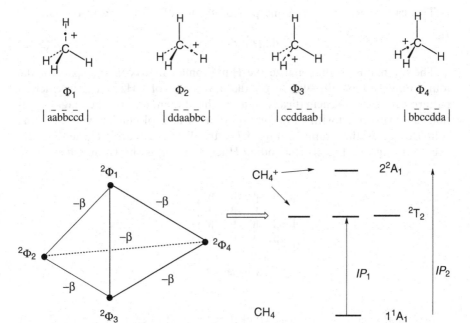

FIGURE 5.7 Generation of the $^2T_2$ and 2^2A_1 states of $CH_4{}^+$, by VB mixing of the four localized structures. The matrix elements between the structures, shown graphically, lead to the three-below-one splitting of the states, and to the observations of two ionization potential peaks in the PES spectrum.

the four forms, such that the wave function does not distinguish the four LBOs that are mutually related by symmetry. The corresponding VB picture, more specifically an FO–VB picture, is illustrated in Fig. 5.7, which enumerates the VB structures and their respective determinants. Each VB structure involves a localized 1e bond situation, while the other bonds are described by doubly occupied LBOs. To make life easier, we can use LBOs that derive from a unitary transformation of the canonical MOs. As such, these LBOs would be orthogonal to each other, and one can calculate the Hamiltonian matrix element between two such VB structures, by simply setting in the VB expressions all overlaps to zero. Thus, to calculate the Φ_1–Φ_2 interaction matrix element between Φ_1 and Φ_2 VB structures (which in this case are simple determinants) one first puts the orbitals of both determinants in maximal correspondence, by means of a transposition in Φ_2. The two so-transformed determinants differ by only one spin-orbital, $\bar{c} \neq \bar{d}$, so that their matrix element is simply β. Going back to the original Φ_1 and Φ_2 VB structures, one finds that their matrix element is negatively signed (Eq. 5.11):

$$\langle \Phi_1 | H^{\mathit{eff}} | \Phi_2 \rangle = \langle |a\bar{a}b\bar{b}c\bar{c}d| | H^{\mathit{eff}} | |d\bar{d}a\bar{a}b\bar{b}c| \rangle = -\langle |a\bar{a}b\bar{b}c\bar{c}d| | H^{\mathit{eff}} | |a\bar{a}b\bar{b}c\bar{d}d| \rangle = -\beta$$

$$(5.11)$$

This can be generalized to any pair of $\Phi_i - \Phi_j$ VB structures in Fig. 5.7:

$$\langle \Phi_i | H^{\text{eff}} | \Phi_j \rangle = -\beta \qquad (5.12)$$

There remains to diagonalize the Hamiltonian matrix in the space of the four configurations, $\Phi_1 - \Phi_4$, to get the four states of CH_4^+. The interaction patterns are shown schematically in a graphical manner, and the diagonalization can be executed with a Hückel program for a "molecule" having the same connectivity as the graph in Fig. 5.7 with all the β matrix elements being negatively signed. The corresponding Hückel matrix is shown in Scheme 5.2.

$$\begin{vmatrix} -E & -\beta & -\beta & -\beta \\ -\beta & -E & -\beta & -\beta \\ -\beta & -\beta & -E & -\beta \\ -\beta & -\beta & -\beta & -E \end{vmatrix} = 0$$

Scheme 5.2

Diagonalization of the above Hückel matrix, with negatively signed β leads to the final states of CH_4^+, shown alongside the interaction graph in Fig. 5.7. These cationic states exhibit a three-below-one splitting, a low lying triply degenerate $^2T_2$ state and above it a $^2A_1$ state. The importance of the sign of the matrix element can be appreciated by diagonalizing the above Hückel matrix using a positively signed β. Doing that would have reversed the state ordering to one-below-three, which is of course incorrect. Thus, simple VB theory predicts correctly that methane will have two ionization peaks, one (IP_1) at lower energy corresponding to transition to degenerate $^2T_2$ states and one (IP_2) at a higher energy corresponding to transition to the $^2A_1$ state. The same result could have been found even without any calculation, for example, by use of symmetry projection operators of the T_d point group on the localized cationic structures $\Phi_1 - \Phi_4$ in Fig. 5.7. The facility of making this prediction and its correspondence to experiment highlight once more that in this story too, the "failure" of VB theory originates more in a myth that caught on due to the naivety of the initial argument.

5.6   THE VALENCE IONIZATION SPECTRUM OF H_2O AND THE "RABBIT-EAR" LONE PAIRS

Another argument that has often been invoked against VB theory and the hybridization concept is the fact that the water molecule has two experimentally measurable ionization potentials, in apparent contradiction with the classical representation of water with its lone electron pairs located in two equivalent

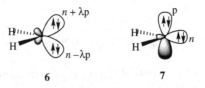

6   7

Scheme 5.3

sp^3 hybrid orbitals, the so-called "rabbit-ears", **6** in Scheme 5.3. This latter picture is popular among chemists, as it readily explains, for example, the anomeric effect, or the structure of ice, with each water molecule being the site of four hydrogen bonds from neighboring molecules arranged along tetrahedral directions with respect to the oxygen atoms, and so on. On the other hand, the two canonical MOs that represent the lone pairs (**7**) are non equivalent in shape and in energy. One of them is a pure p orbital perpendicular to the H_2O plane, while the other, referred to as n in what follows, is a non bonding orbital lying in the molecular plane. As these n and p canonical lone pairs have different energies, the two different IPs are readily explained by use of Koopmans' theorem. In contrast, the description in terms of equivalent hybridized lone pairs might naively be thought to lead to the prediction of a single common IP value.

First, let us express the polyelectronic wave functions, limited to the lone pairs, for the two apparently different representations. In the canonical MO representation, the polyelectronic wave function, Ψ_{MO}, is made of the doubly occupied n and p orbitals:

$$\Psi_{MO} = |n\bar{n}p\bar{p}| \tag{5.13}$$

In the VB-hybridized representation, the equivalent hybrids are $n + \lambda p$ and $n - \lambda p$ (recall Exercise 3.7), leading to the polyelectronic wave function Ψ_{VB} (omitting normalization):

$$\Psi_{VB} = |(n + \lambda p)(\bar{n} + \lambda \bar{p})(n - \lambda p)(\bar{n} - \lambda \bar{p})| \tag{5.14}$$

The wave functions Ψ_{MO} and Ψ_{VB} can be compared with each other, by a simple expansion of Ψ_{VB} in terms of elementary determinants as we did in Chapter 3 for the covalent and ionic parts of a bond wave function:

$$\Psi_{VB} = \lambda^2 |n\bar{n}p\bar{p}| - \lambda^2 |n\bar{p}p\bar{n}| - \lambda^2 |p\bar{n}n\bar{p}| + \lambda^2 |p\bar{p}n\bar{n}| + \cdots |n\bar{n}n\bar{p}| \cdots + \cdots \tag{5.15}$$

After eliminating all the determinants having two identical columns, the first four terms in Equation 5.15 remain, and can be further rearranged by orbital permutation, leading to a unique determinant, which is precisely the MO wave function, Ψ_{MO}:

$$\Psi_{VB} = |n\bar{n}p\bar{p}| = \Psi_{MO} \qquad (5.16)$$

It follows that the two seemingly different representations of the "lone pairs" of water, in terms of equivalent sp^3 hybrids or nonequivalent canonical MOs, are both correct in that they correspond to the same unique polyelectronic wave functions.

If we now remove one electron, the "rabbit-ear" representation leads to two degenerate VB structures, differing from each other by the location of the unpaired electron, in the sp^3 hybrid lying above or below the molecular plane. As these two VB structures differ by the replacement of one spin-orbital, they interact, leading to two states of different energies, as shown in Fig. 5.8. According to the rules outlined in Section 3.A.2, the leading term of the Hamiltonian matrix element between the two VB structures is -β, so that the lowest combination, $\Psi_{VB}^+(^2B_1)$, is the negative one, while the positive combination is an excited ionized state, $\Psi_{VB}^+(^2A_1)$. Removing an electron from the neutral ground state thus leads to two possible ionized states, and therefore to two distinct ionization potentials IP_1 and IP_2 (Fig. 5.8), in agreement with experiment.

In the same manner as we just compared above the ground state of water, expressed in the hybridized VB and canonical MO representations, we can now compare the ionized states in terms of these two representations. The wave function for the lowest energy ionized VB state $\Psi_{VB}^+(^2B_1)$ is expressed by Equation 5.17:

$$\Psi_{VB}^+(^2B_1) = |(n - \lambda p)(\bar{n} - \lambda\bar{p})(n + \lambda p)| - |(n + \lambda p)(\bar{n} + \lambda\bar{p})(n - \lambda p)| \qquad (5.17)$$

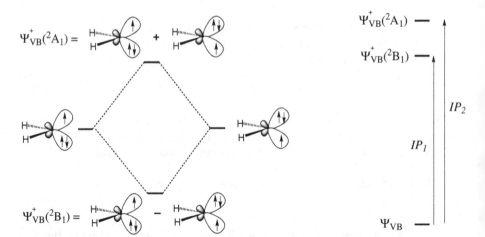

FIGURE 5.8 Generation of the $^2B_1$ and $^2A_1$ states of H_2O^+, by VB mixing of the two localized ionized structures.

As was done for the ground state (Eq. 5.15), here too, Equation 5.17 can be expanded in terms of elementary determinants (see Exercise 5.3), leading to a unique determinant that is nothing else but the wave function for the lowest energy ionized state in the canonical MO representation, with two electrons in n and one electron in the pure p orbital:

$$\Psi_{VB}^{+}(^{2}B_{1}) = |n\bar{n}p| \tag{5.18}$$

Similarly, expanding the excited ionized state, Ψ_{VB}^{+} $(^{2}A_{1})$, leads to a unique determinant that in MO terms represents the second ionized state, in which the electron is excited from the lowest lying lone pair:

$$\Psi_{VB}^{+}(^{2}A_{1}) = |p\bar{p}n| \tag{5.19}$$

It follows from the above analysis that the "rabbit-ears" and canonical MO representations of the water's lone pairs are both perfectly correct, as they lead to equivalent wave functions for the ground state of water, as well as for its two ionized states. Both representations account for the two ionization potentials that are observed experimentally. This example illustrates the well-known fact that, while the polyelectronic wave function for a given state is unique, the orbitals from which it is constructed are not unique, and this holds true even in the MO framework within which a standard localization procedure generates the rabbit-ear lone pairs while leaving the total wave function unchanged. Thus, the question "what are the true lone-pair orbitals of water?" is not very meaningful.

5.7 A SUMMARY

This chapter was dedicated to demonstrations that all the so-called "failures" of VB theory are in fact not real. It was shown that in each such "failure", one could use a simple VB theory, based on the principles outlined in Chapter 3, and arrive at the correct predictions—results. In so doing, this chapter also provided the reader with an opportunity to apply qualitative VB theory to some classical problems in bonding. Having done so, the reader is now more prepared for the material in Chapter 6, where VB theory is applied to chemical reactivity.

REFERENCES

1. W. A. Goddard, III, T. H. Dunning, Jr., W. J. Hunt, P. J. Hay, *Acc. Chem. Res.* **6**, 368 (1973). Generalized Valence Bond Description of Bonding in Low-Lying States of Molecules.
2. R. McWeeny, *J. Mol. Struct. (THEOCHEM)* **229**, 29 (1991). On the Nature of the Oxygen Double Bond.

3. R. D. Harcourt, *J. Phys. Chem.* **96**, 7616 (1992). Valence Bond Studies of O_2 and O_2^-: A Note on One-Electron and Two-Electron Transfer Resonances.

4. S. S. Shaik, P. C. Hiberty, *Adv. Quant. Chem.* **26**, 100 (1995). Valence Bond Mixing and Curve Crossing Diagrams in Chemical Reactivity and Bonding.

5. S. S. Shaik, in *New Theoretical Concepts for Understanding Organic Reactions*, J. Bertran and I. G. Csizmadia, Eds., NATO ASI Series, C267, Kluwer Academic Publishers, 1989, pp. 165–217. A Qualitative Valence Bond Model for Organic Reactions.

6. H. Fischer, J. N. Murrell, *Theor. Chim. Acta (Berlin)* **1**, 463 (1963). The Interpretation of the Stability of Aromatic Hydrocarbon Ions by Valence Bond Theory.

7. N. D. Epiotis, *Nouv. J. Chim.* **8**, 421 (1984). How to "Think" at the Level of MO – CI Theory Using Hückel MO Information!

8. E. Heilbronner, *Tetrahedron Lett.* **5**, 1923 (1964). Hückel Molecular Orbitals of Möbius-Type Conformations of Annulenes.

9. A. A. Frost, B. Musulin, *J. Chem. Phys.* **21**, 572 (1953). A Mnemonic Device for Molecular Orbital Energies.

10. A. F. Voter, W. A. Goddard, III, *J. Am. Chem. Soc.* **108**, 2830 (1986). The Generalized Resonating Valence Bond Description of Cyclobutadiene.

11. S. C. Wright, D. L. Cooper, J. Gerratt, M. Raimondi, *J. Phys. Chem.* **96**, 7943 (1992). Spin-Coupled Description of Cyclobutadiene and 2,4-Dimethylenecyclobutane-1,3-diyl: Antipairs.

12. P. B. Karadakov, J. Gerratt, D. L. Cooper, M. Raimondi, *J. Phys. Chem.* **99**, 10186 (1995). The Electronic Structure of Cyclooctatetratene and the Modern Valence Bond Understanding of Antiaromaticity.

13. S. Shaik, A. Shurki, D. Danovich, P. C. Hiberty, *Chem. Rev.* **101**, 1501 (2001). A Different Story of π-Delocalization—The Distortivity of the π-Electrons and Its Chemical Manifestations.

14. G. W. Wheland, *J. Chem. Phys.* **2**, 474 (1934). The Quantum Mechanics of Unsaturated and Aromatic Molecules: A Comparison of Two Methods of Treatment.

15. G. W. Wheland, *Proc. R. Soc. London, Ser.* A **164**, 397 (1938). The Electronic Structure of Some Polyenes and Aromatic Molecules. V—A Comparison of Molecular Orbital and Valence Bond Methods.

16. J. P. Malrieu, D. Maynau, *J. Am. Chem. Soc.* **104**, 3021 (1982). A Valence Bond Effective Hamiltonian for Neutral States of π-Systems. 1. Method.

17. P. C. Hiberty, D. Danovich, A. Shurki, S. Shaik, *J. Am. Chem. Soc.* **117**, 7760 (1995). Why Does Benzene Possess a D_{6h} Symmetry? A Quasiclassical State Approach for Probing π-Bonding and Delocalization Energies.

18. S. Shaik, S. Zilberg, Y. Haas. *Acc. Chem. Res.* **29**, 211 (1996). A Kekulé Crossing Model for the Anomalous Behavior of the b_{2u} Modes of Polyaromatic Hydrocarbons in the Lowest Excited $^1B_{2u}$ state.

19. R. McWeeny, *Theor. Chim. Acta* **73**, 115 (1988). Classical Structures in Modern Valence Bond Theory.

20. P. C. Hiberty, *J. Mol. Struct. (THEOCHEM)* **451**, 237 (1998). Thinking and Computing Valence Bond in Organic Chemistry.

21. S. Zilberg, Y. Haas, *J. Phys. Chem.* A **102**, 10843 (1998). Two-State Model of Antiaromaticity: the Low Lying Singlet States.

22. S. Shaik, A. Shurki, D. Danovich, P. C. Hiberty, *J. Am. Chem. Soc.* **118**, 666 (1996). Origins of the Exalted b_{2u} Frequency in the First Excited State of Benzene.

23. W. Wu, D. Danovich, A. Shurki, S. Shaik, *J. Phys. Chem. A* **104**, 8744 (2000). Using Valence Bond Theory to Understand Electronic Excited States. Applications to the Hidden Excited State (2^1A_g) of $C_{2n}H_{2n+2}$ $(n = 2-14)$ Polyenes.

24. A. Shurki, P. C. Hiberty, F. Dijkstra, S. Shaik, *J. Phys. Org. Chem.* **16**, 731 (2003). Aromaticity and Antiaromaticity: What Role Do Ionic Configurations Play in Delocalization and Induction of Magnetic Properties?

25. E. Heilbronner, H. Bock, *The HMO Model and its Applications*, Wiley, New York, 1976.

26. E. Honegger, E. Heilbronner, in *Theoretical Models of Chemical Bonding*, Vol. 3, Z. B. Maksic, Ed., Springer Verlag, Berlin- Heidelberg, 1991, pp. 100–151. The Equivalent Orbital Model and the Interpretation of PE Spectra.

27. S. Shaik, P. C. Hiberty, *Helv. Chim. Acta*, **86**, 1063 (2003). Myth and Reality in the Attitude Toward Valence Bond (VB) Theory. Are its "Failures" Real?

EXERCISES

5.1. Let π_x, π_y be the bonding, and π_x^*, π_y^* the antibonding combinations, respectively, of the π atomic orbitals x_1, x_2, y_1, and y_2 of dioxygen (see Fig. 5.1). Ignoring the σ bonds and lone pairs, write the MO wave function $\Psi_{MO}(1)$ that corresponds to the distribution of π electrons schematized in **1** (Fig. 5.1), with a total spin component $S_z = 0$. Expand $\Psi_{MO}(1)$ into its AO-based constituent determinants, and show that it is equivalent to the VB wave function $\Psi_{VB}(1)$ expressed in Fig. 5.1

5.2. Using the guidelines for VB matrix elements (see Appendix 3.A.2), express the leading Hamiltonian matrix element between the VB structures $^1\Phi_1$ and $^1\Phi_2$ of the triangular $C_3H_3^+$ cation in the singlet state (see Fig. 5.3). Do the same for the VB structures $^3\Phi_1$ and $^3\Phi_2$ of the triplet state. Repeat the same for the singlet and triplet states of the triangular $C_3H_3^-$ anion (see Fig. 3.4)

5.3. Let $|n\bar{n}p|$ and $|p\bar{p}n|$ be the wave functions for the lowest and first excited states, respectively, of ionized H_2O in the CMO representation. Prove that the two determinants are equivalent, respectively, to the VB expressions for the same states, $\Psi_{VB}^+(^2B_1)$ and $\Psi_{VB}^+(^2A_1)$.

5.4. It is experimentally known that benzene exhibits diamagnetic ring currents, while cyclobutadiene (CBD) does not. In this exercise, we will consider the wave functions of benzene and CBD as primarily made of their spin-alternant determinants Ω_{QC} and $\tilde{\Omega}_{QC}$, and we will model ring currents by the passage from one spin-alternant determinant to the other $(\Omega_{QC} \rightarrow \tilde{\Omega}_{QC})$, which is associated with a collective circular flow of electrons.

    a. Using the VB mixing rules in Appendix 3.A.2 show that the matrix element between Ω_{QC} and $\tilde{\Omega}_{QC}$ is always small, which prevents any direct transition between the two spin-alternant determinants.

    b. Devise a series of monoionic structures that might possibly mediate the $\Omega_{QC} \to \tilde{\Omega}_{QC}$ transition by successive 1e jumps, in both the benzene and CBD cases.

    c. By using the symmetry of the ionic states in Figs. 5.5 and 5.6, show that the electron flow around the circumference of benzene suffers no interruption, while the electron flow is interrupted in CBD.

Answers

Exercise 5.1 Dropping normalization constants, the π MOs and MO wave functions read

$$\pi_x = x_1 + x_2; \quad \pi_x^* = x_1 - x_2$$
$$\pi_y = y_1 + y_2; \quad \pi_y^* = y_1 - y_2$$
$$\Psi_{MO}(1) = |\pi_x \overline{\pi_x} \pi_x^* \overline{\pi_y} \pi_y \overline{\pi_y^*}|$$

To expand $\Psi_{MO}(1)$ into its constituent AO-based determinants, one can proceed separately first with the MOs of the π_x plane, then with those of the π_y plane:

$$|\pi_x \overline{\pi_x} \pi_x^*| = |(x_1 + x_2)(\overline{x_1} + \overline{x_2})(x_1 - x_2)|$$
$$= |(-x_1 \overline{x_1} x_2 + x_1 \overline{x_2} x_1 - x_1 \overline{x_2} x_2 + x_1 \overline{x_2} x_1)| \propto |(x_1 \overline{x_1} x_2 + x_1 \overline{x_2} x_2)|$$

Similarly,

$$|\overline{\pi_y} \pi_y \overline{\pi_y^*}| \propto |(\overline{y_1} y_1 \overline{y_2} + \overline{y_1} y_2 \overline{y_2})|$$

Combining the two π subsystems leads to:

$$|\pi_x \overline{\pi_x} \pi_x^* \overline{\pi_y} \pi_y \overline{\pi_y^*}| = |(x_1 \overline{x_1} x_2 + x_1 \overline{x_2} x_2)(\overline{y_1} y_1 \overline{y_2} + \overline{y_1} y_2 \overline{y_2})| = \Psi_{VB}(1)$$

Exercise 5.2 The wave functions for the various VB structures are expressed in the captions of Figs. 5.3 and 5.4.

Singlet state structures of triangular $C_3H_3^+$:

$$^1\Phi_1 = 2^{-1/2}(|a\overline{b}| - |\overline{a}b|)$$
$$^1\Phi_2 = 2^{-1/2}(|b\overline{c}| - |\overline{b}c|)$$

To get the leading terms of $\langle ^1\Phi_1|H|^1\Phi_2\rangle$, we will consider only the matrix elements between the determinants that differ by only one orbital, after permutations to put them in maximum spin-orbital correspondence, that is, $\langle |a\overline{b}||H||c\overline{b}|\rangle$ and $\langle |\overline{a}b||H||\overline{c}b|\rangle$. The terms that are to the power of two, βS or S^2, are neglected.

$$\langle |a\overline{b}||H||c\overline{b}|\rangle = \beta_{ac}$$
$$\langle |\overline{a}b||H||\overline{c}b|\rangle = \beta_{ac}$$

$$\langle {}^1\Phi_1|H|{}^1\Phi_2\rangle = \tfrac{1}{2}\langle |a\bar{b}| - |\bar{a}b||H||c\bar{b}| - |\bar{c}b|\rangle = \beta_{ac}$$

Triplet state structures of triangular $C_3H_3{}^+$:

$$^3\Phi_1 = |ab|$$

$$^3\Phi_2 = |bc|$$

$$\langle {}^3\Phi_1|H|{}^3\Phi_2\rangle = -\langle |ab||H||cb|\rangle = -\beta_{ac}$$

Singlet state structures of triangular $C_3H_3{}^-$:

$$^1\Phi_1 = 2^{-1/2}(|a\bar{b}c\bar{c}| - |\bar{a}bc\bar{c}|)$$

$$^1\Phi_2 = 2^{-1/2}(|b\bar{c}a\bar{a}| - |\bar{b}ca\bar{a}|)$$

Here, the first determinant of $^1\Phi_1$ and the second determinant of $^1\Phi_2$, as well as the first one of $^1\Phi_2$ and the second one of $^1\Phi_1$, differ by only one spin-orbital. After some permutations, we get:

$$\langle |a\bar{b}c\bar{c}||H| - |a\bar{b}c\bar{a}|\rangle = -\beta_{ac}$$

$$\langle -|\bar{a}bc\bar{c}||H||\bar{a}ba\bar{c}|\rangle = -\beta_{ac}$$

$$\langle {}^1\Phi_1|H|{}^1\Phi_2\rangle = -\beta_{ac}$$

Triplet state structures of triangular $C_3H_3{}^-$:

$$^3\Phi_1 = |abc\bar{c}|$$

$$^3\Phi_1 = |bca\bar{a}|$$

$$\langle {}^3\Phi_1|H|{}^3\Phi_2\rangle = \langle |abc\bar{c}||H||abc\bar{a}|\rangle = \beta_{ac}$$

Exercise 5.3 Dropping normalization constants, the VB ionized states can be expanded as follows:

$$\Psi^+_{VB}(^2B_1) = |(n - \lambda p)(\bar{n} - \lambda\bar{p})(n + \lambda p)| - |(n + \lambda p)(\bar{n} + \lambda\bar{p})(n - \lambda p)|$$

$$\Psi^+_{VB}(^2A_1) = |(n - \lambda p)(\bar{n} - \lambda\bar{p})(n + \lambda p)| + |(n + \lambda p)(\bar{n} + \lambda\bar{p})(n - \lambda p)|$$

$$|(n - \lambda p)(\bar{n} - \lambda\bar{p})(n + \lambda p)| = \lambda|n\bar{n}p| - \lambda^2|np\bar{p}| - \lambda|p\bar{n}n| + \lambda^2|p\bar{p}n|$$

$$|(n + \lambda p)(\bar{n} + \lambda\bar{p})(n - \lambda p)| = -\lambda|n\bar{n}p| - \lambda^2|n\bar{p}p| + \lambda|p\bar{n}n| + \lambda^2|p\bar{p}n|$$

By taking the negative and positive combinations of the two preceding equations, we have

$$\Psi^+_{VB}(^2B_1) = |n\bar{n}p|$$
$$\Psi^+_{VB}(^2A_1) = |p\bar{p}n|$$

Exercise 5.4 The spin-alternant states along with their symmetry assignment are shown in Scheme 5.Ans.1(a). Each state is composed of two determinants and one can think about the passage of one determinant to the other as a collective circular mode of the electrons as shown in (b). As in electron-transfer theory, here too, the propensity of this movement depends on the matrix element between the two spin-alternant determinants. Using the VB mixing rules in Appendix 3.A.2, these matrix elements are very small since the determinants differ by a change of many spin-orbitals (six in benzene and four in CBD). Therefore, the "electronic motion" will have to be driven by the ionic structures, which can mix into the spin-alternant determinants and mediate the circular motion (note that this is analogous to super-exchange in electron transfer where bridging groups provide states that mediate the electron transfer from one site to a remote one). Part (c) of

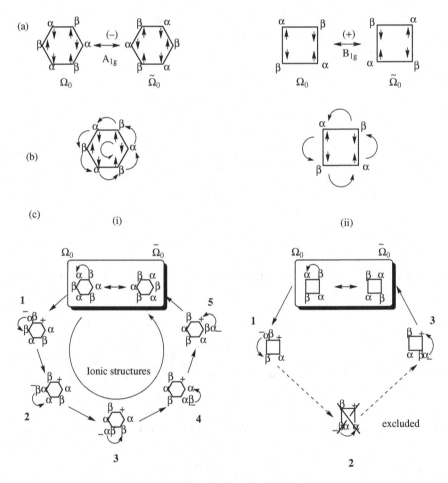

Scheme 5.Ans.1

the drawing shows this mediation. Thus in benzene [case (i)] one starts from Ω_0 and by shifting one β electron, one creates the ionic structure **1**, and subsequently by shifting an α electron one creates **2**, and so on, until one ends with $\tilde{\Omega}_0$. We note that some of the ionic structures are diagonal types. Since the A_{1g} representation of the ionic structures for benzene includes all the monoionic types (Fig. 5.5) including the diagonal ones, the electron flow and delocalization around the circumference of benzene suffers no interruption; complete delocalization of the electrons is achieved and the electrons can flow freely in either direction. In the case of CBD (ii), the diagonal ionics are excluded from mixing with the spin-alternant determinants (Fig. 5.6), and therefore the electron flow around the circumference will be interrupted.

In the presence of a magnetic field, the electrons of benzene will flow around the perimeter in a preferred direction that creates diamagnetic ring currents. One might say that in aromatic species the disposition for ring current exists already in the electronic structure of the ground state; the magnetic field simply sets the preferred direction of flow that repels the external field (hence, diamagnetic currents). In CBD and other antiaromatic systems, where the diagonal ionics are excluded, a diamagnetic ring current cannot be elicited. Instead, one can show using van Vleck's magnetic theory that the electronic flow in antiaromatic species will be paramagnetic and will arise by mixing of excited states into the ground state in the presence of the magnetic field. The interested reader may consult Ref. (24) for details of these paramagnetic ring currents.

6 Valence Bond Diagrams for Chemical Reactivity

6.1 INTRODUCTION

There are two fundamental questions that a model of chemical reactivity has to answer: What are the origins of barriers? What are the factors that determine reaction mechanisms? Since chemical reactivity involves bond breaking and making, VB theory with its focus on the bond is able to provide a lucid model that answers the two questions in a unified manner. The centerpiece of the VB model is the VB correlation diagram that traces the energy of the VB configurations along the reaction coordinate. The subsequent configuration mixing reveals the cause of the barrier, the nature of the transition state, and the reasons for occurrence of intermediates.

The VB diagram that is discussed in this chapter was initially derived by projecting MO-based wave functions onto VB structures (1). This projection was done in the manner that was described in Chapter 4, by taking MO and MO−CI wave functions and projecting them along the entire reaction coordinate. Using this transformation created a bridge between the VB and MO descriptions of chemical reactivity, and provided the means to import into the VB model, the feature of orbital symmetry through the use of the FO−VB representation (see Section 3.1.3). The goal of this chapter is to provide a brief review of the method and its application to reactivity problems, such as barrier heights, stereo- and regioselectivities, and mechanistic alternatives. Exhaustive treatments can be found in a few reviews in the primary literature (2–7). The historical chapter briefly discusses the early VB-based methods developed by London, Evans, Polanyi, Eyring, and co-workers to compute potential energy surfaces for chemical reactions.

6.2 TWO ARCHETYPAL VALENCE BOND DIAGRAMS

For a chemist, the most obvious answer to the question, "What happens to molecules as they react?" (1) is that during the reaction, bonds and electron

---

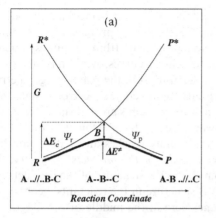
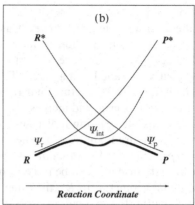

FIGURE 6.1 The VB diagrams for chemical reactivity: (a) The valence bond state correlation diagram (VBSCD) showing the mechanism of barrier formation by avoided crossing of two state curves, Ψ_r and Ψ_p, of reactants and products types. (b) The valence bond configuration mixing diagrams (VBCMD) showing the formation of a reaction intermediate, by avoided crossing of a third curve, Ψ_{int}. The final (adiabatic) states for the thermal reactions are shown by the thick curve.

pairs interchange their location and thereby cause a change in the molecular structures. These changes are native to the VB method that deals with localized bonds and electron pairs distributed in VB configurations in ways that characterize particular bonding patterns in molecules. The VB diagram shows these changes in a vivid manner by focusing on the "active bonds", those that are being broken or made during the reaction. An entire gamut of reactivity phenomena requires just two generic diagrams, which are drawn schematically in Fig. 6.1, and that enable a systematic view of reactivity.

Figure 6.1a depicts a diagram of two interacting states, called a VBSCD, which describes the formation of a barrier in a single chemical step due to avoided crossing or resonance mixing of the VB states for reactants and products. The second is a three-curve diagram (or generally a many-curve diagram), called a VBCMD, which describes a stepwise mechanism derived from the VB mixing of the three curves or more, one of which defines a reaction intermediate (5–7).

6.3   THE VALENCE BOND STATE CORRELATION DIAGRAM MODEL AND ITS GENERAL OUTLOOK ON REACTIVITY

Later in this chapter we construct VBSCDs in a systematic manner, but for the moment let us accept the VBSCD as drawn in Fig. 6.1a. This diagram applies to the general category of reactions that can be described as the interplay of two major VB structures, that of the reactants (e.g., A/B−C in Fig. 6.1a) and that of the products (e.g., A−B/C). It displays the ground-state energy profile for the reacting system (bold curve), as well as the energy profiles for individual VB structures as a function of the reaction coordinate (thinner curves); these latter curves sometimes are also called "diabatic" curves, while the full state energy curve is called "adiabatic". Thus, starting from the reactants geometry on the left, the VB structure that represents the reactant's electronic state, R, has the lowest energy and merges with the supersystem's ground state. Then, as one deforms the reacting molecules toward the products' geometry, the latter VB structure gradually rises in energy and finally reaches an excited state P^* that *represents the VB structure of the reactants in the products' geometry*. A similar diabatic curve can be traced from P, the VB structure of the products in its optimal geometry, to R^*, the same VB structure in the reactants' geometry. Consequently, the two curves cross somewhere in the middle of the diagram. The crossing is of course avoided in the adiabatic ground state, owing to the resonance energy B that results from the mixing of the two VB structures. The barrier is thus interpreted as arising from avoided crossing between two diabatic curves, which represent the energy profiles of the VB state curves of the reactants and products. Thus the diagram shows that when molecules react, the bonding schemes of the reactants and products interchange and mix along the reaction pathway.

The simplest rigorous expression for the barrier, based on Fig. 6.1a, is given by Equation 6.1,

$$\Delta E^{\neq} = \Delta E_c - B \qquad (6.1)$$

in terms of the energy required to reach the crossing point, ΔE_c in Fig. 6.1a, minus B, the resonance energy of the transition state due to the mixing of the VB forms. One can further relate the height of the crossing point to another fundamental factor in the diagram in Fig. 6.1a, as follows:

$$\Delta E_c = fG \quad f < 1 \qquad (6.2)$$

Here, f is some fraction of G, the promotion energy required to convert the electronic structure of R to that of R^*, the vertical state of R that has the same spin pairing as in the product state, P. A useful way of understanding this gap is as *a promotion energy* that is required in order to enable the bonds of the reactants to be broken before they can be replaced by the bonds of the products. As discussed later, this expression forms a basis for structure–reactivity relationships based on the VBSCD.

A very simple demonstration of the VB crossing diagram is shown in Output 6.1 (placed at the end of this chapter), which summarizes the VBSCF/STO-3G calculations (using XMVB) (8) of the two HL configurations for the hydrogen-exchange reaction, $H_{(2)}\text{-}H_{(1)} + H_{(3)}{}^{\bullet} \rightarrow H_{(2)}{}^{\bullet} + H_{(1)}\text{-}H_{(3)}$, for three different geometries of the supersystem $[H_{(2)}\text{—}H_{(1)}\text{—}H_{(3)}]$; one geometry is a short $H_{(2)}\text{—}H_{(1)}$ bond and a long $H_{(1)}\text{—}H_{(3)}$, the second one has identical $H_{(2)}\text{—}H_{(1)}$

and $H_{(1)}$—$H_{(3)}$ distances (0.923 Å) and corresponds to the transition state geometry, and the third one has a long $H_{(2)}$—$H_{(1)}$ and a short $H_{(1)}$—$H_{(3)}$. By inspection of the wave function or the diagonal energies in the Hamiltonian matrix, we can see that the two HL structures interchange along the exchange pathway. Initially, the structures $[H_{(2)}-H_{(1)}/H_{(3)}{}^\bullet]$ and $[H_{(3)}-H_{(1)}/H_{(2)}{}^\bullet]$, noted [1 2 3] and [3 1 2] in the output, have the respective energies, -2.991318 and -2.771714 hartrees, corresponding to an energy gap of 138.0 kcal/mol in the first geometry (closer to R). Then in the transition state the structures attain the same energy, and beyond it, at the geometry closer to P, the HL structure that was higher in energy initially becomes the lowest one. If we further inspect the wave function and total energy of the adiabatic state at the point corresponding to transition state geometry, we can see that the wave function is an equal mixture of the two structures, and the lowest state lies below the energy of an individual structure (see **II.a** in Output 6.1) by the amount of 38.1 kcal/mol, which for this crude application can be associated with the quantity B in Fig. 6.1a. Of course, an accurate calculation of the full VBSCD would require more VB structures (see below), a better basis set, and a more sophisticated VB method. Nevertheless, this qualitative picture of crossing and avoided crossing will not change.

6.4   CONSTRUCTION OF VALENCE BOND STATE CORRELATION DIAGRAMS FOR ELEMENTARY PROCESSES

In this section, we show how to construct VBSCDs by VB following and mixing along the reaction coordinate. We will focus on polar-covalent bonds, where the primary VB structures are the two HL forms. The VBSCDs for cases involving ionic bonds were discussed in the most recent review on VB diagrams (5). The VBSCD for bond heterolysis in solution is given as an exercise (Exercise 6.2).

6.4.1   Valence Bond State Correlation Diagrams for Radical Exchange Reactions

Consider a reaction that involves cleavage of a bond A-Y by a radical $X^\bullet$ (X, A, Y = any atom or molecular fragment):

$$X^\bullet + A - Y \rightarrow X - A + Y^\bullet \tag{6.3}$$

Scheme 6.1 shows the VB structure set for the "active bonds" that interchange during the reaction; there are eight structures that distribute the three electrons among the three fragments in all possible ways (assuming each fragment has a single active orbital). The first three structures (1_R–3_R) are the HL and ionic forms that describe the polar–covalent A—Y bond and the $X^\bullet$ radical in the reactants, while the second trio (1_P–3_P) describes the A—X bond

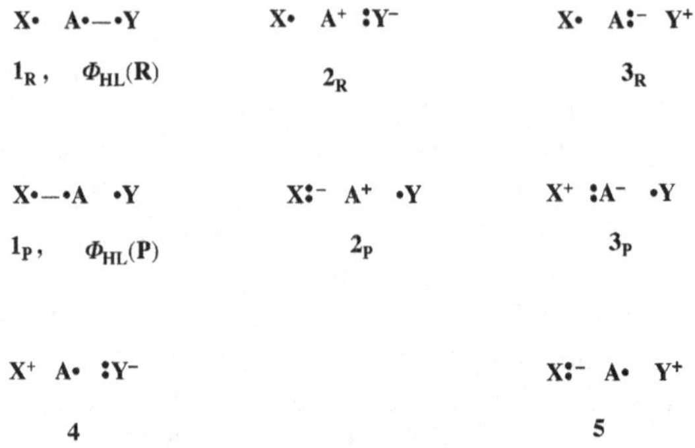

X• A•—•Y   X• A⁺ :Y⁻   X• A:⁻ Y⁺

1_R, $\Phi_{HL}(R)$ 2_R 3_R

X•—•A •Y   X:⁻ A⁺ •Y   X⁺ :A⁻ •Y

1_P, $\Phi_{HL}(P)$ 2_P 3_P

X⁺ A• :Y⁻   X:⁻ A• Y⁺

4   5

Scheme 6.1

and the Y• radical in the products. There are two additional structures, labeled as **4** and **5**, which do not belong to reactant or product bonding, but will mix into the transition state for the reaction.

Figure 6.2 describes the VB correlation diagram for the primary HL structures. Beginning with the HL structure of the reactants (1_R, Scheme 6.1), which is labeled initially as **R**, we can see that along the reaction coordinate the A•—•Y bond of **R** undergoes breaking, while at the same time a repulsive interaction (50% of a triplet repulsion) builds up between the nonbonded X• and A• fragments (consult Section 3.5.4). Consequently, **R** rises in energy and correlates with the excited state **P***. Similarly, starting from the HL structure of the products (1_P, Scheme 6.1), which is initially labeled as **P** on the right-hand side of the diagram, the structure suffers along the reaction coordinate A•—•X bond breaking and triplet repulsion between A• and Y•, and hence, it correlates eventually to the excited state **R***.

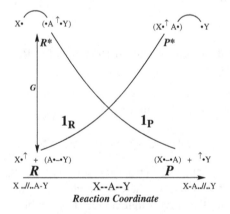

FIGURE 6.2 The state correlation for the radical exchange reaction, X• + A—Y → X—A + Y•, using the HL covalent structures (refer to Scheme 6.1 for the structure set).

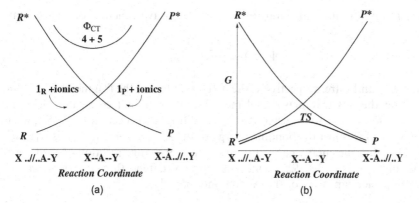

FIGURE 6.3 (a) The state correlation for the radical exchange reaction, $X^\bullet + A{-}Y \rightarrow$ $X{-}A + Y^\bullet$, using the Lewis state curves (refer to Scheme 6.1 for the structure set). (b) The VBSCD showing the avoided crossing of the state curves (augmented by the mixing of Φ_{CT}).

The two intersecting HL curves can be converted to state curves, in Fig. 6.3a, by mixing into them the ionic structures (from Scheme 6.1) to make bond wave functions. Thus mixing the two ionic configurations, 2_R and 3_R into 1_R, will convert the HL A$\bullet{-}\bullet$Y bond into a Lewis A$-$Y bond, and similarly, mixing of 2_P and 3_P into 1_P will generate the Lewis X$-$A bond. So now, we have two Lewis states that can be followed along the reaction coordinate. Based on what we have learnt so far on the bond wave function, the mixing of the ionic structures into the HL configuration is strong when the HL bond is short, for example in the ground states (R and P), and gradually diminishes as the HL bond gets longer. As such, starting with the Lewis state R in Fig. 6.3a and following it along the reaction coordinate where the A$-$Y distance becomes infinitely long, the state curve will gradually lose its ionic components and correlate to P^*, which is a pure HL representation of the reactant bonding in the geometry of the product. Likewise, starting with the Lewis curve at the P extreme and following it along the reaction coordinate will end with the pure HL excited state R^* at the other extreme. Finally, structures **4** and **5**, which do not belong to reactant or product bonding, can be mixed with each other to generate the charge-transfer excited state (Φ_{CT}), which involves a single electron transfer from the radical to the molecule at the two extremes of the diagram. In the subsequent step, we can allow the state curves to mix. Consequently, the crossing will be avoided leading to a transition state, and a barrier on the ground-state energy curve, as shown in Fig. 6.3b.

As we already argued, the origin of the barrier is G. Since R^* in Fig. 6.3 is just the VB image of the product HL wave function in the geometry of the reactants, this excited state displays a covalent-bond coupling between the infinitely separated fragments X and A, and an uncoupled fragment Y$^\bullet$ in the vicinity of A. Dropping normalization factors, the VB wave function of such a state reads

$$\Psi(\boldsymbol{R}^*) = |x\bar{a}y| - |\bar{x}ay| \qquad (6.4)$$

where x, a, and y are, respectively, the active orbitals of the fragments X, A, and Y. Using the VB formulas for elementary interactions (see Section 3.5), the energy of $\boldsymbol{R}^*$ relative to the separated X, A, Y fragments becomes $-\beta_{ay}S_{ay}$, while the energy of $\boldsymbol{R}$ is just the bonding energy of the A−Y fragment, that is, $2\beta_{ay}S_{ay}$. It follows that the energy gap G for any radical exchange reaction of the type in Equation 6.3 is $-3\beta_{ay}S_{ay}$, which is approximately three-quarters of the singlet−triplet gap ΔE_{ST} of the A−Y bond, namely,

$$G \approx 0.75\,\Delta E_{ST}(A - Y) \qquad (6.5)$$

The state $\boldsymbol{R}^*$ in Equation 6.4 strictly keeps the HL wave function of the product $\boldsymbol{P}$, and is hence a quasi/spectroscopic state that has a finite overlap with $\boldsymbol{R}$. If one orthogonalizes the pair of states $\boldsymbol{R}$ and $\boldsymbol{R}^*$, by, for example, a Graham−Schmidt procedure (see Exercise 6.3), the excited state becomes a pure spectroscopic state in which the A−Y is in a triplet state and is coupled to X$^\bullet$ to yield a doublet state. In such an event, one could simply use, instead of Equation 6.5, the spectroscopic gap G_S in Equation 6.6 that is simply the singlet−triplet energy gap of the A−Y bond:

$$G_S = \Delta E_{ST}(A - Y) \qquad (6.6)$$

Each formulation of the state $\boldsymbol{R}^*$ has its own advantages (6,9), Equation 6.6 has the merit of simplicity and is easiest to apply when several bonds are broken−made in a reaction. On the other hand, Equation 6.5, which is more faithful to the sense of the VB correlation, should be preferred when subtle effects are searched for (see Exercises 6.5−6.7). What is essential for the moment is that both expressions use a gap that is either the singlet-to-triplet excitation of the bond that is broken during the reaction, or the same quantity scaled by approximately a constant 0.75. As mentioned above, a useful way of understanding this gap is as a promotion energy that is required in order to enable the A−Y bond to be broken and be replaced by another bond, X−A.

6.4.2 Valence Bond State Correlation Diagrams for Reactions between Nucleophiles and Electrophiles

Scheme 6.2 illustrates the VB structure set for describing an S_N2 reaction where the nucleophile, X$^-$, shifts an electron to the A−Y electrophile and forms a new X−A bond, while the leaving group Y departs with the negative charge, Equation 6.7:

$$X:^- + A - Y \rightarrow X - A + :Y^- \qquad (6.7)$$

$$X:^- \quad A\bullet\!-\!\bullet Y \qquad\qquad X\bullet\!-\!\bullet A \quad :Y^- \qquad\qquad X\bullet \quad A:^- \quad \bullet Y$$

$$6_R, \quad \Phi_{HL}(R) \qquad\qquad 6_P, \quad \Phi_{HL}(P) \qquad\qquad 7_{LB}$$

$$X:^- \quad A^+ \quad :Y^- \qquad\qquad X:^- \quad A:^- \quad Y^+ \qquad\qquad X^+ \quad A:^- \quad :Y^-$$

$$8_{TI} \qquad\qquad\qquad\qquad 9 \qquad\qquad\qquad\qquad 10$$

Scheme 6.2

The principal structures are the two HL forms describing reactants and products, 6_R and 6_P. The long-bond structure, 7_{LB}, is another covalently coupled configuration that involves coupling between the $X\bullet$ and $Y\bullet$ fragments and an anionic fragment $A:^-$. Structures 8–10 are ionic structures, each of which has two anions and one cationic fragment; the most important among these is the alternate triple ion structure, 8_{TI}.

Figure 6.4a illustrates the formation of the VB correlation diagram for this reaction using the two HL structures 6_R and 6_P, while disregarding electrostatic interactions between the anion and the molecule, and focusing on the bonding changes. Starting at the R extreme, the most stable structure is 6_R, which possesses an anion $X:^-$ and a HL $A\bullet\!-\!\bullet Y$ bond. Along the reaction coordinate, the HL bond is broken and a repulsive 3e interaction (see Section 3.5.2) builds up between $X:^-$ and $A\bullet$; hence, the HL structure of the reactants correlates to an excited structure P^* of the products. The HL structure of the products, 6_P, behaves in an analogous manner and correlates with R^* along the reverse reaction coordinate.

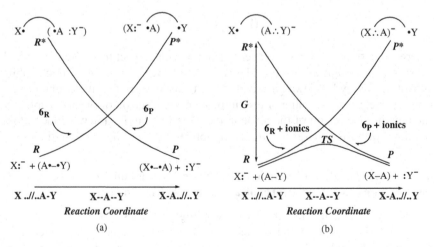

FIGURE 6.4 (a) The state correlation for a reaction between a nucleophile and an electrophile, $X:^- + A\!-\!Y \rightarrow X\!-\!A + :Y^-$, using the HL structures (refer to Scheme 6.2 for the structure set). (b) The VBSCD showing the avoided crossing of the state curves. Note that R^* and P^* are spectroscopic charge-transfer states.

In the same manner as in the previous reaction, here too, the HL curves can be converted to Lewis state curves by mixing in the secondary configurations. Thus, mixing the ionic structures 8_{TI} and **9** into the HL curve 6_R will convert the ground state to the X:$^-$ ion and a Lewis A–Y bond and generate the **R** state curve. Similarly, mixing structures 8_{TI} and **10** into the HL curve 6_P will generate the Lewis state X–A/Y:$^-$ and the corresponding **P** curve. Along the reaction coordinate, these Lewis curves will correlate with the HL excited structures shown already in Fig. 6.4a. However, at these points, the long-bond structure, 7_{LB}, will mix in and will convert the 3e moieties, (A$^\bullet$:Y$^-$) and (A$^\bullet$:X$^-$), to delocalized 3e species, (A∴Y)$^-$ and (A∴X)$^-$, respectively, as shown in Fig. 6.4b. Finally, the two state curves can be allowed to mix, and thereby generate a ground-state curve with a transition state and a barrier for the ground-state reaction.

Now, let us derive an expression for G, by simply examining the nature of the excited state R^* relative to the corresponding ground state in Fig. 6.4b. In R^*, A and Y are geometrically close to each other (as in the ground state, **R**) and separated from X by a long distance. The X fragment, which is neutral in the product **P**, must remain neutral in R^* and therefore carries a single active electron. Consequently, the negative charge is located on the A...Y complex, so that the R^* state is the result of a charge transfer from the nucleophile (X:$^-$) to the electrophile (A-Y), and is depicted in Fig. 6.4b as X$^\bullet$/(A∴Y)$^-$. It follows that the promotion from **R** to R^* is made of two terms: an electron detachment from the nucleophile, X:$^-$, and an electron attachment to the electrophile, A–Y. The promotion energy G is therefore the difference between the vertical ionization potential of the nucleophile, $I^*_{X:}$, and the vertical electron affinity of the electrophile, A^*_{A-Y}, given by Equation 6.8,

$$G = I^*_{X:} - A^*_{A-Y} \qquad (6.8)$$

where the asterisk denotes a vertical quantity with respect to molecular, as well as solvent, configurations (10,11). Thus, the mechanism of a nucleophilic substitution may be viewed as an electron transfer from the nucleophile to the electrophile, and a coupling of the supplementary electron of the electrophile to the remaining electron of the nucleophile, thereby making a new bond. In fact, the description is general for any electrophile–nucleophile combination.

6.4.3   Generalization of Valence Bond State Correlation Diagrams for Reactions Involving Reorganization of Covalent Bonds

As becomes evident from the two preceding examples, the VBSCD has the shape it has because of the crossing of the two HL structures that describe reactant and product bonding. In fact, as long as the bonds that break and form in the reaction are covalent (polar-covalent), this interchange of the HL structures will always occur along the reaction coordinate. This is a simple fundamental rule of VB theory:

Rule 1. For any reaction, that involves bond breaking and making, the HL structure describing the bonds in the reactants will interchange along the reaction coordinate with the corresponding HL structure that describes the bonds in the products. This interchange does not at all depend on the stereochemical course of the reaction; it is an outcome of the different bonding characteristics of reactants and products.

Furthermore, the two examples we worked out in detail represent archetypes of electronic reorganization that are common to all chemical reactions. In each case, the promotion energy G can be deduced from the nature of R^*, namely, by inspecting the HL structure of the products at the geometry of the reactants. Thus, during the radical exchange process in Fig. 6.2, all the fragments keep their electrons and the bond exchange occurs due to a different mode of pairing up the spins into bonds. In the nucleophile–electrophile reaction in Fig. 6.4a, one-electron transfer (from the nucleophile to the electrophile) attends the bond exchange. Accordingly, we can state the following generalities about the promotion energy G we are likely to encounter in chemical reactions:

Rule 2. The R^* state is an electronic image of the product state P, for any reaction (and so is P^* related to R). The promotion energy G involves two types of elementary excitations depending on the type of the reaction: (a) In reactions between nucleophiles and electrophiles, the excitation is a vertical charge-transfer energy corresponding to an electron transfer from the nucleophile to the electrophile. (b) In reactions that involve only bond exchange, a triplet excitation is required for each bond that is broken during the reaction.

Scheme 6.3 applies these rules by showing the HL structures for two cycloaddition reactions; 11_R and 11_P are the structures for the reactants and products of the Woodward–Hoffmann forbidden 2 + 2 reaction, while 12_R and 12_P are the structures for the Woodward–Hoffmann allowed Diels–Alder reaction. In both cases, the difference in the HL structure is only the mode of spin coupling, and therefore the promotion energy G will involve only singlet–triplet excitations. In accord, we drew in Fig. 6.5 the corresponding

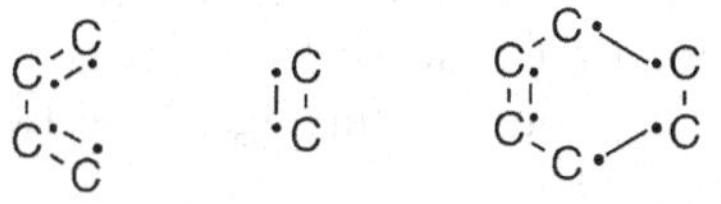

Scheme 6.3

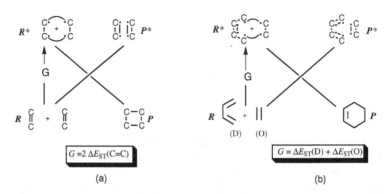

FIGURE 6.5 (a) The state correlation for (a) a 2 + 2 cycloaddition reaction, and (b) a Diels–Alder 4 + 2 reaction. The promotion gap expressions are shown beneath the respective diagrams.

VBSCDs without the VB mixing of the state curves. It is seen that in both cases, the promotion gap is given by the sum of the singlet-to-triplet excitation of the two reactants, while pairing up all the spins into a total singlet spin (see later for a discussion of the meaning of forbiddenness vs. allowedness in the Woodward–Hoffmann sense, in terms of the VB diagram).

The application of the rules is straightforward, and a few more examples are illustrated in the following sections and in Exercises 6.7 and 6.8.

6.5   BARRIER EXPRESSIONS BASED ON THE VALENCE BOND STATE CORRELATION DIAGRAM MODEL

As we argued at the outset of this chapter, the barrier $\Delta E^{\neq}$ of a reaction can be expressed as a function of fundamental parameters of the diagram. With reference to Fig. 6.1a and Equation 6.2, the simplest expression is Equation 6.9:

$$\Delta E^{\neq} = fG - B \qquad (6.9)$$

Here, G is the promotion energy, and f is some fraction <1. This f parameter is associated with the curvature of the diabatic curves, large upward curvatures meaning large values of f, and vice versa for small upward curvatures. The curvature depends on the descent of R^* and P^* toward the crossing point and on the relative pull of the ground states, R and P, so that f incorporates various repulsive and attractive interactions, which are typical of the VB structures of the individual curves along the reaction coordinate. The last quantity is the resonance energy B arising from the mixing of the two VB structures in the geometry of the crossing point.

A similar expression can be given for the barrier of the reverse reaction as a function of the product's gap and its corresponding f factor. Based on Fig. 6.6, one then distinguishes between the reactant's and product's gaps, G_r and G_p,

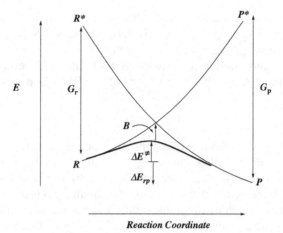

FIGURE 6.6 A generic VBSCD with the key factors that affect the barrier.

their corresponding f factors f_r and f_p, and the energy difference between reactants and products, ΔE_{rp}, that is, the reaction endo−exothermicity. By taking all these factors into a single equation, an expression for the barrier can be derived as (12):

$$\Delta E^{\neq} = f_0 G_0 + (G_p/2G_0)\Delta E_{rp} + (1/2G_0)\Delta E_{rp}{}^2 - B;$$
$$f_0 = 0.5(f_r + f_p), G_0 = 0.5(G_r + G_p) \tag{6.10}$$

Here, G_0 and f_0 are average quantities.

A more compact expression can be given by neglecting the quadratic term in 6.10 and taking $G_p/2G_0$ as $\sim 1/2$:

$$\Delta E^{\neq} \approx f_0 G_0 - B + 0.5\Delta E_{rp}; \quad G_0 = 0.5(G_r + G_p) \tag{6.11}$$

Here, the first two terms jointly define an intrinsic barrier quantity that is determined by the averaged f and G quantities and by the resonance energy of the transition state, due to the avoided crossing, while the third term gives the effect of the reaction thermodynamics (taken only to first order). Equations 6.10 and 6.11 also form a bridge to the popular Marcus equation that is used in physical organic chemistry (13) to analyze the barrier in terms of an intrinsic barrier, $\Delta E_0^{\neq}$, and a thermodynamic attenuation factor:

$$\Delta E^{\neq} = \Delta E_0^{\neq} + 0.5\Delta E_{rp} + (\Delta E_{rp})^2/16\Delta E_0^{\neq} \tag{6.12}$$

Note that both Equations 6.10 and 6.12 contain quadratic terms, albeit with different weighing coefficients of the $(\Delta E_{rp})^2$ term.

The intrinsic barrier in the Marcus equation plays an important role, but is essentially unknown and has to be determined either by averaging the barriers of pairs of identity reactions (when the reaction series possesses identity processes), or by assuming that it is a constant in a series. Now, simplifying the Marcus equation by neglecting the quadratic term, as we did to derive Equation 6.11, and comparing Equations 6.11 and 6.12 reveals an explicit expression of the intrinsic barrier in terms of the VBSCD parameters:

$$\Delta E_0{}^{\neq} = f_0 G_0 - B \tag{6.13}$$

Taken together the barrier expressions describe the interplay of three effects: (a) The $f_0 G_0$ factor describes the energy cost due to undoing the pairing of bonding electrons in order to make new bonds. (b) The ΔE_{rp} factor accounts for the classical rate-equilibrium effect. (c) The quantity B is the transition state resonance energy, which involves information about the electronic structure of the transition state and the preferred stereochemistry of the reaction.

6.5.1   Some Guidelines for Quantitative Applications of the Valence Bond State Correlation Diagram Model

A quantitative application of the VB diagram requires calculations of ΔE_c and B (see Eq. 6.1), or of f, G, and B (remember that f in Eq. 6.9 incorporates the scaling effect by the reaction energy, ΔE_{rp}; see below Fig. 6.7c). The energy gap factor, G, is straightforward to obtain for any kind of process. The height of the crossing point incorporates effects of bond deformations (bond stretching, angular changes, etc.) in the reactants and nonbonded repulsions between them at the geometry corresponding to the crossing point of the lowest energy on the seam of crossing between the two state curves (Fig. 6.1a). This, in turn, can be computed by means of *ab initio* calculations, for example, straightforwardly by use of a VB method (14–17), or with various methods for getting VB quantities from MO-based procedures (18) (see Chapter 9). Alternatively, the height of the crossing point can be computed by molecular mechanical means (19,20). Except for VB theory that calculates B explicitly, in all other methods this quantity is obtained as the difference between the energy of the transition state and the computed height of the crossing point. In a few cases, it is possible to use analytical formulas to derive expressions for the parameters f and B (2,5,16,21). Thus, in principle the VBSCD is a computable qualitative model.

6.6   MAKING QUALITATIVE REACTIVITY PREDICTIONS WITH THE VALENCE BOND STATE CORRELATION DIAGRAM

The VBSCD is a portable tool for making predictions by deducing the magnitudes of barriers from the properties of reactants. The purpose of the following sections is to teach an effective way of using the VBSCD qualitatively as a system of thought about chemical reactivity. Figure 6.7 pictorially outlines the impact of the various factors on the barrier in line with the quantitative

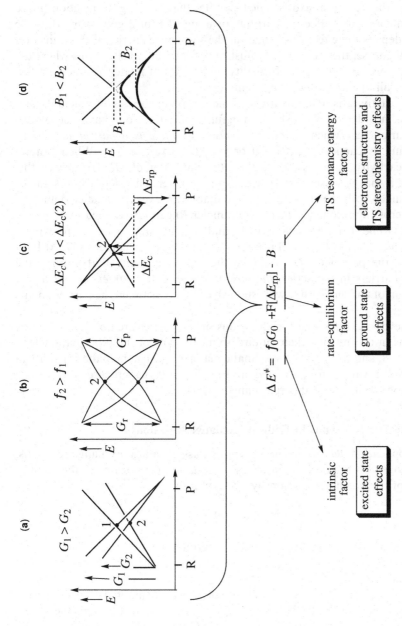

FIGURE 6.7 Pictorial illustration of the factors that control the variation of the barrier heights.

expression in Equation 6.11. Part (a) shows the effect of increasing the promotion energy gap, G, on the height of the crossing point, part (b) shows the effect of changes in the curvature factor, f, part (c) shows the impact of changing the reaction exothermicity (note that making a reaction more exothermic amounts effectively to the lowering of f for a given gap G), while part (d) depicts the effect of changes in the resonance energy of the transition states. Of course, these effects are illustrated for idealized situations where the factors change one at a time. In reality, in a series of related reactions, a few factors or all of them may change simultaneously.

Since reactivity is a multivariable function of properties of reactants and products, in order to establish meaningful correlations, one has to start with the G parameter, which is *the root cause why there is a barrier at all* (try considering a reaction with $G = 0$ or wait for Exercise 6.1). When can we expect reactivity patterns that follow the variation of G? To answer this question, it makes sense to start with families of related reactions; in some of these families both the curvatures of the diabatic curves (parameter f) and the avoided crossing resonance energy (parameter B) can be assumed to be nearly constant, while in other such series, f and B (and ΔE_{rp} in the more explicit expressions, e.g., Eq. 6.11) will usually vary in the same manner as G. At both instances, the parameter G would be the crucial quantity that governs the reaction barriers in the series: *the larger the gap G, the larger the barrier.* A correlation of barriers with G will establish causality between the barrier and its origins.

Whenever one finds that the variations of the barrier are not dominated by variations in G, one may then inquire about the role of the other parameters, notably; f, B, and ΔE_{rp}. In this systematic manner, one can map the reactivity in terms of the variations predicted in Fig. 6.7. Let us proceed with a few applications to illustrate this reasoning.

6.6.1   Reactivity Trends in Radical Exchange Reactions

As an application, let us consider a typical class of radical exchange reactions, the hydrogen abstractions from alkanes. Equation 6.14 describes the identity process of hydrogen abstraction by an alkyl radical:

$$R^{\bullet} + H - R \rightarrow R - H + R^{\bullet} \tag{6.14}$$

Identity reactions proceed without a thermodynamic driving force, and therefore project the role of promotion energy as the origin of the barrier (recall, the intrinsic barrier above).

The barriers for a series of radicals have been computed (22), and were found to increase as the $R-H$ bond energy D increases; the barrier is the largest for $R = CH_3$ and the smallest for $R = C(CH_3)_3$. This trend has been interpreted by Pross et al. (23) using the VBSCD model. The promotion gap G that is the origin of the barrier involves the singlet–triplet excitation of the

R$-$H bond (Eqs. 6.5 and 6.6). Now, according to Equations 3.37 and 3.38, this singlet$-$triplet gap is proportional to the bonding energy of the R$-$H bond, that is, $\Delta E_{ST} \approx 2D$. Therefore, the correlation of the barrier with the bond strength is equivalent to a correlation with the singlet$-$triplet promotion energy (Eq. 6.6), thereby reflecting the electronic reorganization that is required during the reaction. In fact, the computed barriers for the entire series can be fitted (6,7) to the barrier expression in Equation 6.15:

$$\Delta E^{\neq} = 0.3481G - 50\,\text{kcal/mol}, \quad G = 2D_{RH} \tag{6.15}$$

This correlation indicates that this is a reaction family with constants $f = 0.3481$ and $B = 50\,\text{kcal/mol}$. If this is indeed the case (can be checked by VB computations), this kind of correlation enables one to "measure" the resonance energy of the transition state of a chemical reaction.

Recently, computations demonstrated that the ΔE_{ST} quantity that varies as the corresponding bond energy (5–7;15–17) also governs the trends in the barriers for the hydrogen exchange identity reaction, $X^{\bullet} + XH \to XH + {}^{\bullet}X$, when X varies down the column of the periodic table, that is, $X = CH_3$, SiH_3, GeH_3, SnH_3, and PbH_3. Thus, in this series the G quantity, and hence also D_{RH}, decrease as X varies down the column of the periodic table. However, in the same series, B also decreases down the column and varies as $0.5D_{XH}$, while f is approximately a constant ($f = 1/3$). Since both G and B vary in the same manner, the barrier correlates with the size of the bond energy. A similar correlation was shown (24) to govern the trends for the series $X^{\bullet} + HX \to XH + {}^{\bullet}X$, X = halogen, where the barriers for the linear reaction decrease from X = F to X = I, in accordance with the decrease of the $\Delta E_{ST}(H-X)$ quantity and the corresponding bond energy, D_{XH}. We may expect the same trend for other $X^{\bullet} + HX$ series, where X varies down the column in the periodic table.

In contrast to the above series, for changes along a row of the periodic table, for example, in $X^{\bullet} + HX \to XH + {}^{\bullet}X$, where $X = CH_3$, NH_2, OH, and F, the barriers do not correlate with the G quantity (25). When the simple application of the VBSCD does not lead to a correct prediction, this is not because the model fails, since as explained above, the VB crossing of the HL structures underlying the VBSCD is rigorous. Rather, this is because the reactions that are considered are not sufficiently similar to each other to constitute a "family" of reactions, in the sense defined above. More specifically, what fails for the present series is the assumption that the variation of G dominates the variation of the barrier, since the other parameters do not vary in an adverse manner. As argued above, this assumption will be valid, if B is itself a constant (see Eq. 6.15) or when B varies in proportion to the X$-$H bond energy by a constant proportionality factor for the entire series. However, in series where B does not behave in these manners, the correlation of the barrier with the bond energy will often break down. Indeed, as was shown in the computational study, when the terminal groups of the X—H—X transition state carry lone pairs, the

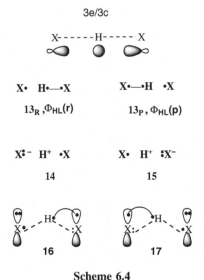

Scheme 6.4

transition state is bent away from linearity (25). This bending lowers the barrier, but not uniformly, for example, by 4.7 kcal/mol for X = F, by 3.8 kcal/mol for X = OH (data calculated for Ref. 26), 1.3 kcal/mol for X = Cl, and so on (24). One can see that whenever X bears lone pairs, the bending of the transition state (TS) will involve in addition to the "normal" structures expected from a 3e/3c process, also some new secondary VB structures, in which the lone pairs participate in the electronic reorganization, as shown in Scheme 6.4. Thus, in addition to the normal HL (13_R and 13_P) and ionic (14 and 15) structures, in which three electrons are distributed in the orbitals along the XHX axis, bending causes the appearance of new covalent configurations 16 and 17 in which the electrons are distributed also in the p_π atomic orbitals that were perpendicular to the X−X axis in the linear geometry. These extra configurations not only reinforce the covalent X−H bonding, but they also induce (due to the resonance between 16 and 17) some 3e bonding between the p_π orbitals that approach each other upon bending. The mixing of the latter VB structures into the 3e/3c transition state will lower the barrier upon bending of the TS. This increase, however, will not be uniform in a given series where the number of lone pairs is not constant, and which therefore do not constitute a family of reactions among which a simple application of the VBSCD model is possible.

6.6.2 Reactivity Trends in Allowed and Forbidden Reactions

Despite the above qualifications, reaction series where G dominates reactivity abound (2,4,5). Thus, in a series of Woodward−Hoffmann forbidden 2 + 2 dimerization reactions, the promotion gap is proportional to the sum of the $\Delta E_{ST}(\pi\pi^*)$ quantities of the two reactants (if needed, see Fig. 6.5a). Consequently, the barrier decreases from 42.2 kcal/mol for the dimerization of ethylene where $\Sigma \Delta E_{ST}(\pi\pi^*)$ is 198 kcal/mol down to <10 kcal/mol for the dimerization of disilene for which $\Sigma \Delta E_{ST}(\pi\pi^*)$ is 80 kcal/mol. Indeed, with differences of 118 kcal/mol in G, and assuming a factor f of 0.3, the barrier difference is predicted to be of the order of 35 kcal/mol. Note that in this comparison, the reduction in the value of G is accompanied by a concomitant change in ΔE_{rp}, namely, the reaction also becomes more exothermic; both factors will therefore reinforce one another. Thus, down the column of the periodic table, the barriers of forbidden reactions decrease markedly (note, however, that the mechanism may still be stepwise).

Of course, formally allowed and forbidden reactions, in the Woodward–Hoffmann sense, must be considered separately, as distinct reaction families, when correlating barriers to the G parameter. This happens because, as a rule, allowed reactions have comparatively larger resonance energies B relative to forbidden reactions, as will be shown below.

As depicted in Fig. 6.5b, the promotion gap for an allowed Diels–Alder reaction is given also by $\Sigma \Delta E_{ST}(\pi\pi^*)$ (5). Indeed, comparison of the Diels–Alder reaction to the trimerization of acetylene, which is also formally allowed, shows that the barrier jumps from 22 kcal/mol for the Diels–Alder reaction (Fig. 6.5b), where $\Sigma \Delta E_{ST}(\pi\pi^*)$ is ca. 173 kcal/mol, to 62 kcal/mol for the trimerization of acetylene, where $\Sigma \Delta E_{ST}(\pi\pi^*)$ is much larger (375 kcal/mol). Thus, with differences of 200 kcal/mol in G, the model predicts that the Diels–Alder reaction should have a lower barrier than the trimerization of acetylene. Note that this occurs despite the much greater exothermicity of the latter reaction, thus further highlighting the importance of G as the fundamental quantity of reactivity. This last comparison is interesting from another angle, because it demonstrates that by linking the two double bonds together as in butadiene, one reduces the value of G to a significant extent, compared with the termolecular reaction where the three double bonds are separated and each has to be promoted in the R^* state (see Exercise 6.8 for the trimerization of ethylene). Generally speaking, G varies linearly with the number of bonds that reacts, and *therefore the chemical transformation involving covalent bonds is normally limited to very few bonds*, unless the bonds are conjugated and lead to intramolecular advantage via the reduction of G. Chemical reactivity is localized.

6.6.3 Reactivity Trends in Oxidative–Addition Reactions

The simplest oxidative–addition reaction in organic chemistry is the bond activation reaction of, for example, a C–H by a carbenoid reagent X_2E:, for example, E = C, Si, Ge, Sn, Pb, while X = H or any other monovalent groups. Scheme 6.5a shows the R and R^* states for this bond activation process, using both electron-dot pairing schemes as well as FO–VB bond-pairing diagrams.

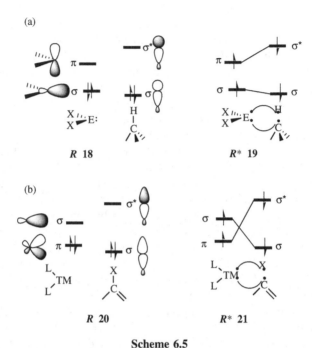

Scheme 6.5

Thus, **18** depicts the **R** state having a singlet paired X_2E:, represented by a doubly filled σ FO and a vacant π-type FO, as well as a molecule undergoing C–H activation, represented by a doubly occupied σ orbital and a vacant σ* orbital. The **R*** state in **19** involves electronic monoexcitations of the two reactants while coupling the spins of the electrons to a total singlet by creating two bond pairs, so as to match the nodal properties of the FOs. Based on that we can draw the state correlation in the VBSCD and assign the G value as the sum of the singlet–triplet promotion energies of the two reactants, as shown in Fig. 6.8a.

An analogous process is the oxidative–addition of a d^{10}–L_2TM (TM = Ni, Pd, Pt) transition metal complex to a C–X bond (X = halogen), which for TM = Pd is a key step in the catalytic cycles of, for example, the Heck reaction and Suzuki coupling (27). To simplify the argument, let us consider a case where the L_2 ligands are parts of a single bidentate ligand, so that the L_2TM has a bent geometry. If the ligands are of the "L" type, that is, they each bring a lone pair to the coordination sphere and are linked to the metal by a dative bond (e.g., NR_3, PR_3), then we know from the "isolobal analogy" (28) that the FOs of the L_2TM: complex resemble those of a carbenoid species (in reverse order), as shown in Scheme 6.5b. These FOs are obtained in the usual manner, shown by Hoffmann, by starting from a square planar L_2TMX_2 complex and cutting the two TM–X bonds, leaving behind the L_2TM fragment with two

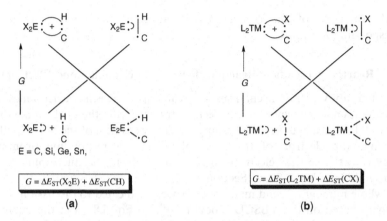

FIGURE 6.8 The state correlations in the VBSCDs of oxidative–additions of (a) a carbenoid reagent into a C–H bond, and (b) a transition metal complex into a C–X bond.

hybrids at the missing coordination sites. Drawing **20** in the scheme depicts the **R** state, where now the singlet paired L_2TM: is represented by a doubly filled π-type FO and a vacant σ type FO, while the C–X bond is represented by a doubly occupied σ orbital. The R^* state, shown in **21** involves unpairing of both closed-shell moieties to monoexcited states and coupling the electrons to a total singlet state with two bond pairs in symmetry matched FOs. The corresponding state correlation in the VBSCD is depicted in Fig. 6.8b, along with the expression of G. The analogy between the two VBSCDs and the dependence of the corresponding barriers on the same G quantity are apparent.

These oxidative–addition reactions have been treated extensively by Su et al. (29–31), using the VBSCD model. In all cases, a good correlation was obtained between the computed barriers of the reaction and the respective ΔE_{ST} quantities (which enter into the expression of G), including the relative reactivity of carbenoids, and of PtL_2 versus PdL_2 (29–31). Another treatment led to the same reactivity patterns for C–F bond activation reactions by $Rh(PR_3)_2X$ and $Ir(PR_3)_2X$ d^8 complexes, which are isolobal to carbenoids (30). A similar extended correlation was found recently for C–Cl activation by d^{10}-PdL_2 (32), and is dealt with in Exercise 6.9.

Despite the similarity between the processes in Figs. 6.8a and b, there is one immediately notable difference, and this is the stereochemistry of the reaction. Since stereochemical issues are treated later, we mention this here in passing. Thus, inspecting the FO–VB bond diagrams in **19** versus **21** (Scheme 6.5), it is apparent that the oxidative cleavage by a carbenoid will lead to a tetrahedral bond insertion product, while the one formed by use of the transition metal complex will be square planar. These different stereochemistries are well known since a d^8 tetracoordinated organometallic complex is expected to be square planar. Still it is interesting to note that these differences are dictated by nodal properties of the orbitals participating in the bond pairs in the corresponding R^* states.

6.6.4   Reactivity Trends in Reactions between Nucleophiles and Electrophiles

As noted above, the mechanism of reactions between nucleophiles and electrophiles may be viewed as an electron transfer from the nucleophile to the electrophile, and a coupling of the supplementary electron of the electrophile to the remaining electron of the nucleophile. The corresponding promotion energy G is the vertical electron-transfer energy from the nucleophile to the electrophile (see, Eq. 6.8 and Section 6.4.2).

A whole monograph and many reviews were dedicated to discussion of S_N2 reactivity based on the VBSCD model (Fig. 6.4, Eq. 6.8), and the interested reader may consult these (2,33–35). In brief, in S_N2 reactions all the reactivity factors change, for example, if we consider the $X:^-/CH_3X$ series, where X is a halogen, both G and B change and increase as the bond energy, D_{CX}, increase, but they vary differently. Consequently, the barrier for $X = F$ is slightly smaller than for $X = Cl$ in the gas phase, while in the rest of the series the barrier decreases as G decreases from $X = Cl$ to $X = I$, namely: $\Delta E^{\neq}$ $(Cl) > \Delta E^{\neq}$ $(Br) > \Delta E^{\neq}$ (I) (5). For the same reactions in a solvent, the vertical charge-transfer states have the same solvent orientations as the respective ground states, and hence are poorly solvated. Therefore, the G factor in solution is augmented by solvation terms (by roughly the solvation energy of the anion X^-), and this increase renders G the dominant factor, leading to the "normal" order of barriers, $\Delta E^{\neq}$ $(F) > \Delta E^{\neq}$ $(Cl) > \Delta E^{\neq}$ $(Br) > \Delta E^{\neq}$ (I) (10,36).

One important feature that emerges from the extensive treatments of S_N2 reactivity is the insight into variations of f. Thus, whether or not the odd electrons in the R^* state are easily accessible to couple to a bond determines the size of the f factor; the easier the bond coupling along the reaction coordinate, the smaller the f and vice versa. For example, in S_N2, with delocalized nucleophiles (e.g., acetate and phenyl thiolate), the active electron is not 100% located on the atom that is going to be eventually linked to the fragment A in the reaction in Fig. 6.4. So the diabatic curve will slowly descend from R^* to P and one may expect a large f factor. On the other hand, localized nucleophiles will correspond to smaller f factors. Of course, the same distinction can be made between localized and delocalized electrophiles, leading to the same prediction regarding the magnitude of f.

In general, since all reactions between closed-shell electrophiles and nucleophiles subscribe to the same diagram type (5) with vertical charge transfer R^* and P^* states, we expect to see the same influence of electron delocalization on the f factor. An example is the nucleophilic cleavage of an ester where the rate-determining step is known (37,38) to involve the formation of a tetrahedral intermediate, as depicted in Fig. 6.9. The promotion energy for the rate-determining step is, accordingly, the difference between the vertical ionization potential of the nucleophile and the electron

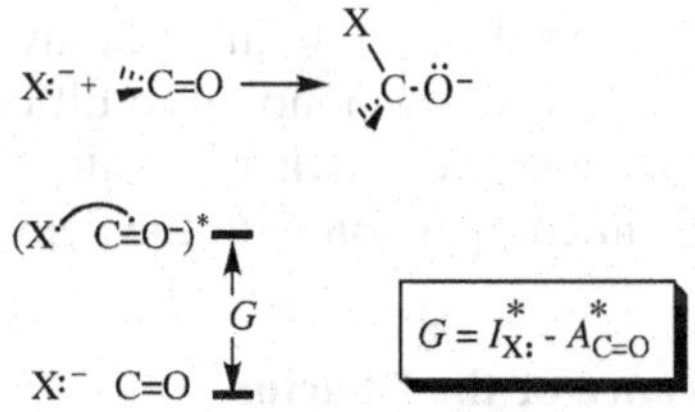

$$X{:}^- + \overset{\shortparallel}{}C{=}O \longrightarrow \overset{X}{\underset{}{\diagdown}}C\cdot\ddot{O}^-$$

$$(X \overset{\frown}{} C{=}O^-)^* \!\!\uparrow$$

$$\underset{X{:}^- \quad C{=}O}{}\,\Big\downarrow G \qquad \boxed{G = I_{X{:}}^* - A_{C=O}^*}$$

FIGURE 6.9 The ground and vertical charge-transfer states in the VBSCD that describes a nucleophilic attack on a carbonyl group.

attachment energy of the carbonyl group. The barriers are given, therefore, by Equation 6.16:

$$\Delta E^{\neq} = f[I_X^* - A_{C=O}^*] - B \tag{6.16}$$

The quantity $A_{C=O}^*$ is a constant for a given ester, and therefore the correlation of barriers with the promotion energy becomes a correlation with the vertical ionization energy of the nucleophiles, I_X^*. Figure 6.10 shows a structure-reactivity correlation for the nucleophilic cleavage of a specific ester based on the VBSCD analysis of Buncel et al (39). It is seen that the free energies of activation correlate with the vertical ionization energies of the nucleophile in the reaction solvent. Furthermore, localized and delocalized nucleophiles appear to generate correlation lines of different slopes. The two correlation lines obtained

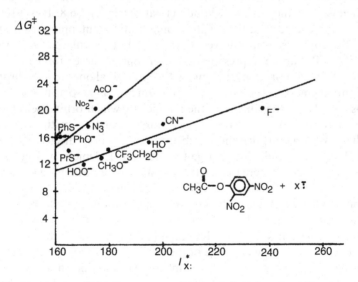

FIGURE 6.10 A plot of the free energy of activation for nucleophilic cleavage of an ester versus the vertical ionization potential of the nucleophile. (Adapted with permission from Ref. 39.)

for the experimental data in Fig. 6.10 are readily understood based on Equation 6.11 (and Fig. 6.7) as corresponding to different f values, where the localized nucleophiles possess the smaller f value, and hence a smaller structure–reactivity slope in comparison with the delocalized nucleophiles.

6.6.5   Chemical Significance of the f Factor

The f factor defines the intrinsic selectivity of the reaction series to a change in the vertical gap (2,5,33), that is,

$$f = \partial(\Delta E^{\neq})/\partial G \qquad (6.17)$$

As we just argued, for reactions of electrophiles and nucleophiles f increases as the nucleophile becomes more delocalized. Thus, the series of delocalized nucleophiles, in Fig. 6.10, is more selective to changes (of any kind) that affect the gap, G, compared with the series of localized nucleophiles. This would be general for other processes as well; delocalization of the single electrons in the R^* states of the diagram results in higher f values, and vice versa. Such trends abound in electrophile–nucleophile combinations; they were analyzed also for radical addition to olefins (40), and are likely to be a general feature of reactivity.

6.6.6   Making Stereochemical Predictions with the VBSCD Model

Making stereochemical predictions is rather facile if we use FO–VB configurations (1,5,41). We already saw that, when we briefly addressed the stereochemistry of the oxidative bond activations in Fig. 6.8. It is essential to recognize at the outset that the original configurational mapping from MO–CI to FO–VB configurations showed that the selection rule derived from the simple FO–VB diagram representations of R and R^* is reinforced, in fact, by other FO–VB configurations that mix with the original ones so as to provide an optimal mixing matrix element (by rehybridizing the FOs so as to increase their overlaps) (1). To further illustrate the manner by which orbital selection rules can be derived (5), let us take a simple example with well–known stereochemistry, the nucleophilic substitution reaction discussed before in Fig. 6.4. The corresponding R^* charge-transfer state is depicted in Fig. 6.11 in its FO–VB formulation, where the nucleophile appears here in its 1e reduced form $X^{\bullet}$, with a single electron in φ_X, while the substrate has an extra electron in its σ_{AY}^* orbital. The two single electrons are coupled into a φ_X–σ_{AY}^* bond pair.

The R^* state correlates to product, $X–A/:Y^-$, since it contains a $\varphi_X - \sigma_{AY}^*$ bond pair that becomes the $X–A$ bond, and at the same time the occupancy of the σ_{AY}^* orbital causes the cleavage of the $A–Y$ bond to release the $:Y^-$ anion. Furthermore, the R^* state contains information about the stereochemical pathway. Since the bond pair involves a φ_X–σ_{AY}^* overlap, due to the nodal properties of the σ_{AY}^* orbital the bond pair will be optimized when the $X^{\bullet}$ is coupled to the substrate in a collinear $X–A–Y$ fashion. Thus, the steepest

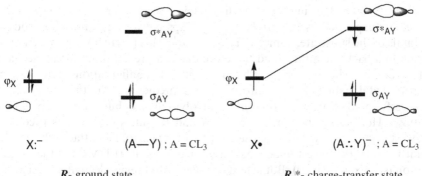

X:⁻   (A—Y) ; A = CL₃   X•   (A∴Y)⁻ ; A = CL₃

R- ground state   **R***- charge-transfer state

FIGURE 6.11   The FO–VB representations of the ground (**R**) and charge-transfer (**R***) states in the VBSCD of the S_N2 reaction, $X:^- + A-Y \rightarrow X-A + :Y^-$.

descent of the **R*** state, and the lowest crossing point will occur along a backside trajectory of the nucleophile toward the substrate.

If we assume that the charge-transfer state remains the leading configuration of **R*** near the crossing point, then we can make predictions about the resonance energy *B*. Indeed, the size of *B* is dominated by the matrix element between **R** and **R***. Since these two VB configurations differ by the occupancy of one spin–orbital (φ_X in **R** is replaced by σ_{AY}^* in **R***) then following the qualitative rules for matrix elements (see Appendix 3.A.2), the resonance energy of the TS will be proportional to the overlap of these orbitals, that is,

$$B \propto \langle \varphi_X | \sigma_{AY}^* \rangle \qquad (6.18)$$

Therefore, it follows that in a backside trajectory, we obtain both the lowest crossing point as well as the largest TS resonance energy. Computationally, the backside barrier is smaller by ~10–20 kcal/mol compared with a front side attack (42). Equation 6.18 defines an orbital selection rule for an S_N2 reaction. Working out this rather trivial prediction is nevertheless necessary since it constitutes a prototypical example for deriving orbital selection rules in other reactions, using FO–VB configurations. Thus, a simple rule may be stated as follows:

> **Rule 3**. The intrinsic bonding features of **R*** provide information about the reaction trajectory, while the $\langle R | R^* \rangle$ overlap provides information about the geometric dependence of the resonance energy of the TS.

By using this approach, it is possible to derive orbital selection rules for cases that are ambiguous in qualitative MO theory. Let us take the radical cleavage of σ bonds as an example. If φ_R is the singly occupied orbital of the attacking radical, the reactants state, **R**, is the determinant $|\varphi_R \sigma \bar{\sigma}|$. The **R*** state, with a triplet $\sigma^1 \sigma^{*1}$ configuration on the substrate, has the following determinant: $|\bar{\varphi}_R \sigma \sigma^*|$. Reordering the orbitals of the latter we obtain $|\sigma^* \sigma \bar{\varphi}_R|$

for the R^* state. By using the rules displayed in Section 3.2, the overlap $\langle R|R^* \rangle$ is readily expressed as the product of overlaps between φ_R and the σ and σ^* orbitals of the substrate, namely, $\langle \varphi_R|\sigma \rangle \langle \varphi_R|\sigma^* \rangle$. This product is optimized once again in a backside attack, and therefore one can predict that radical cleavage of σ-bonds will proceed with inversion of configuration. All known experimental data (43–48) conform to this prediction. Another area where successful predictions have been made involves nucleophilic attacks on radical cations. Here using the corresponding R and R^* states (49), it was predicted that stereoselectivity and regioselectivity of nucleophilic attack should be controlled by the lowest unoccupied molecular orbitals (LUMO) of the radical cation (see Exercise 6.10). Both stereospecificity and regioselectivity predictions were verified by experiment (50,51) and computational means (42). For a more in-depth discussion, the reader may consult the most recent review of the VBSCD and VBCMD models (5).

6.6.7 Predicting Transition State Structures with the Valence Bond State Correlation Diagram Model

Another way to exploit the insights of the VB–FO representation (5,41) is to focus directly on the bonding in the TS. To illustrate this approach, consider a linear X---H---X TS for a hydrogen-abstraction reaction that was discussed above (see Scheme 6.4). Figure 6.12a shows the bond diagram for the linear TS using now the 1s orbital of the central hydrogen (labeled as σ), and the fragment orbitals of the X---X fragment. It is seen that in order to have a bond pair between the X---X and the H, the electrons in the X---X fragment must be in a triplet situation, and be paired to the 1s electron to yield a total doublet state. As such, the X---H---X TS has a delocalized bond pair. This picture reveals a few important features: (a) the bonding in the TS can be regarded as a union between a triplet X---X moiety with the central H (41). As shown recently (25) this bonding interaction is very significant, of the order of the X–H bond energy in the ground-state molecules. (b) As a result of this electronic structure, there is triplet repulsion between the X moieties in the X---X fragment. This triplet repulsion tends to drive the TS toward a linear structure, which is the preferred geometry for TSs, where X = H, CH_3, SiH_3, and so on.

In cases where X = halogen, OH, NH_2, and so on, the X---X fragment has also π-lone-pair orbitals, orthogonal to the X---X axis. One of these orbitals, the symmetric combination, is shown in Fig. 6.12. If the TS will bend, this π-orbital will overlap with the 1s(H) orbital, as shown in Fig. 6.12b, using symmetry labels with respect to the plane bisecting the XHX angle. As such, we can add a secondary bond-diagram structure labeled as $\Psi_2^{\neq}$, which can mix with the primary structure $\Psi_1^{\neq}$, and increase the bonding in the TS. Such an effect happens in TSs, where X = halogen, OH, NH_2, and so on, and the bending lowers the energy of the TS by a few kilocalories per mol, as was noted in Section 6.6.1. The importance of bending the 3e/3c TS was discussed recently by Isborn et al (25).

(a) $(\mathbf{X}\text{-----}\mathbf{H}\text{-----}\mathbf{X})^{\neq}$

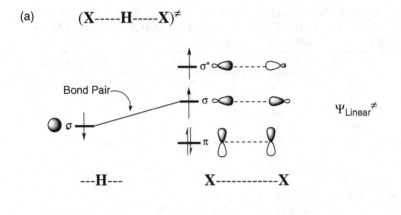

(b) $(\mathbf{X}^{\cdots\overset{\mathbf{H}}{\cdots}\cdots}\mathbf{X})^{\neq}$ $\Psi_{Bent}^{\neq} = \Psi_1^{\neq} + \lambda\,\Psi_2^{\neq}$

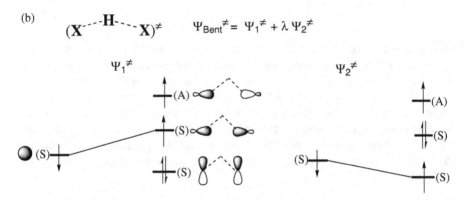

FIGURE 6.12 The FO–VB based bond diagrams showing the spin pairing in the XHX transition state of the $X^{\bullet} + H\text{--}X \rightarrow X\text{--}H + X^{\bullet}$ reaction, for (a) a case with a linear TS, (b) a case of a bent TS.

Most of the above predictions can be made by directly inspecting the VB–AO wave function (e.g., the triplet relationship of the X---X fragment). However, The VB–FO formulation offers a simple pictorial aid to reach these predictions by importing the orbital symmetry into the VB representation.

6.6.8   Trends in Transition State Resonance Energies

Generally speaking, B is related to the strength of the bonding in the TS, and therefore there is often a direct relationship between the bond energy in the ground state and the resonance energy in the TS (41,52,53). Table 6.1 shows the BOVB computed B values (kcal/mol) for 3e/3c TSs for hydrogen atom transfer between MH_3 groups (M = C, Si, Ge, Sn, Pb) (12,16). These values are presented along with the M–H bond energies, D_{M-H} in the ground state,

TABLE 6.1   Trends in Transition State Resonance Energies, B, in XHX Transition States of the $X^{\bullet} + H-X \rightarrow X-H + X^{\bullet}$ Reaction and in the XCH_3X^- Transition States for the $X:^- + CH_3-X \rightarrow X-CH_3 + :X^-$ Reaction[a]

Group	$(X-H-X)^{\neq}$		Group	$(X-CH_3-X)^{-\neq}$	
X	B^b	$D_{HX}{}^c$	X	B^b	$B_{semi}{}^d$
CH_3	51	100	F	29	28
SiH_3	42	85	Cl	21	23
GeH_3	39	78	Br	21	21
SnH_3	33	70	I	20	19
PbH_3	32	64			

[a]All B and D data in kcal/mol.
[b]Calculated by the BOVB *ab initio* method.
[c]The $H-X$ bond energies calculated by the BOVB *ab initio* method.
[d]Calculated using Equation 6.20.

and it is seen that as the bond becomes weaker the TS possesses a smaller resonance energy value. More specifically, for hydrogen-transfer reactions, B has been demonstrated to amount to one quarter of the singlet−triplet excitation energy, $\Delta E'_{ST}$, of the $M-H$ bond at the geometry of the transition state (12,16,52,53) (see Exercise 6.5). The parameter $\Delta E'_{ST}$ is, of course, smaller than ΔE_{ST}, the singlet−triplet gap of the $M-H$ bond in its equilibrium geometry. Now, as it is generally observed that $\Delta E_{ST} > 2D_{M-H}$, we may set $\Delta E'_{ST} \approx 2D_{M-H}$, and a very simple relationship can be derived in this case for B as a function of bond energies:

$$B \approx 0.5 D_{M-H} \qquad (6.19)$$

This relationship predicts pretty well the trends in resonance energies of delocalized 3e/3c TSs in Table 6.1. Of course, this value is subject to small variations depending on the deviation of the TS from linearity (see Section 6.6.7).

Another relationship for 4e/3c TSs was derived for S_N2 reactions (5,36,52), using the same semiempirical approach and is given in Equation 6.20.

$$B \approx 0.5 D(1 - Q_A) \qquad (6.20)$$

Here, D is the $C-X$ bond energy, while Q_A is the charge on the central group in the $(X-A-X)^-$ TS. This charge is associated with the contribution of alternant triple ionic structure $X:^- A^+ :X^-$ to the wave function of the TS (Scheme 6.2). The values, calculated by VB and those estimated from the simple expression in Equation 6.20 are displayed in Table 6.1 for some identity S_N2 reactions, and the fit is seen to be good (B_{semi} data are estimated with Eq. 6.20 based on BOVB charges). It is seen that the B values are condensed within a narrow range of 9 kcal/mol, which reflects the interplay of two opposing effects on the resonance energy: As the electronegativity of X increases from X = I to X = F, not only the bond strength (D) increases, but the charge on the central group also increases, thus roughly canceling each other. In a polar solvent, the central charge will increase and slightly further reduce the B values, more so for the more electronegative X, thus further condensing the B values. This leveling effect of the charge on the central carbon is the root cause why reactivity trends in solvents usually obey the C$-$X bond strength; the stronger the bond the higher the barrier (this is a G effect, see Section 6.6.4).

Why does the charge on the central alkyl group reduce the resonance energy of S_N2 transition states? The answer is straightforward, once it is realized that the X:$^-$ A$^+$:X$^-$ structure commonly contributes to the two Lewis structures. Thus, the higher the contribution of the triple ionic structure, the more similar the two Lewis structures, and the lower the resonance energy becomes. In the theoretical limit, where the X:$^-$ A$^+$:X$^-$ structure becomes the dominant structure in the TS, say 100% of the TS wave function, then the resonance energy should go to zero. It is seen that the semiempirical expression in Equation 6.20 mimics this limit. Other semiempirical expressions make similar predictions (5,53), but Equation 6.20 is the simplest one.

Finally, B is related to the stereochemistry of the reaction and the stereochemistry can be reasoned most simply using *Rule 3*, based on the FO$-$VB representations of R and R^*:

$$B \propto \langle R|R^* \rangle \qquad (6.21)$$

Application of the equation for cycloaddition reactions shows that the B factor is simply proportional to the product of the frontier molecular orbital (HOMO-LUMO) overlaps between the two reactants (5). As such, the following relationship holds for the "allowed" versus the "forbidden" pathways:

$$B(\text{allowed}) > B(\text{forbidden}) \qquad (6.22)$$

Equation 6.22 explains why formally allowed and forbidden reactions must be considered as forming distinct families for simple applications of the VBSCD model. This of course means that comparisons of the same reaction in the two pathways, for example, conrotatory versus disrotatory closure of butadiene to cyclobutene, will always obey the Woodward$-$Hoffmann rules (54). In contrast, trying to correlate barriers to G factors in a series of reactions involving a mixture of allowed and forbidden reactions would be meaningless. For example, consider the dimerization of disylenes, with $G = 80 \, \text{kcal/mol}$, versus the Diels$-$Alder reaction with $G = 173 \, \text{kcal/mol}$; the latter reaction has a much larger barrier even though it is "allowed", whereas the former is "forbidden". Nevertheless, the "forbidden" reaction will still prefer a stepwise mechanism, in which the B value is larger than in the corresponding concerted reaction. As such, the Woodward$-$Hoffmann rules and the VBSCD model are complementary: the former theory classifies the reaction families, while the second rationalizes the magnitudes of the barriers within each family.

For a more in-depth discussion the reader may consult the most recent review of the VBSCD model (5).

6.7   VALENCE BOND CONFIGURATION MIXING DIAGRAMS: GENERAL FEATURES

The VBCMD is an alternative and a complementary diagram to the VBSCD (2–7), typified by more than two curves as shown above in Fig. 6.1b. A few examples are discussed below, while a more in-depth description can be found in a recent review (5).

Figure 6.1b shows the generic VBCMD that features two fundamental curves, labeled as Ψ_r and Ψ_p, and an intermediate curve denoted by Ψ_{int}. In those situations where the intermediate curve lies higher than the crossing point of the fundamental curves, the VB mixing will be prone to generate a single transition state that has a mixed character of the fundamental and intermediate VB structures (5,33). However, the diagram in Fig. 6.1 describes a situation where the intermediate curve, being significantly more stable than the crossing point of the fundamental curves, will generate, though not always (36), an intermediate state in a stepwise mechanism. This intermediate structure provides a low energy pathway that mediates the transformation of *R* to *P* ($\Psi_r \rightarrow \Psi_p$). There are two types of intermediate curves: (a) the intermediate curve is an ionic structure (i.e., native to the Lewis structures in *R* and *P*), and (b) the intermediate curve is a third state that differs from the *R-R** and *P-P** state curves. We refer to this latter state a "foreign state" to underscore the fact that it is not associated with reactants and products.

6.8   VALENCE BOND CONFIGURATION MIXING DIAGRAM WITH IONIC INTERMEDIATE CURVES

Any two-state VBSCD can be transformed into a VBCMD, where the Heitler–London (HL) and ionic VB structures are plotted explicitly as independent curves, instead of being combined into state curves (5). As a rule, ionic structures, which are the secondary VB configurations of polar-covalent bonds, lie above the covalent HL structures at the reactant and product geometries, and generally they cross the two HL structures above their own crossing point. In many cases, the ionic curve is low enough in energy in the hypercoordinated geometry near the transition state, so that solvation can further stabilize the ionic situation and cause it to cross the HL-curves significantly below their own crossing point; in such an event a stepwise mechanism mediated by an ionic intermediate will transpire (e.g., S_N1 mechanism in the reactants of Eq. 6.7). The following examples serve to illustrate the impact of ionic VB structures on the reactivity of covalent bonds.

6.8.1   Valence Bond Configuration Mixing Diagram for Proton-Transfer Processes

The small size of H^+ enables very tight ion-pair geometries with large electrostatic energies. Consequently, the triple ionic structure $X{:}^-$ $H^+{:}X^-$ in a proton-transfer process will usually possess a deep energy minimum (55). An analysis of the case of the $(FHF)^-$ anion, which is a stable symmetric hydrogen bond, can illustrate the importance of the ionic–covalent crossing in this and analogous cases. It should be emphasized that most other hydrogen-bonded dihalide anions are nearly symmetric and feature double well minima separated by a tiny barrier for the proton transfer (5,55,56).

Figure 6.13a depicts the HL and ionic structures for a proton-transfer process between bases, $X{:}^-$, which have moderate or low stability as anions (e.g., carbanions with significant pK_a). In such a case, the ionic structure lies above the HL state (the mixture of the two HL structures), and the avoided crossing will lead generally to a single TS separating the hydrogen-bonded clusters. Nevertheless, the ionic structure is seen to have a deep minimum near the crossing point of the HL curves, and as such the TS will be expected to possess a significant triple ion character. As the anion $X{:}^-$ gets increasingly more stable, so will the ionic structure descends more and more in energy and may dominate the region near the transition state. This is seen in Fig. 6.13b that depicts the computed (5,55) VB configurations for the F^- exchange along the reaction coordinate. At the diagram onset, the ionic structure lies above the HL structure by a moderate energy gap of only $\sim 24\,\mathrm{kcal/mol}$. However, at the symmetric $F{-}H{-}F$ geometry, the lowest VB curve is the ionic structure that

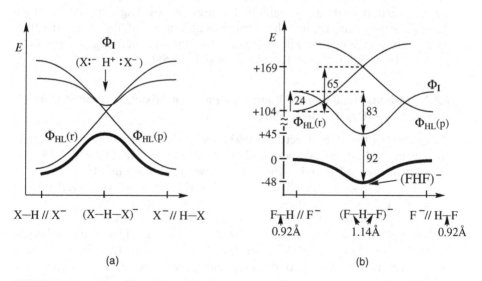

(a)   (b)

FIGURE 6.13   The effect of the triple ionic structure, Φ_I, in proton-transfer processes between (a) very strong $X{:}^-$ bases (e.g., $X = CH_3$), and (b) $F{:}^-$ anions; the latter is computed by VB theory (Adapted with permission from Ref. 5.) Energies are in kcal/mol.

undergoes 83 kcal/mol of stabilization relative to its onset at the reactant geometry. The origins of this remarkable stability of the ionic structure is, as already noted, the small size of H^+ that leads to short $F^- H^+$ distances at the cluster geometry, of $(FHF)^-$, and thereby to very large electrostatic stabilization. This electrostatic stability along with its low initial energy makes the ionic structure the dominant configuration at the cluster geometry.

The short $H-F$ distance is associated with yet another outcome and this is the inception of very large resonance energy due to the mixing of the resonating HL state (which by itself is a resonating mixture of the two HL structures) with the ionic structure. This ionic–covalent resonance energy is seen to be $\sim$90 kcal/mol, which contributes a significant fraction to the bonding in $(FHF)^-$. Thus, the symmetric hydrogen-bonded species is neither fully ionic nor fully covalent; it is virtually a resonating mixture of the two structures (5).

The question whether or not the symmetric $(FHF)^-$ species will be a minimum on the adiabatic (bold) curve, is a question of balance between the difference in electrostatic stabilization and ionic–covalent resonance energies at the cluster geometry relative to the reactant and product geometries. It is seen, from Fig. 6.13b, that the ionic–covalent resonance energy is largest at the reactant–product geometries. It follows, therefore, that the crucial reason why the symmetric $(FHF)^-$ species is so stable, is the electrostatic stabilization that lowers the ionic structure well below the energies of the HL structures at the extremes of the VBSCD. It is this difference that causes the final adiabatic state-profile (in bold) to retain the shape of the ionic curve, and to exhibit a minimum. The relatively small size of the F^- anion is also important for the electrostatic stabilization, and we may expect that, as the anion increases in size (e.g., I^-) or becomes delocalized (e.g., aspartate), the intrinsic stabilization of the ionic structure at the cluster geometry will decrease, and the symmetric geometry may cease to be a minimum of the energy profile (5,55).

6.8.2 Insights from Valence Bond Configuration Mixing Diagrams: One Electron Less–One Electron More

The impact of the ionic structure is fleshed out by comparison of $(FHF)^-$ with the corresponding radical species, $(FHF)^•$. The corresponding VBCMDs are depicted in Figs. 6.14a and b; Fig. 6.14a shows again the VBCMD of $(FHF)^-$, while Fig. 6.14b shows the VBCMD for $F^• + HF$. Thus, with one electron less in the $(FHF)^•$ species, the triple ionic structure is replaced by the $F^- H^+ {}^•F$ and $F^• H^+ F^-$ structures, which lose at least one-half of the electrostatic stabilization, and therefore rise above the HL curves, correlating to excited charge-transfer states. This loss of electrostatic stabilization has a tremendous impact on the reaction profile, and the deep energy well of $(FHF)^-$ becomes an $(FHF)^•$ transition state 18–20 kcal/mol above the reactants (24). For the same reason, it is expected therefore that $(XHX)^•$ species will generally be transition states for the hydrogen-abstraction process with a barrier significantly larger

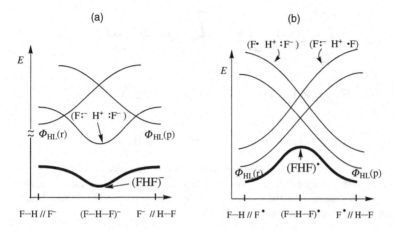

FIGURE 6.14 Comparison of the VBCMDs for the (a) proton transfer process, $F:^- + H-F$, and (b) hydrogen-abstraction process, $F^{\bullet} + H-F$. The curves in bold are adiabatic state curves.

than the corresponding proton-transfer process via the $(XHX)^-$ species. Experimental data show that this is indeed the case (57).

6.8.3 Nucleophilic Substitution on Silicon: Stable Hypercoordinated Species

Another demonstration of the role of ionic structures is the nucleophilic substitution on Si, which proceeds via pentacoordinated intermediates (58,59), in contrast to the situation in carbon where the pentacoordinated species is a transition state. Recent VB calculations (60–62) of C–X and Si–X bonds (X = H, F, Cl) led to an interesting observation that *the minimum of the ionic curve for* Si^+X^- *is significantly shorter than the corresponding minimum for* C^+X^-. By contrast, the minima of the HL curves for Si–X were found to be longer than the corresponding C–X minima. An example of this trend is shown in Fig. 6.15, which depicts the ionic structures, calculated (62) by VB theory, for the C–Cl and Si–Cl bonds. Since Cl^- is common for the ionic structures, these differences mean that SiH_3^+ has a smaller ionic radius compared with CH_3^+; as seen in the figure the value for CH_3^+ is 0.67 Å, compared with 0.31 Å for SiH_3^+ (these values were determined assuming 1.81 Å as the radius of Cl^-) (60,61). Note that the covalent radii behave as expected and the SiH_3 radical is larger than CH_3. Further inspection of Fig. 6.15 shows that the origin of these effective ionic sizes is the charge distribution of the corresponding ions. In CH_3^+ and generally in CL_3^+ (L- a ligand), the charge is distributed over the ligands, while the central carbon possesses a relatively small positive charge (60). Consequently, the minimum distance of approach of an anion X^- toward CL_3^+ will be relatively long and the

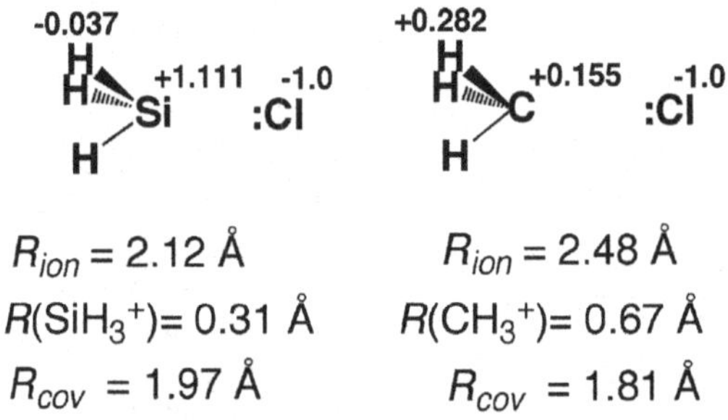

-0.037
H
H
Si
+1.111
-1.0
:Cl
H
+0.282
H
H
C
+0.155
-1.0
:Cl
H
R_ion = 2.12 Å
R(SiH3+)= 0.31 Å
R_cov = 1.97 Å
R_ion = 2.48 Å
R(CH3+)= 0.67 Å
R_cov = 1.81 Å

FIGURE 6.15 Coulson–Chirgwin charges in the ionic structures of $H_3Si{-}Cl$ and $H_3C{-}Cl$, the corresponding minima of these ionic structures, the ionic radii of CH_3^+ and SiH_3^+, and the minima of the corresponding covalent structures of these bonds. The ionic radii were determined assuming $R(Cl^-) = 1.81$ Å.

electrostatic energy will be small. In contrast, in SiL_3^+, the charge is fully localized on Si, and consequently the minimum distance of approach of an X^- anion will be relatively short and the electrostatic stabilization large. Indeed, the depth of the ionic curve H_3Si^+ X^- was found to exceed the depth of H_3C^+ X^-, by >50 kcal/mol. In conclusion, therefore, the silicenium ion L_3Si^+ is expected to behave more like the small proton, whereas the corresponding carbenium ion CL_3^+ will be bulky.

Based on these differences, it is possible to represent the VBCMDs for typical nucleophilic substitution reactions on Si versus C as shown in Figs. 6.16a and b. In Fig. 6.16a, the ionic curve $X{:}^-$ $SiL_3^+{:}X^-$ is very stable in the pentacoordinated geometry due to the electrostatic energy of the triple ion structure, much like the case of the $(FHF)^-$ species discussed before. Consequently, the pentacoordinated $(XSiL_3X)^-$ species will generally be a

(a)
E
ΦHL(r)
ΦHL(p)
X:⁻ Si⁺ :X⁻
R
P
X—Si—X
L
L L
X:⁻ / L3Si—X
X—SiL3 / :X⁻
Reaction Coordinate
(b)
X:⁻ C⁺ :X⁻
E
ΦHL(r)
ΦHL(p)
R
P
X—C—X
L
L L
X:⁻ / L3C—X
X—CL3 / :X⁻
Reaction Coordinate

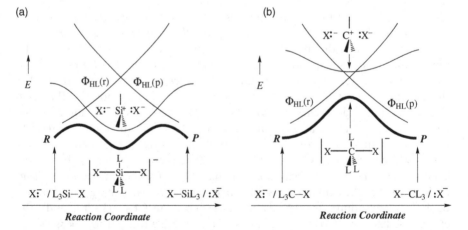

FIGURE 6.16 The VBCMDs comparing $S_N2(Si)$ in (a) versus $S_N2(C)$ in (b). The curves in bold are adiabatic state curves (Adapted with permission from Ref. 5.)

stable entity. By the same analogy to the $(FHF)^-$ species, the pentacoordinated silicon species will be neither ionic nor covalent, but rather a resonating mixture of the two structures with bonding augmented by significant ionic–covalent resonance.

Figure 6.16b shows the typical situation for the carbon analogue where the ionic structure is relatively high in energy and the VB mixing leads to a single-step reaction with a pentacoordinated TS. In a polar solvent, we might expect that the ionic structure will cross slightly below the HL state and give rise to a transient triple-ion intermediate species (or will proceed by the related S_N1 mechanisms). However, even in such an event the pentacoordinated species of carbon will be significantly different than the corresponding silicon species. Thus, the larger size of the CL_3^+ species will lead to smaller ionic–covalent resonance energy compared with the silicon analogue, because the ionic–covalent matrix element, being proportional to the overlap between the active orbitals of the fragments, is a distance-dependent quantity (60–62). Consequently, should a pentacoordinated intermediate of carbon become stable, it will generally be highly ionic with loose bonds.

6.9 VALENCE BOND CONFIGURATION MIXING DIAGRAM WITH INTERMEDIATES NASCENT FROM "FOREIGN STATES"

Every reaction system possesses, in addition to the promoted excited states, R^* and P^*, which are localized in the active bonds, numerous foreign excited states that involve electronic excitations in orbitals and bonds that do not belong to the active bonds (5). Some of these "foreign" states are high in energy, but some are of low energy and can become accessible along the reaction coordinate. As already stated, mixing of foreign states provides means by which complex molecules find low energy pathways for otherwise difficult transformations. To elucidate this mechanistic significance of the foreign states, we have chosen three examples; others can be found in a recent review (5).

6.9.1   The Mechanism of Nucleophilic Substitution of Esters

In order to learn how to generate a complete mechanistic sequence for a complex reaction, we consider now a nucleophilic substitution reaction of esters, amides and so on, shown in Equation 6.23,

$$X:^- + R'(RO)C{=}O \rightarrow R'(X)C{=}O + RO:^- \qquad (6.23)$$

in which the nucleophile substitutes a leaving group, for example, RO^- in the case of an ester (5). Figures 6.9 and 6.10 considered one step of the mechanism, whereas the entire mechanism requires a VBCMD, as presented in Fig. 6.17.

First, let us consider the R^* and P^* states for the overall process. Since the net process is a nucleophilic displacement reaction, then by complete analogy

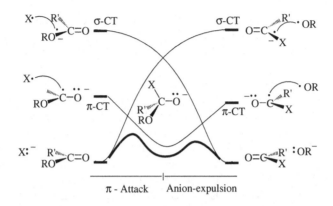

FIGURE 6.17 The VBCMD for the nucleophilic displacement process, $X:^- + R'(RO)$ $C=O \rightarrow R'(X)C=O + RO:^-$ (Adapted with permission from Ref. 5.)

to the S_N2 reaction and following **_Rule 2_**, the R^* and P^* states are σ charge-transfer states, where an electron from the anion is transferred to the bond that undergoes cleavage in the given direction; in the R^* state, the electron is transferred from $X:^-$ to the C—OR bond, while in the P^* state the transfer is from $RO:^-$ to the C—X bond. If this were the only option for VB crossing, the reaction would have proceeded in a manner analogous to the S_N2 reaction, which in this case would have resulted in a high barrier (large promotion energies to the σ-CT state due to the strength of the bonds to an sp^2 carbon, and since the backside attack of the nucleophile on the bond to the leaving group is sterically difficult). The π-CT states are mediated by the transformation, which are significantly lower in energy compared with the σ-CT states, and as such, the π-CT states cut through the high ridge of the σ-CT states and provide a low energy path for the transformation. In fact, as shown in Fig. 6.17, the π-CT states also split the reaction coordinate into two phases; a π-attack phase that results in a tetrahedral intermediate, and an anion-expulsion phase that results in substitution. The VBCMD in Fig. 6.17 is in fact common to other nucleophilic substitution processes of unsaturated systems, such as nucleophilic vinylic substitution and aromatic nucleophilic susbsitution; in all of these cases the π-CT states cut through the σ-CT states and provide a low energy pathway in a stepwise mechanism (5).

6.9.2   The $S_{RN}2$ and $S_{RN}2^c$ Mechanisms

Low energy foreign excited states can lead to novel reaction mechanisms. An interesting strategy to achieve such low energy states is to modify classical reactions by substituting a radical center adjacent to the reaction centers (63). A case in point is the recent proposition of new nucleophilic substitution mechanisms (63), termed $S_{RN}2$ and $S_{RN}2^c$, and shown in Scheme 6.6, for the identity reactions of β-chloro ethyl radical with chloride anion. These

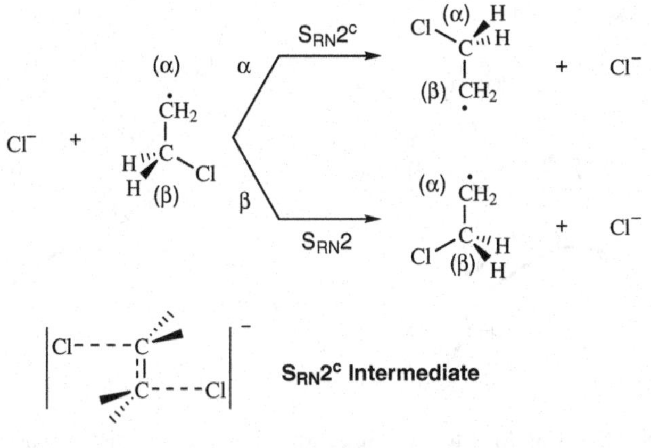

Scheme 6.6

nucleophilic reactions involve Cl$^-$ exchange via attack on either the β or α positions of the radical. It was found (63) that the S$_{RN}$2 process occurs in a single-step reaction with a TS that is very similar to the corresponding S$_N$2 TS for the reaction of Cl$^-$ with ethyl chloride. At the same time, the S$_{RN}$2 barrier was shown to be lower by ~11 kcal/mol in comparison with the S$_N$2 barrier. Even more intriguing were the findings (63) that the S$_{RN}$2^c mechanism proceeds in fact in a stepwise manner via a C_{2h} intermediate that is ~3 kcal/mol lower than the TS of the S$_{RN}$2 mechanism. Thus, the adjacent radical center on the one hand, considerably lowers the barriers for the Cl$^-$ exchange reaction, and on the other, leads to a novel intermediate species.

Scheme 6.7 shows the HL structures, **21** and **22**, of reactants and products for the reaction of Cl$^-$ with β-chloro ethyl radical, and the intermediate states **23** and **24** generated by the presence of the radical adjacent to the reaction centers. It is seen that since we now have three odd electrons in the covalent

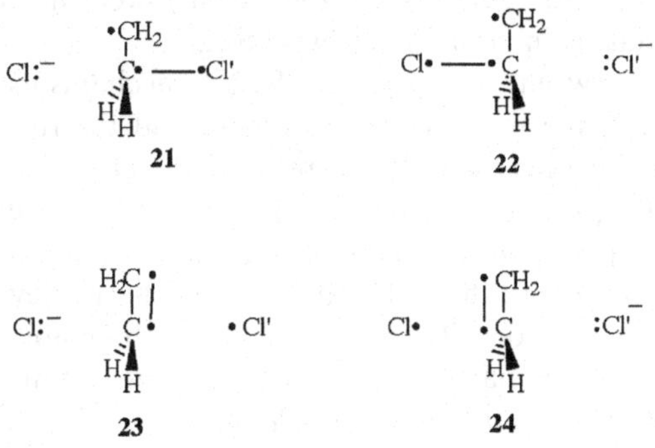

Scheme 6.7

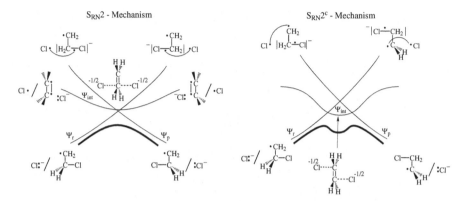

FIGURE 6.18 The VBCMD for the $S_{RN}2$ and $S_{RN}2^c$ mechanisms. The lower Ψ_{int} in the latter mechanism is due to the less repulsive electrostatic interactions between the negatively charged chlorines (Adapted with permission from Ref. 5.)

structures there are alternative ways to pair up the spins. Thus, we can either pair up the central carbon to Cl′ and leave the radical on the methylene group, as in **21**, or pair the central carbon to the methylene and leave the radical on Cl′ as in **23**. In the same manner, we can generate **22** and **24** by pairing up the central carbon to Cl or to the methylene group, respectively. Thus, the presence of the radical on the methylene group generates a new state, not associated with reactants and products (hence, a "foreign" state), and certainly low enough to impact the mechanism and the reaction barriers. The new intermediate state will be a linear combination of **23** and **24**.

Figure 6.18 shows the VBCMDs for the the $S_{RN}2$ and the $S_{RN}2^c$ mechanisms. In both diagrams, there exist two fundamental curves identical to those of the classical S_N2 reaction (Fig. 6.4), and a low lying intermediate curve, Ψ_{int}, in which the C_2H_4 moiety is π-bonded and corresponding to the mixture of **23** and **24**. It is the mixing of this intermediate structure into the fundamental curves that accounts for the much lower energetic of the $S_{RN}2$ and $S_{RN}2^c$ mechanisms in comparison with S_N2.

The difference between the $S_{RN}2$ and $S_{RN}2^c$ mechanisms is rooted in the relationship between the intermediate structure and the fundamental curves. Thus, the intermediate structure Ψ_{int} can lower the electrostatic repulsions of the chlorines in the $S_{RN}2^c$ mechanism and define a lower energy state compared with the crossing point of the HL configurations. Indeed, based on the computed C=C distance in the C_2H_4 moiety, the charge development on this moiety, and on the spin density development on the chlorine moieties, it is apparent that the intermediate-state character is more dominant in the $S_{RN}2^c$ mechanism. Thus, $S_{RN}2^c$ is *a stepwise mechanism mediated by a low energy state due to strong electronic coupling with the accessory radical centre*. Zipse (63) has generalized the conclusions and showed that a radical center adjacent to the reaction centre is a novel strategy to generate low energy pathways via intermediate states, in a variety of processes including damage mechanisms of DNA bases.

6.10 VALENCE BOND STATE CORRELATION DIAGRAM: A GENERAL MODEL FOR ELECTRONIC DELOCALIZATION IN CLUSTERS

The VBSCD serves also as a model for understanding the status of electronic delocalization in isoelectronic series. Consider, for example, the following exchange process between monovalent atoms, which exchange a bond while passing through an $X_3{}^{\bullet}$ cluster in which three electrons are delocalized over three centers.

$$X^{\bullet} + X - X \rightarrow [X_3^{\bullet}] \rightarrow X - X +{}^{\bullet} X \qquad (6.24)$$

We can imagine a variety of such species, for example, X = H, F, Cl, Li, Na, Cu, and ask ourselves the following question: When do we expect the $X_3{}^{\bullet}$ species to be a transition state for the exchange process, and when will it be a stable cluster, an intermediate en route to exchange? In fact, the answer to this question comes from the VBSCD model that describes all these process in a single diagram where G is given by Equation 6.5, that is, $G \approx 0.75 \ \Delta E_{ST}(X-X)$. Thus, as shown in Fig. 6.19 a very large triplet promotion energy for X = H results in an $H_3{}^{\bullet}$ transition state, while the small promotion energy for X = Li results in a stable $Li_3{}^{\bullet}$ cluster. The VB computations of Maître et al. (15) in

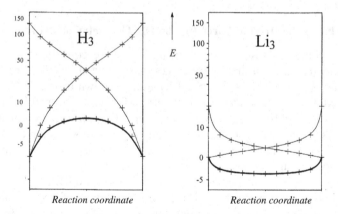

FIGURE 6.19 *Ab initio* computations of VBSCDs for the exchange reactions $X^{\bullet} + X-X \rightarrow [X_3^{\bullet}] \rightarrow X-X + {}^{\bullet}X$ for X = H (left-hand side) and X = Li (right-hand side). The abscissa is the reaction coordinate defined as RC = $0.5 \ (n_1 - n_2 + 1)$, where n_1 and n_2 are the bond orders in the X---X---X species. Energies are in kcal/mol. (Adapted with permission from Ref. 15.)

Fig. 6.19 show that, as the promotion gap drops drastically, the avoided crossing state changes from a transition state for $H_3^\bullet$ to a stable cluster for $Li_3^\bullet$. Moreover, this transition from a barrier to an intermediate can in fact be predicted quantitatively from the barrier equation, by deriving explicit expressions for G, f, and B (see Exercise 6.6) (16,64,65).

The spectacular relationship between the nature of the $X_3^\bullet$ species and the promotion energy shows that the VBSCD is in fact a general model of the pseudo-Jahn-Teller effect (PJTE). A qualitative application of PJTE would predict all the $X_3^\bullet$ species to be transition-state structures that relax to the distorted $X^\bullet$---X-X and X-X---$^\bullet X$ entities. The VBSCD makes a distinction between strong binders, which form transition states, and weak binders that form stable intermediate clusters. Thus, the VBSCD model is in tune with the general observation that as one moves in the periodic table from strong binders to weak ones (e.g., metallic) matter changes from discrete molecules to extended delocalized clusters and/or lattices. The delocalized clusters of the strong binders are the transition states for chemical reactions.

The variable nature of the $X_3^\bullet$ species in the isoelectronic series, form a general model for electronic delocalization, enabling one to classify the species either as distortive or as stable ones. Using the isoelectronic analogy, one might naturally ask about the isoelectronic π-species in allyl radical; does it behave by itself like $H_3^\bullet$ or like $Li_3^\bullet$? Moreover, the same two extreme VBSCDs for $X_3^\bullet$ in Fig. 6.19 can be shown for X_3^+, X_3^-, X_4 and X_6 species (64). Likewise one might wonder about the status of the corresponding isoelectronic π-components in allyl cation, anion, cyclobutadiene, or benzene. These questions were answered in detail elsewhere and the reader is advised to consult a recent review (64). Below we briefly discuss the problem of one of our molecular icons, benzene.

6.10.1   What is the Driving Force for the D_{6h} Geometry of Benzene, σ or π?

The regular hexagonal structure of benzene can be considered as a stable intermediate in a reaction that interchanges two distorted Kekulé-type isomers, each displaying alternating C–C bond lengths as shown in Fig. 6.20.

It is well known that the D_{6h} geometry of benzene is stable against a Kekuléan distortion (of b_{2u} symmetry), but one may still wonder which one of the two sets of bonds, σ or π, is responsible for this resistance to a b_{2u} distortion. The σ frame, which is just a set of identical single bonds, is by nature symmetrizing and prefers a regular geometry with equal C–C bond lengths. It is not obvious whether the π electronic component, by itself, is also symmetrizing or on the contrary distortive, with a weak force constant that would be overwhelmed by the symmetrizing driving force of the σ frame. To answer this question, consider in Fig. 6.20 the VBSCD that represents the interchange of Kekulé structures along the b_{2u} reaction coordinate; the middle of the b_{2u} coordinate corresponds to the D_{6h} structure, while its two extremes correspond to the bond-alternated mirror image Kekulé geometries. Part (a) of

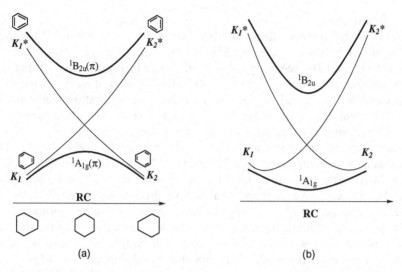

FIGURE 6.20   The VBSCDs showing the crossing and avoided crossing of the Kekulé structures of benzene along the bond alternating mode, b_{2u} for: (a) π-only curves, (b) full $\sigma+\pi$ curves.

the figure considers π *energies only*. Starting from the left-hand side, Kekulé structure K_1 correlates to the excited state K_2^* in which the π bonds of the initial K_1 structure are elongated, while the repulsive nonbonding interactions between the π bonds are reinforced. The same argument applies if we start from the right-hand side, with structure K_2 and follow it along the b_{2u} coordinate; K_2 will then rise and correlate to K_1^*. To get an estimate for the gap, we can extrapolate the Kekulé geometries to a complete distortion, in which the π bonds of K_1 and K_2 would be completely separated (which in practice is prevented by the σ frame that limits the distortion). At this asymptote the promotion energy, $K_i \rightarrow K_i^* (i = 1, 2)$, is due to the unpairing of three π bonds in *the ground state*, K_i, and replacing them, in K_i^*, by three nonbonding interactions. As we recall the latter are repulsive by a quantity that amounts to half the size of a triplet repulsion. The fact that such a distortion can never be reached is of no concern. What matters is that this constitutes an asymptotic estimate of the energy gap G that correlates the two Kekulé structures, and that eventually determines if their mixing results in a barrier or in a stable situation, in the style of the X_3 problem in Fig. 6.19. According to the VB rules, G is given by Equation 6.25:

$$G(K \rightarrow K^*) = 3[0.75\Delta E_{ST}(C{=}C)] = 9/4\Delta E_{ST}(C{=}C) \qquad (6.25)$$

Since the ΔE_{ST} value for an isolated π bond is of the order of 100 kcal/mol, Equation 6.25 places the π electronic system in the region of large gaps. Consequently, the π-component of benzene is predicted by the VBSCD model to be an *unstable transition state*, $^1A_{1g}(\pi)$, as illustrated in Fig. 6.20a. This "π-transition state" prefers a distorted Kekuléan geometry with bond alternation, but is forced by the σ frame, with its strong symmetrizing driving force, to adopt the regular D_{6h} geometry. This prediction, which was derived at the time based on qualitative considerations of G in the VBSCD of isoelectronic series (65), was later confirmed by rigorous *ab initio* $\sigma-\pi$ separation methods (66–69). The prediction was further linked (70) to experimental data associated with the vibrational frequencies of the excited states of benzene.

The spectroscopic experiments (70–73) show a peculiar phenomenon. This phenomenon is both state specific, to the $^1B_{2u}$ excited state, as well as vibrational mode specific, to the bond-alternating mode, that is, *the Kekulé mode b_{2u}*. Thus, upon excitation from the $^1A_{1g}$ ground state to the $^1B_{2u}$ excited state, with exception of b_{2u} all other vibrational modes behave "normally" and undergo frequency lowering in the excited state, as expected from the decrease in π-bonding and disruption of aromaticity following a $\pi \rightarrow \pi^*$ excitation. By contrast, the Kekulé b_{2u} mode undergoes an upward shift of $257-261$ cm^{-1}. As explained below, this phenomenon is predictable from the VBSCD model and constitutes a critical test of π distortivity in the ground state of benzene.

Indeed, the VBSCD model is able to predict not only on the ground state of an electronic system, but also on a selected excited state. Thus, the mixing of the two Kekulé structures K_1 and K_2 in Fig. 6.20a leads to a pair of resonant and antiresonant states $K_1 \pm K_2$; the $^1A_{1g}$ ground state $K_1 + K_2$ is the resonance-stabilized combination, and the $^1B_{2u}$ excited state $K_1 - K_2$ is the antiresonant mixture [this is the first excited state of benzene (74)]. In fact, the VBSCD in Fig. 6.20a predicts that the curvature of the $^1A_{1g}(\pi)$ ground state (restricted to the π electronic system) is negative, whereas by contrast, that of the $^1B_{2u}(\pi)$ state is positive. Of course, when the energy of the σ frame is added as shown in Fig. 6.20b, the net total driving force for the ground state becomes symmetrizing, with a small positive curvature. By comparison, the $^1B_{2u}$ excited state displays now a steeper curve and is much more symmetrizing than the ground state, having more positive curvature. As such, the VBSCD model predicts that the $^1A_{1g} \rightarrow ^1B_{2u}$ excitation of benzene should result in the reinforcement of the symmetrizing driving force, which will be manifested as a frequency increase of the Kekuléan b_{2u} mode.

In order to show how delicate the balance is between the σ and π opposing tendencies, we recently (64) derived an empirical equation (Eq. 6.26) for $4n + 2$ annulenes:

$$\Delta E_{\pi+\sigma} = 5.0(2n+1) - 5.4(2n) \quad \text{kcal/mol} \tag{6.26}$$

Here, $\Delta E_{\pi+\sigma}$ stands for the total (π and σ) distortion energies, the terms $5.0(2n+1)$ represent the resisting σ effect, which is 5.0 kcal/mol for an adjacent pair of σ-bond, whereas the negative term, $-5.4(2n)$, accounts for the π-distortivity. This expression predicts that for $n = 7$, namely, the $C_{30}H_{30}$ annulene, the $\Delta E_{\pi+\sigma}$ becomes negative and the annulene undergoes bond localization. If we increase the π-distortivity coefficient, by just a tiny bit, namely, to Equation 6.27,

$$\Delta E_{\pi+\sigma} = 5.0(2n+1) - 6.0(2n) \quad \text{kcal/mol} \tag{6.27}$$

the equation now would predict that the annulene with $n = 3$, namely, $C_{14}H_{14}$, will undergo bond localization. This extreme sensitivity, which is predicted to manifest in computations and experimental data of annulenes, is a simple outcome of the VBSCD prediction that the π-component of these species behaves as a transition state with a propensity toward bond alternation.

6.11   VALENCE BOND STATE CORRELATION DIAGRAM: APPLICATION TO PHOTOCHEMICAL REACTIVITY

Photochemistry is an important field for future applications. The pioneering work of van Der Lugt and Oosterhoff (75) and Michl (76) highlighted the importance of avoided crossing regions as decay channels in photochemistry. Köppel and co-workers (77,78) showed that conical intersections, rather than avoided crossing regions, are the most efficient decay channels, from excited to ground states. Indeed, this role of conical intersections in organic photochemistry has been extensively investigated by Robb et al. (79,80) and conical intersections are calculated today on a routine basis using software, such as GAUSSIAN. Bernardi, et al. (79) have further shown that VB notions can be useful for rationalizing the location of conical intersections and their structures.

As was subsequently argued (53), the VBSCD is a very natural model for discussing the relation between thermal and photochemical reactions and between the avoided crossing region and a conical intersection. Thus, the avoided crossing region of the VBSCD leads to the twin-states $\Psi^{\neq}$ and Ψ^* (Fig. 6.21); one corresponds to the resonant state of the VB configurations and the other to the antiresonant state (1,5,53). Since the extent of this VB mixing is a function of geometry, there in principle should exist *specific distortion modes that convert the avoided crossing region into conical intersections*, where the twin-states $\Psi^{\neq}$ and Ψ^* become degenerate, and thereby enable the excited reaction complex to decay into the ground-state surface. In this manner, a conical intersection will be anchored at three structures; two of them are the reactant (R) and product (P_1) of the thermal reaction, and the third is the product (P_2) generated by the distortion mode that causes the degeneracy of the twin-states $\Psi^{\neq}$ and Ψ^*. *The new product (P_2) would therefore be characteristic of the distortion mode that is required to convert the avoided crossing region into a conical intersection.* Assuming that most of the excited species roll down eventually to the Ψ^* funnel, then P_2 would be a major photoproduct. If, however, there exist other excited state funnels near the twin-excited state, Ψ^*, other products will be formed, which are characteristic of these other excited states and can be predicted in a similar manner provided one knows the

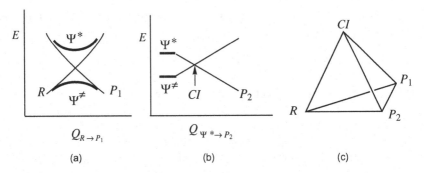

FIGURE 6.21 (a) The VBSCD showing the twin-states formed by avoided crossing along the reaction coordinate of the thermal process leading from reactant R to product P_1. (b) The crossover of the twin-states generates a conical intersection (CI) along a coordinate that stabilizes Ψ^* and leads to product P_2. (c) The conical intersection will be anchored in three minima (or more): R, P_1, and P_2.

identity of these excited states. The following description is restricted to the analysis of the twin-excited states. Another qualification that needs to be made is that once we decide what R and P_1 are, we have to specify the identity of the P_2 product; the latter is by definition the combination of R and P_1 that corresponds to the twin excited state of the transition state for the $R \rightarrow P_1$, process. Since there might be more than one distortion mode that stabilize Ψ^* and destabilize $\Psi^{\neq}$, we may be able to predict a few P_2 products coming from the twin-excited state (see Exercise 6.14).

6.11.1 Photoreactivity in 3e/3c Reactions

A simple example is the celebrated hydrogen-exchange reaction, $H_a + H_b$-$H_c \rightarrow H_a$-$H_b + H_c$, where the transition state has a collinear geometry, H_a-H_b-H_c. In this geometry, the ground state $\Psi^{\neq}$ is the resonating combination of R and P (Eq. 6.28) and the TS for the thermal reaction, while the twin-excited state Ψ^* is the corresponding antiresonating combination in Equation 6.29 (dropping normalization constants):

$$\Psi^{\neq} = R - P \qquad R = |ab\bar{c}| - |a\bar{b}c| \qquad P = |a\bar{b}c| - |\bar{a}bc| \qquad (6.28)$$

$$\Psi^* = R + P = |ab\bar{c}| - |\bar{a}bc| \qquad (6.29)$$

where the orbitals a, b, and c belong to H_a, H_b, and H_c, respectively (see Exercise 6.11 for the demonstration that the negative combination of R and P is the lowest one).

It is clear from Equation 6.29 that Ψ^* involves a bonding interaction between H_a and H_c and will be lowered by the bending mode that brings H_a and H_c together. Furthermore, based on the semiempirical VB approach described in Chapter 3 (neglecting squared overlap terms), the expression for the avoided crossing interaction B in Equation 6.30,

$$B = H_{RP}^{eff} - S_{RP}(H_{RR}^{eff} + H_{PP}^{eff})/2 = -\beta_{ab}S_{ab} - \beta_{bc}S_{bc} + 2\beta_{ac}S_{ac} \qquad (6.30)$$

shows that this quantity shrinks to zero in an equilateral triangular structure, where the H_a-H_c, H_a-H_b, and H_b-H_c interactions are identical (for details of the calculations, see Exercise 6.11). As such, the equilateral triangle defines a conical intersection with a doubly degenerate state, in the crossing point of the VBSCD. This D_{3h} structure will relax to the isosceles triangle with short H_a-H_c distance that will give rise to a "new" product $H_b + H_a$-H_c. The photocycliza-tion of allyl radical to cyclopropyl radical is precisely analogous. The ground state of allyl is the resonating combination of the two Kekulé structures, while the twin-excited state, Ψ^*, is their antiresonating combination with the long bond between the allylic terminals (41). As such, rotation of the two allylic terminals will lower Ψ^*, raise the ground state $\Psi^{\neq}$, and establish a conical intersection that will channel the photoexcited complex to the cyclopropyl radical, and vice versa. This structural dichotomy of the resonant and antiresonant states in the VBSCD accounts for the thermal–photochemical dichotomy as first highlighted in the Woodward–Hoffmann rules (54), and as amply observed.

6.11.2   Photoreactivity in 4e/3c Reactions

Let us consider, as a model system, the nucleophilic reaction $H_a^- + H_b$-$H_c \rightarrow H_a$-$H_b + H_c^-$. As in the previous case, the TS for the thermal reaction has a linear conformation, $[H_a$-H_b-$H_c]^-$, but we are now dealing, in all rigor, with three VB structures: $H_a^-\ H_b\bullet$—$\bullet H_c$, $H_a\bullet$—$\bullet H_b\ H_c^-$, and $\bullet H_a\ H_b^-\ H_c\bullet$, the latter becoming entirely equivalent to the former two structures in the equilateral triangular geometry (see Section 4.1). However, at the linear geometry, the third VB structure is a long-bond structure, which is unbound, and can be neglected. Therefore, we can describe the ground state and the first excited state of the symmetrical linear complex in terms of the two first VB structures, respectively, referred to as R and P (Eq. 6.31). By using the rules of the semiempirical VB theory displayed in Chapter 3, the reduced Hamiltonian-matrix element B between R and P is readily calculated, dropping normal-ization (see Exercise 6.12):

$$R = |a\bar{a}b\bar{c}| - |a\bar{a}\bar{b}c| \qquad P = |c\bar{c}a\bar{b}| - |c\bar{c}\bar{a}b| \qquad (6.31)$$

$$B = 4(\beta_{ab}S_{bc} + \beta_{bc}S_{ab}) - 2\beta_{ac} + \text{higher order terms} \qquad (6.32)$$

Since H_a and H_b are far away from each other in the linear structure, we can neglect β_{ac}. With the remaining two terms, it is clear that B is negative, so that the ground state $\Psi^{\neq}(A^{'})$ is the positive combination of R and P, while the twin-excited state $\Psi^*(A'')$ is their negative combination. Taking the bisecting plane passing through H_b as the unique symmetry element, the state symmetry can be labeled as follows using C_s symmetry:

$$\Psi^{\neq}(A') = R + P \qquad \Psi^*(A'') = R - P \qquad (6.33)$$

If we now start to bend the H_3^- complex to a triangular shape, the β_{ac} term will gradually gain importance as the H_a and H_c atoms approach each other. According to Equation 6.32, the avoided crossing interaction ($|B|$) decreases, and consequently, the gap between $\Psi^{\neq}$ and Ψ^* will diminish. Further bending the complex, the reasoning requires the third VB structure, $^\bullet H_a\, H_b^-\, H_c^\bullet$, which gains importance and mixes with the ground state; this is a bit complex for qualitative reasonings. However, the situation gets simple again for a bending angle of 60°, since we know, by analogy with the cyclopropenyl anion that has been treated in Chapter 5, that the gap between the ground state and first excited state drops to zero for the equilateral triangular structure of H_3^-. This D_{3h} structure is therefore the site of the conical intersection for the H_3^- system, as illustrated in Fig. 6.22a. Now, let us examine the nature of $\Psi^*(A'')$, in order to predict what kind of products are expected for the photochemical reaction. Combining Equations 6.31 and 6.33, we get

$$\Psi^*(A'') = |a\bar{a}b\bar{c}| - |a\bar{a}\bar{b}c| - |c\bar{c}a\bar{b}| + |c\bar{c}\bar{a}b| \qquad (6.34)$$

Rearranging Equation 6.34 to Equation 6.35 reveals a stabilizing 3e bonding interaction between H_a and H_c, of the type ($H_a^\bullet\, :H_c^- \leftrightarrow H_a:^-\, ^\bullet H_c$):

$$\Psi^*(A'') = |a\bar{a}c\bar{b}| + |a\bar{c}c\bar{b}| + |\bar{a}a\bar{c}b| + |\bar{a}c\bar{c}b| \qquad (6.35)$$

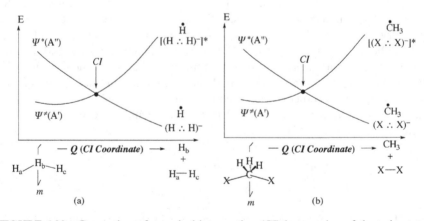

FIGURE 6.22 Generation of a conical intersection (CI) by crossing of the twin states along the bending distortion mode, (a) in H_3^- and (b) in S_N2 systems. Symmetry labels refer to the mirror plan, m.

Therefore, it is expected that the photochemical reaction will generate a radical anion made from the original terminal atoms, that is, $[H_a \therefore H_c]^-$ and a radical on the original central atom, that is, $H_b{}^{\bullet}$.

As a "more chemical" application, let us now consider the photostimulation of S_N2 systems, such as $X^- + A\text{-}Y$ (A = Alkyl), using a VBSCD-based rationale, for predicting the location of conical intersections (53). Here, the transition state for the thermal reaction is the collinear $[X-A-Y]^-$ structure, which is not really analogous to the $H_3{}^-$ system as the orbital of the central fragment, which is of p type along the axis, is antisymmetric with respect to the bisecting symmetry plane (assuming that X and Y are equivalent). As a consequence, the ground state $\Psi^{\neq}(A')$ of collinear $[X-A-Y]^-$ is the negative combination of the reactants' and products' Lewis structures, while the twin-excited state, $\Psi^*(A'')$, is their positive combination, opposite to the states in the linear $H_3{}^-$ case (see Exercise 6.13). If, however, the collinear $[X-A-Y]^-$ structure undergoes bending of the $X-A-Y$ angle with concomitant rotation of the alkyl group, as shown in Fig. 6.22b, then the orbital of the alkyl fragment is symmetrical relative to the bisecting plane and the states become entirely analogous to those of the $H_3{}^-$ case. As such, the A' ground state is now the positive combination of the reactants' and products' structures, while the A'' excited state is the negative combination, as in Equation 6.36:

$$\Psi^* = (|x\bar{x}a\bar{y}| - |x\bar{x}\bar{a}y|) - (|y\bar{y}x\bar{a}| - |y\bar{y}\bar{x}a|) \tag{6.36}$$

where the orbitals x, a, and y belong to the fragments X, A, and Y, respectively.

By analogy with the $H_3{}^-$ case, we can predict that for some bending angle, the geometry displayed in Fig. 6.22b induces the appearance of a conical intersection, leading to a 3e bonded species as the product of the photochemical reaction. Indeed, rearranging Equation 6.36 to Equation 6.37 reveals a stabilizing 3e bonding interaction between X and Y, of the type $(X^{\bullet} : Y^- \leftrightarrow X:^- {}^{\bullet}Y)$.

$$\Psi^* = (|\bar{x}x\bar{y}a| + |\bar{x}y\bar{y}a|) + (|x\bar{x}\bar{y}a| + |x\bar{y}\bar{y}a|) \tag{6.37}$$

As such, the bending mode that brings the X and Y groups together and rotates the methyl group destabilizes the $[X-A-Y]^-$ structure and stabilizes the twin-excited state, until they establish a conical intersection that correlates down to $X \therefore Y^-$ and $R^{\bullet}$ (for a very simple alternative demonstration based on a FO−VB construction, see Exercise 6.13). This analysis is in accord with experimental observation that the irradiation of the I^-/CH_3I cluster at the charge-transfer band leads to $I_2{}^-$ and $CH_3{}^{\bullet}$, while for or I^-/CH_3Br, such excitation generates IBr^- and $CH_3{}^{\bullet}$(81).

The presence of excited-state minima above the thermal TS is a well-known phenomenon (75–80). The VBSCD model merely gives this ubiquitous phenomenon a simple mechanism in terms of the avoided crossing of VB structures, and hence enables one to make predictions in a systematic manner. Other important applications of the twin-states concern the possibility of spectroscopic probing or accessing the twin-excited state that lies directly above the TS of a thermal reaction. Thus, much like the foregoing story of benzene, any chemical reaction will possess a TS, $\Psi^{\neq}$ and a twin-excited state, Ψ^* (1,5,82). For most cases, albeit not all, the twin-excited state should be stable and hence observable, and its geometry will be almost coincident with the thermal transition state. In addition, the twin-excited state will possess a real and sometimes greatly increased frequency of the reaction coordinate mode, by analogy to the benzene story, where the b_{2u} mode was enhanced in the $^1B_{2u}$ twin-excited state. Thus, since the twin pair has coincident geometries, a spectroscopic characterization of Ψ^* will provide complementary information on the transition state $\Psi^{\neq}$ and will enable resolution of the TS structure.

As a proof of principle, the twin states were characterized for the semibullvalene rearrangement and found to possess virtually identical geometries. As shown in Fig. 6.23, the twin-excited state possesses B_2 symmetry as the symmetry of the reaction coordinate of the thermal process. The TS mode, b_2, which is imaginary for $\Psi^{\neq}(A_1)$, was shown to be real for $\Psi^*(B_2)$ (83). These calculations match the intriguing findings (84,85) in the

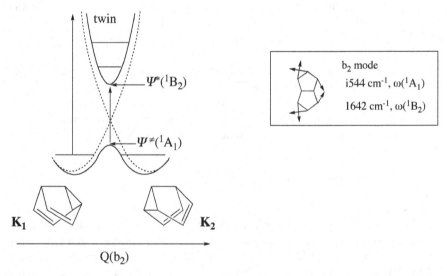

FIGURE 6.23 The twin states along the b_2 reaction coordinate for the semibullvalene rearrangement. When the thermal barrier is NOT much higher than the zero-point levels of the two isomers, the transition state ($\Psi^{\neq}$) region becomes available thermally. Absorption in the transition state region is in the visible, leading to thermochromism at elevated temperatures.

semibullvalene and barbaralene systems. Thus, semibullvalene and barbaralene derivatives in which the barrier for the rearrangement could be lowered drastically, to a point where it almost vanishes, exhibit thermochromism without having a chromophore; they are colorless at low temperatures, but highly colored at 380 K. According to Quast (85), the thermochromism arises due to the low energy transition from the transition state ($\Psi^{\neq}$) to the twin excited state (Ψ^*), Fig. 6.23. Thus, since the thermal barrier is exceptionally low, at elevated temperatures the transition state becomes thermally populated. Since the $\Psi^{\neq} - \Psi^*$ gap is small, one observes color due to absorption within the visible region. However, at low temperatures, the molecules reside at the bottom of the reactant–product wells, where the gap between the ground and excited states is large, and hence, the absorption is in the ultraviolet (UV) region and the color is lost. To quote Quast, "thermochromic ... semibullvalene allow the observation of transition states even with one's naked eye" (85). Of course, identifying appropriate systems where the twin-excited state is observable is required for the eventual "observation" of the TS of thermal reactions.

Coherent control (86) is a new method that makes use of the availability of the twin-excited state to control the course of chemical reactions by laser excitation. Thus, laser excitation from $\Psi^{\neq}$ to Ψ^*(Fig. 6.21a), using two different and complementary photons causes the decay of Ψ^* to occur in a controlled manner either to the reactant or products. In the case where the reactants and products are two enantiomers, the twin-excited state is achiral, and the coherent control approach leads to chiral resolution.

In summary, the twin-excited state plays an important role in photochemistry as well as in thermal chemistry.

6.12   A SUMMARY

This chapter showed how to conceptualize reactivity in a variety of different fields and problems using just two archetypal diagrams shown in Fig. 6.1. Section 6.10 on electronic delocalization and Section 6.11 dealing with photochemical reactions, both prepare us for Chapter 7 which deals with VB descriptions of excited states for a variety of molecules.

REFERENCES

1. S. S. Shaik, *J. Am. Chem. Soc.* **103**, 3692 (1981). What happens to Molecules As They React? A Valence Bond Approach to Reactivity.
2. S. Shaik, P. C. Hiberty, *Adv. Quant. Chem.* **26**, 100 (1995). Valence Bond Mixing and Curve Crossing Diagrams in Chemical Reactivity and Bonding.

3. A. Pross, *Theoretical and Physical Principles of Organic Reactivity*, John Wiley & Sons, Inc., New York, 1995, pp. 83–124; 235–290.

4. S. Shaik, P. C. Hiberty, in *Theoretical Models of Chemical Bonding*, Vol. 4, Z. B. Maksic, Ed., Springer Verlag, Berlin–Heidelberg, 1991, pp, 269–322. Curve Crossing Diagrams as General Models for Chemical Structure and Reactivity.

5. S. Shaik, A. Shurki, *Angew. Chem. Int. Ed. Engl.* **38**, 586 (1999). Valence Bond Diagrams and Chemical Reactivity.

6. S. Shaik, P. C. Hiberty, *Rev. Comput. Chem.*, **20**, 1 (2004). Valence Bond, Its History, Fundamentals and Applications: A Primer.

7. S. Shaik, P. C. Hiberty, The Valence Bond Diagram Approach: A Paradigm for Chemical Reactivity, in *Theory and Applications of Computational Chemistry: The First Fourty Years,* C. Dykstra, Ed., Elsevier, New York, 2005, pp. 635–668.

8. L. Song, Y. Mo, Q. Zhang, W. Wu, *J. Comput. Chem.* **26**, 514 (2005). XMVB: A Program for Ab Initio Nonorthogonal Valence Bond Computations.

9. R. Méreau, M. T. Rayez, J. C. Rayez, P. C. Hiberty, *Phys. Chem. Chem. Phys.* **3**, 3650 (2001). Alkoxyl Radical Decomposition Explained by a Valence Bond Model.

10. S. S. Shaik, *J. Am. Chem. Soc.* **106**, 1227 (1984). Solvent Effect on Reaction Barriers. The S_N2 Reaction. 1. Application to the Identity Exchange.

11. S. S. Shaik, *J. Org. Chem.* **52**, 1563 (1987). Nucleophilicity and Vertical Ionization Potentials in Cation-Anion Recombinations.

12. P. Su, L. Song, W. Wu, P. C. Hiberty, S. Shaik, *J. Am. Chem. Soc.* **126**, 13539 (2004). Valence Bond Calculations of Hydrogen Transfer Reactions: A General Predictive Pattern Derived from VB Theory.

13. L. Eberson, *Electron Transfer Reactions in Organic Chemistry*, Springer-Verlag, Heidelberg, 1987.

14. G. Sini, S. S. Shaik, J.-M. Lefour, G. Ohanessian, P. C. Hiberty, *J. Phys. Chem.* **93**, 5661 (1989). Quantitative Valence Bond Computation of a Curve Crossing Diagram for a Model S_N2 reaction: $H^- + CH_3H' \rightarrow HCH_3 + H'^-$.

15. P. Maître, P. C. Hiberty, G. Ohanessian, S. S. Shaik, *J. Phys. Chem.* **94**, 4089 (1990). Quantitative Valence Bond Computations of Curve-Crossing Diagrams for Model Atom Exchange Reactions.

16. S. Shaik, W. Wu, K. Dong, L. Song, P. C. Hiberty, *J. Phys. Chem. A* **105**, 8226 (2001). Identity Hydrogen Abstraction Reactions, $X^{\bullet} + H\text{-}X' \rightarrow X\text{-}H + X'^{\bullet}$ ($X = X' = CH_3$, SiH_3, GeH_3, SnH_3, PbH_3): A Valence Bond Modeling.

17. L. Song, W. Wu, K. Dong, P. C. Hiberty, S. Shaik, *J. Phys. Chem. A* **106**, 11361 (2002). Valence Bond Modeling of Barriers in the Nonidentity Hydrogen Abstraction Reactions, $X'^{\bullet} + H\text{-}X \rightarrow X'\text{-}H + X^{\bullet}$ ($X' \neq X = CH_3$, SiH_3, GeH_3, SnH_3, PbH_3).

18. Y. Mo, S. D. Peyerimhoff, *J. Chem. Phys.* **109**, 1687 (1998). Theoretical Analysis of Electronic Delocalization.

19. F. Jensen, P.-O. Norrby, *Theor. Chem. Acc.* **109**, 1 (2003). Transition States from Empirical Force Fields.

20. Y. Kim, J. C. Corchado, J. Villa, J. Xing, D. G. Truhlar, *J. Chem. Phys.* **112**, 2718 (2000). Multiconfiguration Molecular Mechanics Algorithm for Potential Energy Surfaces of Chemical Reactions.

21. L. Song, W. Wu, P. C. Hiberty, S. Shaik, *Chem. Eur. J.* **9**, 4540 (2003). An Accurate Barrier for the Hydrogen Exchange Reaction from Valence Bond Theory: Is this Theory Coming of Age?

22. H. Yamataka, S. Nagase, *J. Org. Chem.* **53**, 3232 (1988). *Ab Initio* Calculations of Hydrogen Transfers. A Computational Test of Variations in the Transition-State Structure and the Coefficient of Rate-Equilibrium Correlation.

23. A. Pross, H. Yamataka, S. Nagase, *J. Phys. Org. Chem.* **4**, 135 (1991). Reactivity in Radical Abstraction Reactions: Application of the Curve Crossing Model.

24. P. C. Hiberty, C. Megret, L. Song, W. Wu, S. Shaik, *J. Am. Chem. Soc.* **128**, 2836 (2006). Barriers for Hydrogen vs. Halogen Exchange—An Experimental Manifestation of Charge-Shift Bonding.

25. C. Isborn, D. A. Hrovat, W. T. Borden, J. M. Mayer, B. K. Carpenter, *J. Am. Chem. Soc.* **127**, 5794 (2005). Factors Controlling the Barriers to Degenerate Hydrogen Atom Transfers.

26. S. Shaik, S. P. de Visser, W. Wu, L. Song, P. C. Hiberty, *J. Phys. Chem. A* **106**, 5043 (2002). Reply to Comment on: Identity Hydrogen Abstraction Reactions, $X^{\bullet} + H\text{-}X' \rightarrow XH + X'^{\bullet}$ ($X = X' = CH_3$, SiH_3, GeH_3, SnH_3, PbH_3): A Valence Bond Modeling.

27. C. Amatore, A. Jutand, *Acc. Chem. Res.* **33**, 314 (2000). Anionic Pd(0) and Pd(II) Intermediates in Palladium Catalyzed Heck and Cross-Coupling Reactions.

28. R. Hoffmann, *Angew. Chem. Int. Ed. Engl.* **21**, 711 (1982). Building Bridges Between Inorganic and Organic Chemistry (Nobel Lecture).

29. M.-D. Su, S.-Y. Chu, *Inorg. Chem.* **37**, 3400 (1998). Theoretical Study of Oxidative Addition and Reductive Elimination of 14-Electron d^{10} ML_2 Complexes: A $ML_2 + CH_4$ (M = Pd, Pt; L = CO, PH_3, $L_2 = PH_2'CH_2CH_2PH_2$) Case Study.

30. M.-D. Su, S.-Y. Chu, *J. Am. Chem. Soc.* **119**, 10178 (1997). C-F Bond Activation by the 14-Electron $M(X)(PH_3)_2$ (M = Rh, Ir; X = CH_3, H, Cl) Complex. A Density Functional Study.

31. M.-D. Su, *Inorg. Chem.* **34**, 3829 (1995). A Configuration Mixing Approach to the Reactivity of Carbene and Carbene Analogs.

32. S. Kozuch, A. Jutand, C. Amatore, S. Shaik, *Organometallics* **24**, 2139 (2005). What Makes for a Good Catalytic Cycle ? A Theoretical Study of the Role of Anionic Palladium(0) in the Cross-Coupling Reaction of An Aryl Halide with an Anionic Nucleophile.

33. S. S. Shaik, *Prog. Phys. Org. Chem.* **15**, 197 (1985). The Collage of S_N2 Reactivity Patterns: A State Correlation Diagram Model.

34. S. S. Shaik, H. B. Schlegel, S. Wolfe, *Theoretical Aspects of Physical Organic Chemistry,* Wiley-Interscience, New York, 1992.

35. S. Shaik, *Acta Chem. Scand.* **44**, 205 (1990). S_N2 Reactivity and its Relation to Electron Transfer Concepts.

36. L. Song, W. Wu, P. C. Hiberty, S. Shaik, *Chem Eur. J.* **12**, 7458 (2006). The Identity S_N2 Reaction $X^- + CH_3X \rightarrow XCH_3 + X^-$ (X=F, Cl, Br, and I) in Vacuum and in Aqueous Soution: A Valence Bond Study.

37. I. M. Kovach, J. P. Elrod, R. L. Schowen, *J. Am. Chem. Soc.* **102**, 7530 (1980). Reaction Progress at the Transition State for Nucleophilic Attack on Esters.

38. D. G. Oakenfull, T. Riley, V. Gold, *J. Chem. Soc., Chem. Comm.* 385 (1966). Nucleophilic and General Base Catalysis by Acetate Ion in Hydrolysis of Aryl Acetates: Substituent Effects, Solvent Isotope Effects, and Entropies of Activation.
39. E. Buncel, S. S. Shaik, I.-H. Um, S. Wolfe, *J. Am. Chem. Soc.* 110, 1275 (1988). A Theoretical Treatment of Nucleophilic Reactivity in Additions to Carbonyl Compounds. Role of the Vertical Ionization Energy.
40. S. S. Shaik, E. Canadell, *J. Am. Chem. Soc.* 112, 1446 (1990). Regioselectivity of Radical Attacks on Substituted Olefins. Application of the State-Correlation-Diagram (SCD) Model.
41. S. S. Shaik, in *New Theoretical Concepts for Understanding Organic Reactions*, J. Bertran and I. G. Csizmadia, Eds., NATO ASI Series, C267, Kluwer Academic Publishers, 1989, pp. 165–217. A Qualitative Valence Bond Model for Organic Reactions.
42. S. S. Shaik, A. C. Reddy, A. Ioffe, J. P. Dinnocenzo, D. Danovich, J. K. Cho, *J. Am. Chem. Soc.* 117, 3205 (1995). Reactivity Paradigms. Transition State Structures, Mechanisms of Barrier Formation, and Stereospecificity of Nucleophilic Substitutions on σ-Cation Radicals.
43. J. H. Incremona, C. J. Upton, *J. Am. Chem. Soc.* 94, 301 (1972). Bimolecular Homolytic Substitution With Inversion, Stereochemical Investigation of an SH_2 Reaction.
44. C. J. Upton, J. H. Incremona, *J. Org. Chem.* 41, 523 (1976). Bimolecular Homolytic Substitution at Carbon. Stereochemical Investigation.
45. B. B. Jarvis, *J. Org. Chem.* 35, 924 (1970). Free Radical Additions to Dibenzotricyclo[3.3.0.02,8]-3,6-octadiene.
46. G. G. Maynes, D. E. Appliquist, *J. Am. Chem. Soc.* 95, 856 (1973). Stereochemistry of Free Radical Ring Cleavage of *cis*-1,2,3-trimethylcyclopropane by Bromine.
47. K. J. Shea, P. S. Skell, *J. Am. Chem. Soc.* 95, 6728 (1973). Photobromination of Alkylcyclopropanes. Stereochemistry of Homolytic Substitution at a Saturated Carbon Atom.
48. M. L. Poutsma, *J. Am. Chem. Soc.* 87, 4293 (1965). Chlorination Studies of Unsaturated Materials in Nonpolar Media. V. Norbornene and Nortricyclene.
49. S. S. Shaik, J. P. Dinnocenzo, *J. Org. Chem.* 55, 3434 (1990). Nucleophilic Cleavages of One-electron σ Bonds are Predicted to Proceed with Stereoinversion.
50. J. P. Dinnocenzo, W. P. Todd, T. R. Simpson, I. R. Gould, *J. Am. Chem. Soc.* 112, 2462 (1990). Nucleophilic Cleavage of One-Electron σ Bonds: Stereochemistry and Cleavage Rates.
51. L. Eberson, R. González-Luque, M. Merchán, F. Radner, B. O. Roos, S. Shaik, *J. Chem. Soc., Perkin Trans 2*, 463 (1997). Radical Cations of Non-Alternant Systems as Probes of the Shaik-Pross VB Configuration Mixing Model.
52. S. S. Shaik, E. Duzy, A. Bartuv, *J. Phys. Chem.* 94, 6574 (1990). The Quantum Mechanical Resonance Energy of Transition States. An Indicator of Transition State Geometry and Electronic Structure.
53. S. Shaik, A. C. Reddy, *J. Chem. Soc., Faraday Trans.* 90, 1631 (1994). Transition States, Avoided Crossing States and Valence-bond Mixing: Fundamental Reactivity Paradigms.

54. R. B. Woodward, R. Hoffmann, *The Conservation of Orbital Symmetry*, Verlag Chemie, Weinheim, 1971.

55. W. Wu, S. Shaik, W. H. Saunders, Jr., *J. Phys. Chem.* A **106**, 11616 (2002). Comparative Study of Identity Proton Transfer Reactions Between Simple Atoms or Groups by VBSCF Methods.

56. B. S. Ault, *Acc. Chem. Res.* **15**, 103 (1982). Matrix Isolation Investigation of the Hydrogen Bihalide Anions.

57. D. M. Neumark, *Acc. Chem. Res.* **26**, 33 (1993). Transition State Spectroscopy via Negative Ion Photodetachment.

58. Y. Apeloig, in *The Chemistry of Organic Silicon Compounds*, S. Patai, Z. Rappoport, Eds., John Wiley & Sons, Inc., Chichester, 1989, pp. 59–225. Theoretical Aspects of Organosilicon Compounds.

59. R. R. Holmes, *Chem. Rev.* **96**, 927 (1996). Comparison of Phosphorous and Silicon: Hypervalency, Sterechemistry, and Reactivity.

60. S. Shaik, D. Danovich, B. Silvi, D. L. Lauvergnat, P. C. Hiberty, *Chem Eur. J.* **11**, 6358 (2005). Charge-Shift Bonding: A Class of Electron-Pair Bonds That Emerges from Valence Bond Theory and Is Supported by the Electron Localization Function Approach.

61. D. Lauvergnat, P. C. Hiberty, D. Danovich, S. Shaik, *J. Phys. Chem.* **100**, 5715 (1996). Comparison of C-Cl and Si-Cl Bonds. A Valence Bond Study.

62. A. Shurki, P. C. Hiberty, S. Shaik, *J. Am. Chem. Soc.*, **121**, 822 (1999). Charge-Shift Bonding in Group IVB Halides: A Valence Bond Study of MH_3-Cl (M = C, Si, Ge, Sn, Pb) Molecules.

63. H. Zipse, *Acc. Chem. Res.* **32**, 571 (1999). The Methylenology Principle: How Radicals Influence the Course of Ionic Reactions.

64. S. Shaik, A. Shurki, D. Danovich, P. C. Hiberty, *Chem. Rev.* **101**, 1501 (2001). A Different Story of π-Delocalization—The Distortivity of the π-Electrons and Its Chemical Manifestations.

65. S. S. Shaik, R. Bar, *Nouv. J. Chim.* **8**, 411 (1984). How Important is Resonance in Organic Species?

66. P. C. Hiberty, S. S. Shaik, J.-M. Lefour, G. Ohanessian, *J. Org. Chem.* **50**, 4657 (1985). Is the Delocalized π-System of Benzene a Stable Electronic System?

67. S. S. Shaik, P. C. Hiberty, J.-M. Lefour, G. Ohanessian, *J. Am. Chem. Soc.* **109**, 363 (1987). Is Delocalization a Driving Force in Chemistry ? Benzene, Allyl Radical, Cyclobutadiene and their Isoelectronic Species.

68. S. S. Shaik, P. C. Hiberty, G. Ohanessian, J.-M. Lefour, *J. Phys. Chem.* **92**, 5086 (1988). When Does Electronic Delocalization Become a Driving Force of Chemical Bonding ?

69. P. C. Hiberty, D. Danovich, A. Shurki, S. Shaik, *J. Am. Chem. Soc.* **117**, 7760 (1995). Why Does Benzene Possess a D_{6h} Symmetry? A Quasiclassical State Approach for Probing π-Bonding and Delocalization Energies.

70. Y. Haas, S. Zilberg, *J. Am. Chem. Soc.* **117**, 5387 (1995). The ν_{14} (b_{2u}) Mode of Benzene in S_0 and S_1 and the Distortive Nature of the π Electron System: Theory and Experiment.

71. L. Wunsch, H. J. Neusser, E. W. Schlag, *Chem. Phys. Lett.* **31**, 433 (1975). Two Photon Excitation Spectrum of Benzene and Benzene-d_6 in the Gas Phase: Assignment of Inducing Modes by Hot Band Analysis.

72. L. Wunsch, F. Metz, H. J. Neusser, E. W. Schlag, *J. Chem. Phys.* **66**, 386 (1977). Two-Photon Spectroscopy in the Gas Phase: Assignments of Molecular Transitions in Benzene.

73. D. M. Friedrich, W. M. McClain, *Chem. Phys. Lett.* **32**, 541 (1975). Polarization and Assignment of the Two-Photon Excitation Spectrum of Benzene Vapor.

74. E. C. da Silva, J. Gerratt, D. L. Cooper, M. Raimondi, *J. Chem. Phys.* **101**, 3866 (1994). Study of the Electronic States of the Benzene Molecule Using Spin-Coupled Valence Bond Theory.

75. W. Th. A. M. Van der Lugt, L. J. Oosterhoff, *J. Am. Chem. Soc.* **91**, 6042 (1969). Symmetry Control and Photoinduced Reactions.

76. J. Michl, *Topics Curr. Chem.* **46**, 1 (1974). Physical Basis of Qualitative MO Arguments in Organic Photochemistry.

77. U. Manthe, H. Köppel, *J. Chem. Phys.* **93**, 1658 (1990). Dynamics on Potential Energy Surfaces with a Conical Intersection: Adiabatic, Intermediate, and Diabatic Behavior.

78. H. Köppel, W. Domcke, L. S. Cederbaum, *Adv. Chem. Phys.* **57**, 59 (1984). Multimode Molecular Dynamics Beyond the Born-Oppenheimer Approximation.

79. F. Bernardi, M. Olivucci, M. Robb, *Isr. J. Chem.* **33**, 265 (1993). Modeling Photochemical Reactivity of Organic Systems. A New Challenge to Quantum Computational Chemistry.

80. M. A. Robb, M. Garavelli, M. Olivucci, F. Bernardi, *Rev. Comput. Chem.* **15**, 87 (2000). A Computational Strategy for Organic Photochemistry.

81. D. M. Cyr, G. A. Bishea, M. G. Scranton, M. A. Johnson, *J. Chem. Phys.* **97**, 5911 (1992). Observation of Charge-Transfer Excited States in the $I^- \cdot CH_3I$, $I^- \cdot CH_3Br$, and $I^- \cdot CH_2Br_2$ S_N2 Reaction Intermediates Using Photofragmentation and Photoelectron Spectroscopies.

82. I. B. Bersuker, *Nouv. J. Chim.* **4**, 139 (1980). Are Activated Complexes of Chemical Reactions Experimentally Observable Ones?

83. S. Zilberg, Y. Haas, D. Danovich, S. Shaik, *Angew. Chem. Int. Ed. Engl.* **37**, 1394 (1998). The Twin-Excited State as a Probe for the Transition State in Concerted Unimolecular Reactions. The Semibullvalene Rearrangement.

84. H. Quast, K. Knoll, E.-M. Peters, K. Peters, H. G. von Schnering, *Chem. Ber.* **126**, 1047 (1993). 2,4,6,8-Tetraphenylbarbaralan—ein Orangeroter, Thermochromer Kohlenwasserstoff ohne Chromophor.

85. H. Quast, M. Seefelder, *Angew. Chem. Int. Ed. Engl.* **38**, 1064 (1999). The Equilibrium between Localized and Delocalized States of Thermochromic Semibullvalenes and Barbaralanes—Direct Observation of Transition States of Degenerate Cope Rearrangements.

86. (a) M. Shapiro, P. Brumer, *Adv. At. Mol. Opt. Phys.* **42**, 287 (2000). Coherent Control of Atomic, Molecular and Electronic Processes. (b) See recent application in bioprocesses: M. Chergui, *Science* **313**, 1246 (2006). Controlling Biological Functions.

Output 6.1

XMVB OUTPUT

(I) Energy of a Reactant Like State

H₂-H₁ ----- H₃

```
H1 -- H2   0.712 A
H1 ---H3   1.253 A
```

```
                   Distance matrix (angstroms):
                     1          2          3
      1   H    0.000000
      2   H    0.712210   0.000000
      3   H    1.252992   1.965202   0.000000
```

```
---------------Input File---------------
H3 Sto-3g
```

Number of Structures: 2

The following structures are used in calculation:

```
1 *****    1  2  3
2 *****    3  1  2
```

Nuclear Repulsion Energy: 1.434612

Total Energy: −1.561460559

 First Excited: −1.011696

```
              ****** MATRIX  OF  HAMILTONIAN ******
              1          2
   1    −2.991318    2.149711
   2     2.149711   −2.771714
```

```
           ****** COEFFICIENTS OF STRUCTURES ******

   1    0.90280  ******    1  2  3
   2   −0.13142  ******    3  1  2
```

```
           ******  WEIGHTS OF STRUCTURES ******
Coulson-Chirgwin Weights

   1    0.89889  ******    1  2  3
   2    0.10111  ******    3  1  2
```

Inverse Weights

```
   1    0.97925  ******    1  2  3
   2    0.02075  ******    3  1  2
```

(II) Energy of the Transition State (TS)

H₂--H₁--H₃

```
H1 -- H2   0.912 A
H1 ---H3   0.912 A

                  Distance matrix (angstroms):
                  1          2          3
    1  H   0.000000
    2  H   0.912340   0.000000
    3  H   0.912340   1.824680   0.000000


Nuclear Repulsion Energy:      1.450055

  Total Energy:     -1.53248216

  First Excited:    -1.140009

             ****** MATRIX  OF  HAMILTONIAN ******

          1          2
    1   -2.921797   2.120110
    2    2.120110  -2.921797

             ****** COEFFICIENTS OF STRUCTURES ******

    1    0.54385 ******   1 2 3
    2   -0.54385 ******   1 3 2


             ****** WEIGHTS OF STRUCTURES ******
  Coulson-Chirgwin Weights
    1    0.50000 ******   1 2 3
    2    0.50000 ******   1 3 2
```

(II.a) Energy of a Single structure at the TS
```
Electronic energy           -2.921797
Nuclear Repulsion Energy:    1.450055
Total Energy:               -1.47174232331
```

(III) Energy of a Product Like State

```
    H₂------H₁-H₃

H2 --- H1 -- H3

H2 --- H1 1.253 A
H1 -- H3 0.712 A

Number of Structures:    2

The following structures are used in calculation:

  1 *****   1 3 2
  2 *****   2 1 3

Nuclear Repulsion Energy:      1.434612

Total Energy:    -1.561460559
```

```
First Excited:      -1.011696

                 ****** MATRIX  OF  HAMILTONIAN ******
                  1          2
     1     -2.991318    2.149711
     2      2.149711   -2.771714

                 ****** COEFFICIENTS OF STRUCTURES ******

     1       0.90280  ******  1  3  2
     2      -0.13142  ******  2  1  3

                 ****** WEIGHTS OF STRUCTURES ******
    Coulson-Chirgwin Weights
     1       0.89889  ******  1  3  2
     2       0.10111  ******  2  1  3

    Inverse Weights
     1       0.97925  ******  1  3  2
     2       0.02075  ******  2  1  3
```

EXERCISES

6.1. Use a VB analysis to show why radical recombination reactions generally do not possess an energy barrier. For simplicity, use the gas phase recombination of an alkyl radical $R^•$ with an electronegative radical, $X^•$.

6.2. (a) Analyze the process of bond dissociation of for example, t-BuCl, in solution. (b) Based on your analysis predict the dependence of the barrier for cation–nucleophile recombination in solution on two fundamental properties of the reactants: the electron affinity of the cation and the ionization potential of the nucleophile. For advanced reading, consult Ref. 11.

6.3. Use semiempirical VB theory (Chapter 3) to derive the expressions for G in the case of the radical exchange process in Equation 6.3. Start from Equation 6.5 and the expression for the ground-state's wave function. Use Graham–Schmidt orthogonalization to show that the spectroscopic R^* state can be formulated as a pure $A^{••}Y$ triplet coupled to $X^•$ to give a total doublet. In so doing, derive the expression in Equation 6.6. Using the pure triplet expression for the R^* state, express the wave function, $\Psi(R^*, P)$ that describes a smooth diabatic curve connecting P to R^* as one goes from the products's geometry to the reactants's one. *Hint*: the wave function involves three determinants and a variable parameter λ.

6.4. Use semiempirical VB arguments to predict trends in the barriers for hydrogen-abstraction reaction $X^• + H–X \rightarrow X–H + {}^•X$; $X = CH_3$, SiH_3, GeH_3, SnH_3, PbH_3, F, Cl, Br, I, OH, and SH. You will need H–X bond energies, which you can extract from the literature (Refs. (12,16,25,26)) or you can use experimental values, or calculate your own values. For simplicity, assume that you have only linear X---H---X transition states.

6.5. Use semiempirical VB theory (described in Chapter 3) to derive the following expression for the avoided crossing term B in the process $X^\bullet + H{-}X \rightarrow X{-}H + {}^\bullet X$:

$$B = 0.25\,\Delta E'_{ST}\,(H{-}X) \qquad (6.\text{Ex}.1)$$

where $\Delta E'_{ST}$ is the singlet–triplet transition energy of the X–H bond at the geometry of the transition state. *Hint*: Use Equation 3.45 and neglect squared overlaps to simplify.

6.6. Use the semiempirical VB theory in Chapter 3, to show why the process $X^\bullet + H{-}X \rightarrow X{-}H + {}^\bullet X$ has a barrier for $X = CH_3$, SiH_3, GeH_3, SnH_3, PbH_3, H, while the Li_3 species in the process $Li^\bullet + Li{-}Li \rightarrow Li{-}Li + {}^\bullet Li$ is a stable intermediate. First, construct a VBSCD with the usual parameters ΔE_{ST}, f, G, and B. For convenience, define the energy of a Lewis bond, for example, H–X (or X–X), relative to the nonbonded quasiclassical reference determinant, as follows:

$$E_S(H{-}X) = -\lambda_S \qquad D(H{-}X) = \lambda_S \qquad (6.\text{Ex}.2)$$

where λ_S is used as a shorthand notation for the $2\beta S/(1 + S^2)$ derived in Chapter 3.

Similarly, denote the energy of the triplet pair $H\uparrow\,\uparrow X$ (or of $X\uparrow\,\uparrow X$) by

$$E_T(H\uparrow\,\uparrow X) = \lambda_T \qquad (6.\text{Ex}.3)$$

where λ_T is the corresponding $-2\beta S/(1 - S^2)$ terms for the triplet repulsion. While λ_S and λ_T are defined at the equilibrium distance of the H–X (X–X) bond, analogous parameters λ'_S and λ'_T are defined for the stretched H–X (X–X) bond corresponding to the geometry of the transition state. Based on these notations derive the following relations and quantities:

(a) Express ΔE_{ST} and G as functions of λ_S and λ_T.

(b) Express the avoided crossing interaction B as a function of λ'_S and λ'_T (use Eq. 6.Ex.1 in the preceding exercise).

(c) Express the energy of the crossing point (relative to the reference quasiclassical determinant) as a function of λ'_S and λ'_T.

(d) To enable yourself to derive a simple expression for the barrier, assume that $\lambda = \lambda'$. Then express ΔE_c, the height of the crossing point relative to the reactants, and derive an expression for f, as a function of α, defined as follows:

$$\alpha = \lambda_T/\lambda_S \qquad (6.\text{Ex}.4)$$

(e) Derive an expression for the barrier $\Delta E^{\neq}$, as a function of ΔE_{ST} and α. Knowing that λ_T is generally larger than λ_S for strong binders, such as X–H (X = CH_3, SiH_3, GeH_3, SnH_3, PbH_3, H) while the opposite is true for weak binders, such as alkali atoms, show that the reaction $X^\bullet + H–X \rightarrow X–H + {}^\bullet X$ has a barrier while the Li_3 species is a stable intermediate in the process $Li^\bullet + Li–Li \rightarrow Li–Li + {}^\bullet Li$.

6.7. Consider an alkoxyl radical that undergoes unimolecular decomposition in the gas phase, by the following reaction, which is known to be endothermic:

$$S_1S_2S_3C–O^\bullet \rightarrow S_1^\bullet + S_2S_3C{=}O \qquad (6.Ex.5)$$

Here, S_1–S_3 are alkyl substituents.

As expected, the barrier can be lowered by stabilizing the products by appropriate changes in S_1–S_3. There is, however, a curious dichotomic relationship between the barrier and the endothermicity. That is, for a given amount of products stabilization, the effect on the barrier depends on whether one affects the product side by stabilizing the leaving radical $S_1^\bullet$ (as in the series S_1 = Me, Et, i-Pr, t-Bu, ...) or by substituting the $S_2S_3C{=}O$ molecule (S_2, S_3 = H, Me,...). Interpret this finding by means of the VBSCD model, following the guidelines in (a−e):

(a) Draw the VB correlation diagram for this reaction, and express the states R^* and P^* according to Equation 6.4.

(b) Derive qualitative predictions on the energy variations of P, R,* and P^* relative to the reactants' ground state R, for cases where one stabilizes the product radical $S_1^\bullet$.

(c) Repeat the procedure for cases where the carbonyl product is stabilized by substitution (e.g., $H_2C{=}O \rightarrow RHC{=}O \rightarrow RR'C{=}O$, R = alkyl)

(d) From the previous results, show that the barrier is lowered about as much as the reaction energy if the products are stabilized through the leaving radical $S_1^\bullet$, and much less if they are stabilized through the substituents on the carbonyl product.

(e) Use Equation 6.Ex.6, which is a developed form of Equation 6.11 in the chapter, to reach the conclusions as in (a−d) by reasoning on the VBSCD parameters G_R, G_P, and ΔE_{rp}.

$$\Delta E^{\neq} \approx f_0(G_R + G_P)/2 - B + 0.5\Delta E_{rp} \qquad (6.Ex.6)$$

6.8. The trimerization of ethylene to cyclohexane is computed to be highly exothermic, −67 kcal/mol, but has a large barrier of 49 kcal/mol. By comparison, the Diels−Alder reaction is much less exothermic, −44 kcal/ mol, but has a much lower barrier of 22 kcal/mol. Both are computed to proceed via a concerted 6-electron/6-centered transition state, namely, both are formally allowed. Why are the barriers so different? Provide a VB-based explanation for this puzzle. For more details, you may consult the original work in, A. Ioffe, S. Shaik, *J. Chem. Soc. Perkin Trans. 2*, 2101 (1992).

6.9. The complex L_2Pd ($L_2 = PPh_2(CH_2)_nPPh_2$) activates C–X bonds of aromatic molecules. It was observed that the barriers for Ph–Cl activation decrease steadily as n varies from $n = 6$ to $n = 2$. Provide an explanation based on VBSCDs. You may consult Ref. (32).

6.10. Use **Rule 3** in the main text (Section 6.6.6) to predict the stereochemistry of nucleophilic cleavage of the one-electron σ-bond in a substituted cyclopropane cation radical (see Fig. 6.Ex.1.). *Hint*: Use the FO–VB representation, i.e. the σ and σ* MOs of the C–C one-electron bond. You may consult Ref. (49) for further reading.

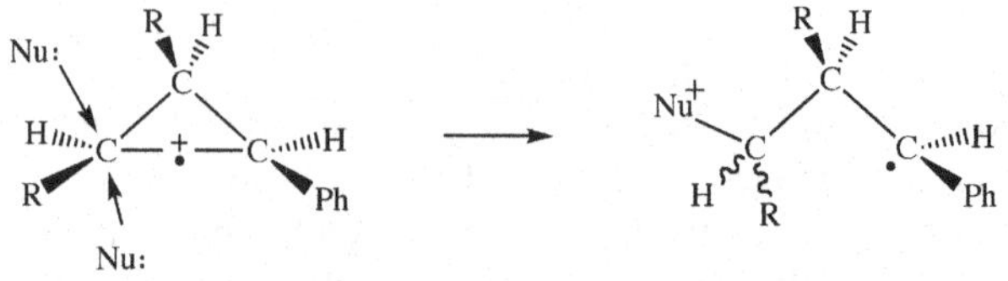

FIGURE 6.Ex.1 Two trajectories for the nucleophilic cleavage of the one-electron σ-bond in a substituted cyclopropane cation radical, leading to retention or inversion of configuration.

6.11. By using the semiempirical VB theory described in Chapter 3, setting the sum of orbital energies to zero (Eq. 3.35) and neglecting squared overlap terms, derive the expression of the reduced Hamiltonian matrix element, in Equation 6.30, between R and P for the 3-orbital, 3-electrons reacting system $[H_a \ldots H_b \ldots H_c]^\bullet$. From the sign of this integral, derive the expressions of $\Psi^{\neq}$ and Ψ^* in Equations 6.28 and 6.29. Show that the reduced Hamiltonian matrix element is largest in the collinear transition state geometry, and drops to zero in the equilateral triangular structure.

6.12. Consider the 3-orbital, 4-electron anionic complex $[H_a \ldots H_b \ldots H_c]^-$, in a linear conformation. The lowest VB structures $H_a^-\ H_b{}^\bullet{-}{}^\bullet H_c$ and $H_a{}^\bullet{-}{}^\bullet H_b\ H_c^-$ are referred to as R and P, respectively. Since the H_a-H_b and H_b-H_c distances are equal, the β_{ab} and β_{bc} integrals will be noted by β, the S_{ab} and S_{bc} overlaps by S, while the long-distance terms β_{ac} and S_{ac} will be neglected. The reduced Hamiltonian matrix element between R and P is defined as follows:

$$B = \langle R|H^{\text{eff}}|P\rangle - \langle R|P\rangle(\langle R|H^{\text{eff}}|R\rangle + \langle P|H^{\text{eff}}|P\rangle)/2 \qquad (6.\text{Ex.}7)$$

(a) By using the rules of semiempirical VB theory (Chapter 3), and neglecting squared overlap terms, express B as a function of β and S, and show that this quantity is stabilizing in the collinear conformation.

(b) Then consider B for a collinear transition state $[X\text{-}A\text{-}Y]^-$ for an S_N2 reaction, where X is the nucleophile, Y the leaving group, and A the central methyl group. The active fragment orbitals are referred to as

x, and y, respectively, for X and Y, and a for the axial p orbital of the central methyl group.

Since the central orbital a overlaps positively with y and negatively with x, we will use the following notations:

$$S_{ay} = -S_{xa} = S \qquad S > 0 \qquad\qquad (6.\text{Ex}.8)$$

$$\beta_{ay} = -\beta_{xa} = \beta \qquad \beta < 0 \qquad\qquad (6.\text{Ex}.9)$$

Denoting the lowest VB structures X⁻ A•—•Y and X•—•A Y⁻ by R and P, respectively, express B as a function of β and S, neglecting β_{xy}. Show that, contrary to the H_3^- system in (a), the ground state is now the $R - P$ combination, of A′ symmetry, while the A″ combination, $R + P$, is the first excited state.

(c) Subsequently, consider the [X–A–Y]⁻ structure as in Fig. 6.22b, where the central fragment is rotated, and the transition state is slightly bent, such that the β_{xy} integral is still negligible. Show that now the ground state is the $R + P$ combination, while the $R - P$ combination is the first excited state, thus making the reacting system analogous to the H_3^- system.

6.13. This exercise is related to the preceding one and it deals with a bent [X...CH₃...X]⁻ transition state, such as the one represented in Fig. 6.22b. The aim is to show in the simplest possible manner why the twin states in Fig. 6.22b behave in the manner they do, by using the FO−VB representation. The system will be considered as the interaction between the *rotated* CH_3 fragment, represented by an orbital a, and the X...X fragment, represented by two MOs σ and σ^*.

(a) By VB coupling the single electron in the orbital a to one of the single electrons in either σ or to σ^*, write the expressions of $\Psi^{\neq}$ and Ψ^* in the FO−VB representation, specify their symmetries and show that $\Psi^{\neq}$ is lower in energy than Ψ^* in the collinear conformation.

(b) Show that upon bending, as the two X fragments get close to each other, while CH_3 moves away, Ψ^* gets lower than $\Psi^{\neq}$, so that the two states cross and lead to the products that are specified in Fig. 6.22b.

6.14. The aim of this exercise is to predict the possible photochemical products by means of a VBSCD prescription for conical intersections. For this purpose, consider the photochemical conversion of butadiene, and follow the steps outlined in (a)−(d):

(a) By considering only the AOs that are directly involved in the reaction, write the VB wave functions for R (butadiene) and P (cyclobutene).

(b) By neglecting all overlaps between AOs, find the sign of the overlap between R and P. Deduce the sign of the corresponding reduced Hamiltonian matrix element.

(c) Write the VB wave functions for the ground state $\Psi^{\neq}$ and the twin excited state Ψ^* in the transition state geometry.

(d) Find the VB structure that corresponds to Ψ^*, and predict the photochemical product(s) that is (are) likely to be observed in addition to cyclobutene (which is the photochemically allowed product).

Answers

Exercise 6.1   Consider two radicals, $R^\bullet$ and $X^\bullet$ (not Na and Cl of course) combining to form a bond. Since the bond is polar–covalent, its wave function will be dominated by the HL structure. Letting r and x represent the singly occupied orbitals of the two radicals, the unnormalized HL wave function is

$$\Phi_{HL} = (|r\bar{x}| - |\bar{r}x|) \qquad (6.Ans.1)$$

The other two structures are the corresponding ionics, $R^+{:}X^-$ and $R{:}^- X^+$. If R is less electronegative than X, we can neglect, for qualitative purposes, the $R{:}^- X^+$ structure. As shown in Figure 6.Ans.1 the ionic structure, Φ_I, lies well above the covalent structure in the gas phase (by the difference between the ionization energy of $R^\bullet$ and the electron affinity of $X^\bullet$). As the R and X moieties approach each other, the HL structure will be stabilized by the bond-pairing energy, which is augmented by the mixing with the ionic structure; the energy will continually go down until the radicals are coupled into a bond. Thus, in radical coupling the same dominant wave function is valid at infinite separation and at the bond equilibrium distance. As such, there is no crossing

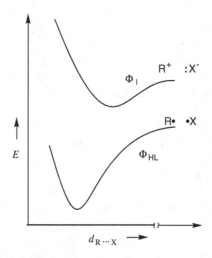

FIGURE 6.Ans.1   Qualitative dissociation energy curves for the R–X bond in the gas phase, using the covalent (Φ_{HL}) and ionic (Φ_I) structures.

of VB structures here and there is no electronic reorganization that leads to an electronic barrier, in other words this is a case with $G = 0$. Of course, there may be small features on the energy profile, like shallow minima due to electrostatic-polarization-dispersion interactions, or small barriers in the case the radicals are heavily delocalized.

Exercise 6.2

(a) Consider again the R–X bond (R = t-Bu) treated in Ex. 6.1, but now in a polar solvent. The major change that happens in solution is illustrated in part (1) of Fig. 6.Ans.2, where it is shown that the ionic curve, which in the gas phase (dashed curve) was above the HL curve, is stabilized relative to the HL curve by the solvation, and is pulled down to become $\Phi_I(s)$ that lies in solution below the HL curve. Now there is crossing of the covalent and ionic VB structures. In part (2) of the Fig. 6.Ans.2, the crossing is avoided by mixing of the two structures, leading to two adiabatic states (bold lines). This leads to a transition state, Ψ^*, on the lower surface (and a "twin excited state", Ψ^*, on the excited-state surface); this is the transition state for bond heterolysis in solution. This treatment of bond heterolysis has appeared for the first time in the pioneering works of Evans and Polanyi, in the 1930s, where chemical reactivity was treated by empirical VB calculations.

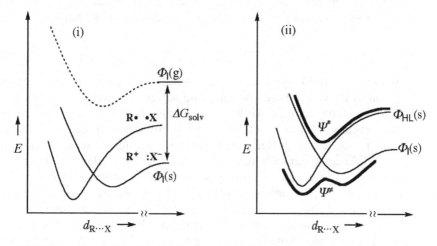

FIGURE 6.Ans.2 Qualitative dissociation energy curves for the R–X bond in a polar solvent. (1) The covalent and ionic curves, $\Phi_{HL}(s)$ and $\Phi_I(s)$, shown in regular lines, while the gas- phase ionic curve [$\Phi_I(g)$] is shown in a dashed line, (2) Covalent and ionic curves in solvent (thin curves), and their avoided crossing leading to the ground-state and twin-excited states (bold curves).

(b) Figure 6.Ans.3 again shows the ionic and covalent curves in solution. The barrier for cation–anion recombination can be expressed as the height of the crossing point, ΔE_c, minus the resonance energy of the transition state,

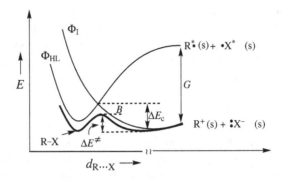

FIGURE 6.Ans.3   Qualitative dissociation energy curves for the R–X bond in a polar solvent, along with the reactivity parameters of the VBSCD for the cation–anion recombination process.

B. Since ΔE_c can be expressed as a fraction of the promotion gap, we may write the following equation for the barrier:

$$\Delta E^{\neq} = \Delta E_c - B = fG - B \qquad (6.Ans.2)$$

The promotion gap is given by the difference between the vertical ionization energy of X:⁻ anions $(I_{X:}^*)$ and the vertical electron affinity of the cation (A_{R+}^*), leading to the following expression for the barrier:

$$\Delta E^{\neq} = f(I_{X:}^* - A_{R+}^*) - B \qquad (6.Ans.3)$$

where the asterisk symbolizes that during the electron transfer from the anion to the cation both geometry and solvent orientation remain frozen. These quantities are available from experiment and can also be evaluated theoretically using a simple Hess cycle. For this purpose, the interested reader should consult Ref. (10) or (11). For example, if we take a reaction series of a single cation with a family of nucleophiles, then the A_{R+}^* term is constant, and one may anticipate that the relative reactivity will be dominated by a single quantity, the $I_{X:}^*$ of the nucleophiles. Such a spectacular correlation is discussed in Ref. (11) for the reactions of Y-pyronin cation with a series of nucleophile, X:. Reference (11) discusses further meanings of the correlation in terms of the potential to quantify the resonance energy of the transition states, B.

Exercise 6.3   Consider the R and R^* states for the radical exchange process, $X^{\bullet} + A\!-\!Y \rightarrow X\!-\!A + {}^{\bullet}Y$, in Equation 6.3. The HL wave functions for R and R^* are the following (dropping normalization constants):

$$\Psi(R) = |xa\bar{y}| - |x\bar{a}y| \qquad (6.Ans.4)$$

$$\Psi(R^*) = |x\bar{a}y| - |\bar{x}ay| \qquad (6.Ans.5)$$

One can use semiempirical VB theory to derive the G value for these states (see text). But one can do so by inspection of the wave function for R^*. It is seen that the AY moiety in this state is described in the second determinant as pure triplet, while in the first determinant AY is neither a triplet nor a singlet, but in fact an average of the singlet and triplet wave functions (with $M_S = 0$). Therefore AY is 75% triplet in R^*. This leads to the following G expression:

$$G = 0.75\Delta E_{ST}(A\text{–}Y) \qquad (6.\text{Ans}.6)$$

which is nothing else but Equation 6.5 in the text.

Note that the wave functions in the above R and R^* states share a common determinant. If we now crudely normalize the wave functions using $2^{-1/2}$ as a normalization factor (this means we are using a zero differential overlap approximation), using the same approximation the overlap will be the following:

$$\langle R | R^* \rangle = -1/2 \qquad (6.\text{Ans}.7)$$

Since the two wave functions overlap, we can form from them two orthogonal wave functions using the Graham–Schmidt orthogonalization procedure, namely:

$$\Psi_1 = \Psi(R) \text{ and } \Psi_2 = \Psi(R^*) - \langle R | R^* \rangle \Psi(R) \qquad (6.\text{Ans}.8)$$

Thus, the new excited state Ψ_2 (which is unnormalized) becomes:

$$\Psi_2 = \Psi(R^*) + 0.5\Psi(R) \qquad (6.\text{Ans}.9)$$

which by using the expressions for R and R^* gives:

$$\Psi_2 = 0.5(|xa\bar{y}| + |x\bar{a}y|) - |\bar{x}ay| \qquad (6.\text{Ans}.10)$$

In this wave function, the last determinant has a pure triplet AY moiety as before in R^*. However, now the first two terms are also a pure triplet AY (with $M_S = 0$). As such, by using the orthogonalization procedure, we generated a spectroscopic promoted state, where $X^\bullet$ is coupled to a triplet AY to a total doublet spin state; the spin pairing is 50% between $X^\bullet$ and $A^\bullet$ and 50% between $X^\bullet$ and $Y^\bullet$. The corresponding G expression becomes then:

$$G = \Delta E_{ST}(A\text{–}Y) \qquad (6.\text{Ans}.11)$$

as in Equation 6.6 in the text.

When the R^* state is defined as a pure triplet connected to a distant radical, it is important to construct a wave function $\Psi(R^*, P)$ that varies smoothly from P to R^*, so that a VBSCD can be calculated and generated. Such a wave function can be chosen as a linear combination of Ψ_2 and $\Psi(P)$, with a variable parameter λ:

$$\Psi(R^*, P) = \lambda|\text{xa}\bar{\text{y}}| + (1 - \lambda)|\text{x}\bar{\text{a}}\text{y}| - |\bar{\text{x}}\text{ay}| \qquad (6.\text{Ans.}12)$$

With $\lambda = 0$ the above wave function corresponds to $\Psi(P)$, while when $\lambda = 0.5$, this wave function becomes the spectroscopic state expressed in Ψ_2 above. Thus, letting λ vary between zero (at the geometry corresponding to P) and 0.5 (at the geometry corresponding to R), the wave function $\Psi(R^*, P)$ will vary smoothly from $\Psi(P)$ to Ψ_2 above.

Exercise 6.4   Based on the analysis of the barrier for radical exchange reactions, we can use the relationship $G = 2D_{HX}$ for the promotion gap. If we assume only linear transition states, we can use the relationship $B = 0.5D_{HX}$ given in Equation 6.19 (16). Combining the two expressions and the value $f = 1/3$, we get the following expression for the barrier (16):

$$\Delta E^{\neq} = fG - B = (1/6)D_{HX} \qquad (6.\text{Ans.}13)$$

Of course, this is a very crude expression, but it should give reasonable trends for reaction families, such as changes of X down the column of the periodic table. Since the H–X bond energy decreases down the column, we may predict based on the above expression that the identity barriers in the halogen exchange series will vary in the order F > Cl > Br > I. For the same reason, we can predict that when the X group is MH_3 (M = C, Si, Ge, Sn, Pb), the barriers will vary in the order $CH_3 > SiH_3 > GeH_3 > SnH_3 > PbH_3$. Finally, for the series X = OH, SH, and so on, we predict the same trends, namely the barriers decrease down the column. The table below shows the barriers estimated with the above equation and computed by the CCSD(T) method. Similar trends are obtained for the other series. The semiempirically estimated barriers are generally higher than ab initio computed ones, but all the series show that the identity barrier for H abstraction decrease as X changes down the column of the periodic table. All these trends (not the absolute values of the barriers) are in accord with experiment.

The H–X Bond Energies and Barriers for $X^{\bullet} + H–X \rightarrow X–H + {}^{\bullet}X$

X	D_{HX} (kcal/mol)[a]	$\Delta E^{\neq}$ (CCSD(T))[a]	$\Delta E^{\neq}$ (semiempirical)[d]
F	122.9 (137.1)[b]	20.9	20.5 (22.9)[b]
Cl	88.8	11.0	14.8
Br	75.4	8.0	12.6
I	68.3[c]		11.4

[a]Data from Ref. (24). Bond energies calculated by BOVB.
[b]A CCSD(T) datum from Ref. (25).
[c]A VBCI datum from Ref. (12).
[d]Using the equation derived above. For similar data see Ref. (16).

Note, however, as discussed in this chapter, that when X has lone pairs, the TS is bent, and the B value increases making the barrier smaller than the simple prediction in Equation 6.Ans.13. The F–H–F transition state is bent with an angle of 132.6° and a barrier of 17.8 kcal/mol (UCCSD(T)/6-311++G(3df,3pd)//RCCSD(T)/6-311++G(3df,3pd) (26). At the same level, for the linear geometry the barrier is 20.9 kcal/mol, in accord with the prediction in the table above. Using a UCCSD(T)/6-311++G** bond energy datum of the H–OH bond, the above equation predicts a barrier of 19.6 kcal/mol for the linear HO...H...OH transition state, the calculated barrier is 15.6 kcal/mol. Similarly, at the same levels, the predicted barrier for X = SH is 15.8 kcal/mol versus 11.0 computed one.

Exercise 6.5   Let us consider the general process $X^\bullet + H\text{–}Y \rightarrow X\text{–}H + {}^\bullet Y$, with $X = Y$. At the transition state geometry, the ground state $\Psi^{\neq}$ is the normalized combination of the reactants' and products' wave functions, Φ_1 and Φ_2. To facilitate the derivation, let us write these two wave functions so that their overlap is positive, that is,

$$\Phi_1 = \frac{1}{\sqrt{2}}(|xh\bar{y}| - |x\bar{h}y|) \qquad (6.\text{Ans}.14)$$

$$\Phi_2 = \frac{1}{\sqrt{2}}(|\bar{x}hy| - |x\bar{h}y|) \qquad (6.\text{Ans}.15)$$

where x, h, and y stand for the active orbitals of the corresponding fragments during the process.

The integrals S_{12}, H_{12}, and the quantity E_{ind} are readily calculated:

$$B = [H_{12} - E_{ind}S_{12}]/(1 + S_{12}) = (2\beta S - 0.5\beta S)/(1 + 0.5) = \beta S \qquad (6.\text{Ans}.16)$$

Since ΔE_{ST} for a bond equals to $4\beta S$ (see Section 6.6.8), the expression for B becomes:

$$B = 0.25 \, \Delta E'_{ST}(H - X) \qquad (6.\text{Ans}.17)$$

where the prime signifies that the value corresponds to the TS geometry.

Exercise 6.6
(a) With the shorthand notations, the singlet–triplet excitation becomes then,

$$\Delta E_{ST}(H\text{–}X) = (\lambda_S + \lambda_T) \qquad (6.\text{Ans}.18)$$

(b) and the expressions for G and B read:

$$G = 0.75 \, \Delta E_{ST}(H\text{–}X) = 0.75(\lambda_S + \lambda_T) \qquad (6.\text{Ans}.19)$$
$$B = 0.25 \, \Delta E'_{ST}(H\text{–}X) = 0.25(\lambda'_S + \lambda'_T) \qquad (6.\text{Ans}.20)$$

The primes appearing in Equation 6.Ans.20 signify that these values correspond to the TS geometry.

(c) At the crossing point the Lewis structure $X^\bullet$ H–X has on one side a bond with energy $-\lambda'_S$ and half a repulsive interaction on the other side with energy $0.5\lambda'_T$. Thus the energy of the crossing point is given by,

$$E_c(X^\bullet H\text{–}X) = -\lambda'_S + 0.5\lambda'_T \qquad (6.\text{Ans}.21)$$

(d) With the crude assumption that $\lambda = \lambda'$, the height of the crossing point becomes Equation 6.Ans.22:

$$\Delta E_c = 0.5\lambda_T \qquad (6.\text{Ans}.22)$$

This expression shows that the height of the crossing point depends on the triplet repulsive interactions between the bonded atom in the center (e.g., H) and the two terminal groups X.

The f-factor becomes then Equation 6.Ans.23:

$$f = \Delta E_c/G = \{2\alpha/[3(1+\alpha)]\} \qquad \alpha = \lambda_T/\lambda_S \qquad (6.\text{Ans}.23)$$

(e) Combining Equations 6.Ans.23, 6.Ans.22, 6.Ans.19, and 6.Ans.20, the barrier expression becomes:

$$\Delta E^{\neq} = 0.25\,\Delta E_{ST}[(\alpha - 1)/(\alpha + 1)] \qquad (6.\text{Ans}.24)$$

It is seen from Equation 6.Ans.24 that the barrier is positive as long as $\alpha > 1$. When $\alpha < 1$, the barrier becomes negative, and the delocalized 3-electron/3-center species XHX (or X–X–X) becomes a stable cluster.

The parameter α is the crucial quantity that determines the transition from a saddle-point species to a stable cluster. From Equation 6.Ans.18 and the expression for D (see Eq. 6.Ex.2) we can show that α determines the ratio of the singlet–triplet excitation of the bond to its bond energy:

$$\Delta E_{ST}/D = (\lambda_S + \lambda_T)/\lambda_S = 1 + \alpha \qquad (6.\text{Ans}.25)$$

All the X–H bonds in the previous exercise are typified by $\Delta E_{ST}/D > 2$, and hence their α values are > 1. These are the "strong binders". In contrast, in a bond, such as Li_2, $\Delta E_{ST} = 32.9$ kcal/mol and $D = 24.6$ kcal/mol, and hence $\alpha = 0.3374$. This is the class of "weak binders", in which the triplet repulsion (λ_T) is significantly shallower than the bonding interaction ($-\lambda_S$). Equation 6.Ans.24 predicts that clusters of "weak binders" will be stable intermediates, in contrast to clusters of "strong binders", which are transition states.

Equation 6.Ans.24 turns out to be useful also for calculating barriers, provided the ΔE_{ST} and α quantities are available. A sample of these calculated barriers is given in Table 6.Ans.1. The interested reader can further consult Ref. (16).

TABLE 6.Ans.1. The ΔE_{ST}, α and Barriers ($\Delta E^{\neq}$) Calculated with Equation 6.Ans.24 for $X^{\bullet} + H{-}X' \to X{-}H + {}^{\bullet}X'$

X	$\Delta E_{ST}{}^{a}$	α	$\Delta E^{\neq}$ (Eq. Ans.24)^{a}	$\Delta E^{\neq}$ [CCSD(T)]^{a}
H	240.3	1.559	13.1	14.8
CH_3	276.4	1.778	19.4	21.4
SiH_3	224.2	1.622	13.3	14.7
GeH_3	212.0	1.708	13.9	11.1
SnH_3	187.5	1.671	11.8	10.2
PbH_3	172.1	1.693	11.1	7.6
Li^{b}	32.9	0.3374	-4.1	-3.8^{c}

^{a}In kcal/mol from Ref. (16).
^{b}This corresponds to $Li + Li{-}Li' \to Li{-}Li + Li'$.
^{c}This is a VB datum from Ref. (15).

Exercise 6.7

(a) The VBSCD for the alkoxyl radical decomposition is displayed in Fig. 6.Ans.4. In the case of P^{*} we show the spin distribution in the carbonyl moiety using the two determinants in the corresponding wave function.

(b) As can be seen from Fig. 6.Ans.5, part (1), substituting the leaving radical S_1 lowers R^{*}, P^{*} and P by the same quantity, relative to R. As a consequence, the energy of the crossing point is lowered almost as much as the product stabilization. The barrier follows the same tendency and will be lowered significantly.

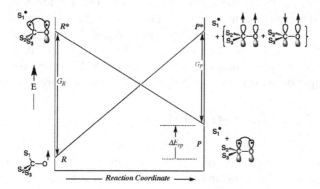

FIGURE 6.Ans.4 The VBSCD for the dissociation of a substituted alkoxyl radical showing only the correlation of the VB structures. In some of the structures, we do not use dots to indicate electrons. Instead we indicate electron spins.

(c) As can be seen from Fig. 6.Ans.5, part (2), stabilizing the carbonyl product, for example, by changing from $H_2C{=}O$ to $HMeC{=}O$ or $Me_2C{=}O$,

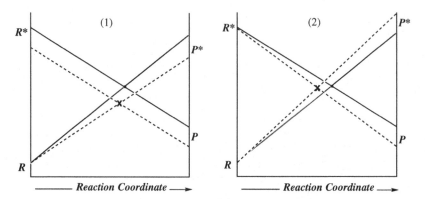

FIGURE 6.Ans.5 Schematic representations of the energetic effect on the crossing point of the diabatic curves, when the reaction endothermicity is diminished by substituent effects. Full lines and a dotted crossing point represent the diabatic curves for the parent reaction. Dashed lines and a cross represent the effect of substitutents on the diabatic curves and their crossing point. (1) Changes due to substitution on the leaving radical S_1. (2) Changes due to substitution on the carbonyl.

stabilizes P owing to the π-donating character of the methyl groups and the C^+–O^- polarity of the π CO bond. On the other hand, a methyl group on the carbonyl moiety cannot stabilize R^*, as its pseudo-π-orbital (highest occupied orbital) overlaps only weakly with the singly occupied orbital of the central carbon atom, which is pyramidal. By contrast, the P^* state is *destabilized* by a methyl group because the pseudo-π-orbitals of the substituent maintain 3e Pauli repulsion with the π system of the carbonyl. This extra Pauli repulsion is not compensated for by hyperconjugation for reasons that will become clear with the example of $HCH_3C{=}O$. In the absence of hyperconjugation, that is, if the methyl group is not involved in conjugation with the CO π bond, the P^* state is expressed as follows:

$$P^* = H\overset{\uparrow\downarrow}{C}H_3{-}\overset{\uparrow}{C}{-}\overset{\uparrow}{O} + H\overset{\uparrow\downarrow}{C}H_3{-}\overset{\downarrow}{C}{-}\overset{\uparrow}{O} \qquad (6.\text{Ans.}26)$$

The first term has a triplet CO moiety, while the second exhibits spin alternation on the carbonyl fragment. Since each interaction between two electrons of the same spin account for one Pauli repulsion, there are two Pauli repulsion in the first term of P^* and only one in the second (the interaction between the spins on CH_3 and O has a longer distance and is discounted). Hyperconjugation would have the effect of delocalizing the pseudo--orbital of methyl on the neighboring carbon, or, equivalently of adding the following two VB structures to P^*:

$$P^*(hyper) = H\overset{\uparrow}{C}H_3{-}\overset{\uparrow\downarrow}{C}{-}\overset{\uparrow}{O} + H\overset{\downarrow}{C}H_3{-}\overset{\uparrow\downarrow}{C}{-}\overset{\uparrow}{O} \qquad (6.\text{Ans.}27)$$

Here there are two Pauli repulsions in each of the two terms. Therefore, hyperconjugation destabilizes the P^* state that can only rise in energy relative to R upon substitution of the carbonyl group.

(d) It follows from the preceding points that substituting the carbonyl group lowers P and raises P^* while leaving R^* unchanged in energy relative to R. As a consequence, as shown in part (2) of Fig. 6.Ans.5, the energy of the crossing point, and consequently the barrier, is hardly lowered as the products are stabilized by carbonyl substitution. In contrast, as demonstrated above, stabilizing the leaving radical S_1 leads to a barrier lowering about as important as the lowering of the products. Hence, the dichotomic relationship between the barrier and the endothermicity. Experimentally, the two different behaviors are clear-cut and cover a number of different reactions. (For more details, see Ref. (9).)

(e) Now, let us show that the use of the effective parameters of the VBSCD (see Fig. 6.7 and Eq. 6.Ex.6) leads to the same predictions as the more detailed considerations in (a–d). Consider Fig. 6.Ans.4 as a VBSCD with a promotion gap, G_R, between R and R^*, a promotion gap G_P between P and P^*, and reaction energy ΔE_{rp}. When the substituent S_1 is made more stable, this weakens the C–S_1 bonding energy in the reactant state, and since the promotion gap is given by $G_R \approx 2D(\text{C–}S_1)$, stabilization of S_1 will result in the lowering of both G_R and ΔE_{rp} (making the reaction less endothermic), thus having a strong barrier lowering effect (consult Fig. 6.7 and consider Eq. 6.Ex.6). In contrast, changes of S_2 and S_3 that are not involved in the bond cleavage will leave G_R unchanged, lower ΔE_{rp} and increase G_P, thereby causing a smaller effect on the barrier lowering.

Exercise 6.8 The answer to this question is rooted in the promotion energy needed to generate the R^* states for the two processes. These states and corresponding promotion energies are illustrated pictorially in Fig. 6.Ans.6. It is seen that in the case of the three ethylene molecules, we need to undo the pairing of all the π-bonds to triplets and pair the three reactants across the linkages leading to products (for simplicity, we use the spectroscopic state). The triplet excitation energy of ethylene, calculated at the recommended paper, is 101 kcal/mol. The total G for this trimerization reaction is then 303 kcal/mol. By comparison, in the Diels–Alder reaction when we excite the diene portion, one double bond is formed at the central C–C of the butadiene and the triplet electrons are located on the termini of the molecule. Since these termini are far away from each other, the triplet repulsion is small and the excitation energy of the diene is only 78 kcal/mol, which together with the excitation of the ethylene gives us a G value of 179 kcal/mol. This comparison of the two reactions is interesting because it demonstrates that one reduces the value of G by a significant extent (78 kcal/mol for butadiene instead of 202 kcal/mol for two ethylenes), by linking the two double bonds together as in butadiene, compared with the termolecular reaction where the three double bonds are separated and each has to be promoted in the R^* state.

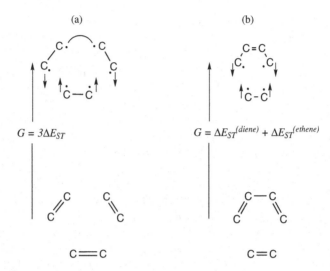

FIGURE 6.Ans.6   Ground and promoted states of the reactants for the trimerization of ethylene (left) and a Diels–Alder reaction (right).

With a difference of 124 kcal/mol in the G values, the Diels–Alder reaction will have a much smaller barrier. Using a value of $f = 1/3$ as in radical exchange, the G factors will contribute to a barrier difference of 41 kcal/mol in favor of the Diels–Alder reaction. However, the reaction energy term ΔE_{rp} will offset some of this difference. Using Equation 6.11, the reaction energy is weighted by a factor of 0.5. Since the reaction energies of the two processes differ by 23 kcal/mol in favor of the trimerization of ethylene, this will lower the energy barrier differences by $\sim$12 kcal/mol, making the net difference only 29 kcal/mol. The ab initio calculated difference is 27 kcal/mol.

Exercise 6.9   Let us assume that our series is a reaction family where the dominant reactivity factor is G. The VBSCD for C–X activation by PdL_2 is shown in Fig. 6.8b. It is seen that in the R^* state both the catalyst and the C–X bond are unpaired to their triplet states and the electrons are newly paired across the Pd–C and Pd–X bonds. In a series where there is a common substrate, Ph–Cl, and a family of catalysts, L_2Pd ($L_2 = PPh_2(CH_2)_nPPh_2$), which vary in the bite angle of the diphosphine ligand, the only variable that matters is the change in the singlet triplet excitation of the catalyst, $\Delta E_{ST}(PdL_2)$. Figure 6.Ans.7 shows a Walsh diagram for a d^{10} PdL_2 complex (the L orbitals are simplified to s-like orbitals). It is apparent that the natural geometry of PdL_2 is linear, in which there is a large HOMO–LUMO energy gap, and hence a large $\Delta E_{ST}(PdL_2)$ excitation energy. As we bend the L–Pd–L angle, the HOMO is destabilized and the LUMO is stabilized, resulting in a decrease of the $\Delta E_{ST}(PdL_2)$ excitation as the angle shrinks. The bidentate

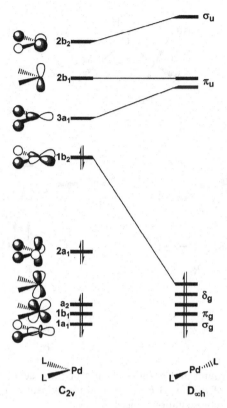

FIGURE 6.Ans.7   A Walsh diagram showing the orbitals of PdL_2 in a bent (C_{2v}) and linear geometries. The orbitals of the L ligands are simplified to s AOs.

ligand $PPh_2(CH_2)_nPPh_2$ enforces on the palladium a bent P–Pd–P angle, and the smaller the $(CH_2)_n$ linker, the smaller the angle gets and the lower becomes the $\Delta E_{ST}(PdL_2)$ excitation energy. Thus, as the linker of the two phosphines gets smaller, the barriers for C–Cl bond activation decreases [by 11 kcal/mol (32) from $n = 6$ to $n = 2$].

Exercise 6.10   The ground and promoted states of the reactants, R and R^* are shown in Fig. 6.Ans.8 below using the FO–VB representation, which is the simplest one for making predictions on stereochemistry. The electronic structure of R^* displays an electron transfer from the nucleophile to the C–C bond. A transfer to the σ-orbital is not relevant, as this would generate a closed-shell cyclopropane that cannot form a bond with the oxidized nucleophile. It is therefore the σ* orbital that accepts the transferred electron, thus generating the triplet σσ* configuration of the C–C bond. Now, since R and R^* differ by one-electron shift from n_{Nu} to the σ* orbital of the cation radical, the corresponding

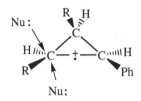

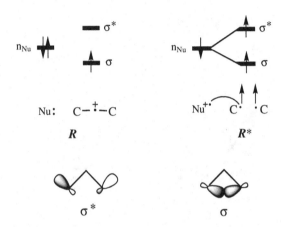

FIGURE 6.Ans.8 Ground and promoted states of the reactants, R and R^*, for the nucleophilic cleavage of the one-electron σ-bond in a substituted cyclopropane cation radical. The FO–VB representation is used.

transition-state resonance energy will be proportional to the overlap of the two orbitals (consult **Rule 3**), that is,

$$B \propto \langle n_{Nu} | \sigma^* \rangle \qquad (6.\text{Ans.}28)$$

Considering the nodal properties of the σ^* orbital it is apparent that the prediction is that nucleophilic cleavage of one-electron bonds will proceed with inversion of configuration. (For more details consult Refs. (5,42,49).)

Exercise 6.11 Neglecting βS^2 terms, the matrix elements between determinants are as follows:

$$\langle |a\bar{b}c|\,|H^{\text{eff}}||ab\bar{c}|\rangle = -2\beta_{bc}S_{bc} \qquad (6.\text{Ans.}29)$$

$$-\langle |a\bar{b}c||H^{\text{eff}}||a\bar{b}c|\rangle = +2\beta_{ac}S_{ac} \qquad (6.\text{Ans.}30)$$

$$-\langle |\bar{a}bc||H^{\text{eff}}||ab\bar{c}|\rangle = +2\beta_{ac}S_{ac} \qquad (6.\text{Ans.}31)$$

$$\langle |\bar{a}bc||H^{\text{eff}}||a\bar{b}c|\rangle) = -2\beta_{ab}S_{ab} \qquad (6.\text{Ans.}32)$$

$$\langle R|H^{\text{eff}}|P\rangle = H^{\text{eff}}_{RP} = -2\beta_{ab}S_{ab} - 2\beta_{bc}S_{bc} + 4\beta_{ac}S_{ac} \qquad (6.\text{Ans.}33)$$

For calculating the reduced Hamiltonian matrix element B, we need the diagonal matrix elements H_{RR}^{eff} and H_{PP}^{eff}, and the overlap between R and P:

$$H_{RR}^{\text{eff}} = -2\beta_{ab}S_{ab} - 2\beta_{ac}S_{ac} + 4\beta_{bc}S_{bc} \tag{6.Ans.34}$$

$$H_{PP}^{\text{eff}} = -2\beta_{ac}S_{ac} - 2\beta_{bc}S_{bc} + 4\beta_{ab}S_{ab} \tag{6.Ans.35}$$

$$S_{RP} = -1 \tag{6.Ans.36}$$

$$B = H_{RP}^{\text{eff}} - S_{RP}(H_{RR}^{\text{eff}} + H_{PP}^{\text{eff}})/2 = -\beta_{ab}S_{ab} - \beta_{bc}S_{bc} + 2\beta_{ac}S_{ac} \tag{6.Ans.37}$$

This term B is largest when the last term is negligible, which will correspond to the collinear geometry. On the other hand, B falls to zero for a bending angle of 60°, where all overlaps are equivalent, as well as all the corresponding β integrals.

Exercise 6.12

(a) Neglecting terms of the type βS^2 or βS^3, the matrix elements between determinants are as follows:

$$\langle|a\bar{a}b\bar{c}||H||c\bar{c}a\bar{b}|\rangle = \beta_{ab}S_{bc} + \beta_{bc}S_{ab} \tag{6.Ans.38}$$

$$-\langle|a\bar{a}b c||H||c\bar{c}a\bar{b}|\rangle = -\beta_{ac} + \beta_{ab}S_{bc} + \beta_{bc}S_{ab} \tag{6.Ans.39}$$

$$-\langle|a\bar{a}\bar{b}c||H||c\bar{c}a\bar{b}|\rangle = -\beta_{ac} + \beta_{ab}S_{bc} + \beta_{bc}S_{ab} \tag{6.Ans.40}$$

$$\langle|a\bar{a}\bar{b}\bar{c}||H||c\bar{c}a\bar{b}|\rangle = \beta_{ab}S_{bc} + \beta_{bc}S_{ab} \tag{6.Ans.41}$$

From the expressions of R and P in Equation 6.31, it follows (dropping normalization constants):

$$\langle R|H^{\text{eff}}|P\rangle = 4(\beta_{ab}S_{bc} + \beta_{bc}S_{ab}) - 2\beta_{ac} \tag{6.Ans.42}$$

To express the reduced Hamiltonian matrix element B, we need the overlap between R and P and their diagonal matrix elements:

$$\langle R|P\rangle = -2S_{ac} \tag{6.Ans.43}$$

$$\langle R|H^{\text{eff}}|R\rangle = -4\beta_{ab}S_{ab} - 4\beta_{ac}S_{ac} + 4\beta_{bc}S_{bc} \tag{6.Ans.44}$$

$$\langle P|H^{\text{eff}}|P\rangle = -4\beta_{ac}S_{ac} - 4\beta_{bc}S_{bc} + 4\beta_{ab}S_{ab} \tag{6.Ans.45}$$

$$B = 4(\beta_{ab}S_{bc} + \beta_{bc}S_{ab}) - 2\beta_{ac} = 8\beta S < 0 \tag{6.Ans.46}$$

It is clear from Equation 6.Ans.46 that B, the reduced matrix element between R and P, is negative in the linear conformation.

(b) Dealing now with the collinear transition state $[X–A–Y]^-$ of an S_N2 reaction, let us get the expression for B by replacing (S_{ab}, β_{ab}) by (S_{ax}, β_{ax}), (S_{bc}, β_{bc}) by (S_{ay}, β_{ay}) and (S_{ac}, β_{ac}) by (S_{xy}, β_{xy}) in Equation 6.Ans.46, leading to 6.Ans.47:

$$B = 4(\beta_{ax}S_{ay} + \beta_{ay}S_{ax}) - 2\beta_{xy} - 16\beta_{xy}S_{xy}^2 \qquad (6.\text{Ans.}47)$$

Recalling Equations 6.Ex.8 and 6.Ex.9, we have,

$$B = -8\beta S > 0 \qquad (6.\text{Ans.}48)$$

Since the matrix element is positive, the $R - P$ combination is the ground state, and the $R + P$ combination is the first excited state.

Under mirror reflection passing through the bisecting plane, R and P are transformed into $-P$ and $-R$, respectively. Therefore, the ground state is symmetrical (A′ symmetry), while the excited state is antisymmetrical (A″ symmetry).

(c) Owing to the rotation of the central fragment, the central a orbital now overlaps positively with both x and y:

$$S_{ay} = S_{xa} = S \quad S > 0 \qquad (6.\text{Ans.}49)$$

$$\beta_{ay} = \beta_{xa} = \beta \quad \beta < 0 \qquad (6.\text{Ans.}50)$$

$$B = +8\beta S < 0 \qquad (6.\text{Ans.}51)$$

Since the matrix element is negative, the $R + P$ combination is the ground state, and the $R - P$ combination is the first excited state, as in the H_3^- case. Under reflection with respect to the symmetry plane, R and P are transformed into $+P$ and $+R$, respectively, leading to an A′ ground state and an A″ excited state. Note that the FO–VB representation in Exercise 6.13 and the corresponding answer leads to these conclusions in a rather immediate manner.

Exercise 6.13

(a) Consider the twin states, $\Psi^{\neq}$ and Ψ^* at the geometry of a slightly bent transition state for the S_N2 reaction, $(X...CH_3...X)^-$, as in Fig. 22b. The simplest way to view the VB wave functions of the two states is to use FO–VB. This is done in the figure below (Fig. 6.Ans.9), by cutting the XCH_3X^- species into two fragments, one the central CH_3, the other the terminal atoms X...X. Two possible spin-pairing modes can be considered: $\Psi^{\neq}$, involves electron pairing of the electrons in the a and the σ FO, while two electrons are located in the σ^* FO. The symmetry of the state with respect to the plane passing through CH_3 is A′ in the C_s point group. On the other hand, Ψ^* involves singlet pairing of the electrons in the a and σ^* FOs and an electron pair in σ; the corresponding state symmetry is A″. Note that since the a and σ^* FOs have different symmetries, the singlet pairing does not stabilize the state, it is merely pairing of the spins to singlet (drawn as such by a dashed line between the respective orbitals). $\Psi^{\neq}$

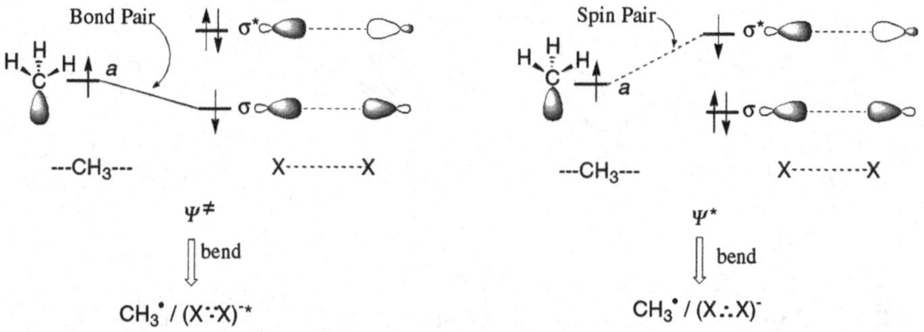

FIGURE 6.Ans.9 The VB–FO representation of the twin states, $\Psi^{\neq}$ and Ψ^* obtained after rotation of the methyl group at the collinear geometry for the S_N2 reaction, $(X\ldots CH_3\ldots X)$.

and Ψ^* are readily written as

$$\Psi^{\neq} = |\sigma^*\bar{\sigma}^*a\bar{\sigma}| - |\sigma^*\bar{\sigma}^*\bar{a}\sigma| \qquad (6.Ans.52)$$
$$\Psi^* = |\sigma\bar{\sigma}a\bar{\sigma}^*| - |\sigma\bar{\sigma}\bar{a}\sigma^*| \qquad (6.Ans.53)$$

Since the σ and σ^* FOs are quasidegenerate in the collinear or slightly bent conformation, the lowest state is the one that involves a stabilizing spin-pairing of a with the $X\ldots X$ FO of the right symmetry, namely, the σ FO. As such, $\Psi^{\neq}$ is the lowest state in the linear conformation, while Ψ^* is the excited state.

(b) Inspection of the $X\ldots X$ moiety in both states reveals that this moiety is a ground-state anion radical (X_2^-) in Ψ^*, and an excited anion radical (X_2^{-*}) in $\Psi^{\neq}$. As we further bend the $X\ldots X\ldots X$ angle this shortens the $X\ldots X$ distance and moves CH_3 away. Consequently, the $\Psi^{\neq}(A')$ state will undergo destabilization and generate $CH_3^{\bullet}/X_2^{-*}$, while $\Psi^*(A'')$ will undergo stabilization and generate $CH_3^{\bullet}$ and X_2^-.

Exercise 6.14

(a) This is a classical problem in photochemistry and is associated with the pioneering study of Oosterhoff who showed for the first time that the important state in the photochemistry of butadiene is not the first excited state, but rather the second excited state (of A_{1g} symmetry, so called the "dark state" that is doubly excited in MO terms). In terms of the VBSCD, this second excited state is nothing else but the "twin" excited state. Based on the atom ordering depicted in the figure below (Fig. 6.Ans.10), the wave functions for **R** and **P** are as follows:

$$R = |a\bar{b}c\bar{d}| - |a\bar{b}\bar{c}d| - |\bar{a}bc\bar{d}| + |\bar{a}b\bar{c}d| \qquad (6.Ans.54)$$
$$P = |ab\bar{c}\bar{d}| - |a\bar{b}c\bar{d}| - |\bar{a}bc\bar{d}| + |\bar{a}\bar{b}cd| \qquad (6.Ans.55)$$

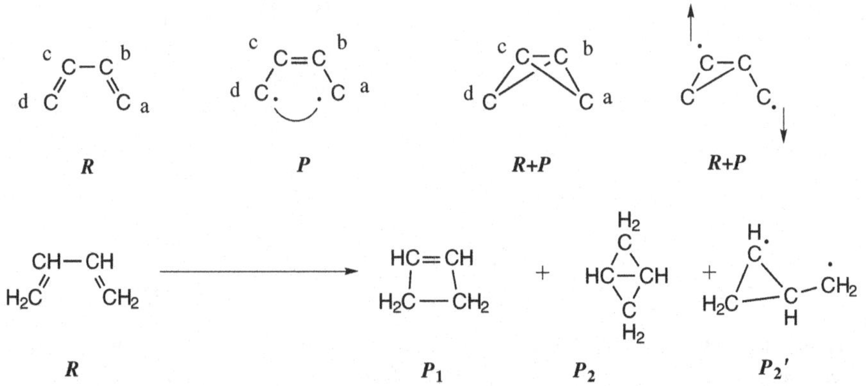

FIGURE 6.Ans.10 Products of the singlet photochemical excitation of butadiene (*R*).

Each one of these wave functions is obtained, as explained in Chapter 3, as a product of the corresponding bond wave functions. Thus, in *R* the bonds are a-b and c-d, while in *P* these are b-c and a-d.

(b) If all orbital overlaps are neglected, the overlap between *R* and *P* is calculated very simply by considering only the determinants that are common to both *R* and *P*. This overlap between the two unnormalized wave functions is −2. Since the overlap is negative, the reduced Hamiltonian matrix element between *R* and *P* will be proportional to −β, and is therefore positive. It follows that the transition state, $\Psi^{\neq}$, is the negative combination *R−P*:

(c) $\Psi^{\neq} = 2|a\bar{b}c\bar{d}| - |a\bar{b}\bar{c}d| - |\bar{a}bc\bar{d}| + 2|\bar{a}b\bar{c}d| - |ab\bar{c}\bar{d}| - |\bar{a}\bar{b}cd|$ (6.Ans.56)

The twin state is the positive *R* + *P* combination, given by:

$\Psi^{*} = -|a\bar{b}\bar{c}d| - |\bar{a}bc\bar{d}| + |ab\bar{c}\bar{d}| + |\bar{a}\bar{b}cd|$ (6.Ans.57)

(d) It can be easily shown that Ψ^{*} involves a-c and b-d couplings. You can reconstruct the wave function that corresponds to the product of the a-c and b-d bond wave functions, and verify that it is identical to the expression of Ψ^{*} above. Therefore, among the products of butadiene irradiation will be cyclobutene and [1.1.0]-bicyclobutane (see *R* + *P* and P_2 in Fig. 6.Ans.10). It is important to recognize that the diagonal bond making of two bonds in butadiene may not be synchronous, and therefore another mode that leads to crossing of the transition state and the twin excited state will generate the methylene cyclopropyl diradical (see P_2') that will lead to a variety of products (by H-abstraction etc.). The products of the photochemical reaction are depicted in Fig. 6.Ans.10. An alternative (but related) way to predict these products can be found in *J. Phys. Chem. A* **103**, 2364 (1999).

7 Using Valence Bond Theory to Compute and Conceptualize Excited States

Computation of excited states is not a simple matter in quantum chemistry. By using MO theory, one has an arsenal of methods, starting from configuration interaction with single excitations (CIS) (1) and going all the way to the more elaborate CASPT2 or CASMP2 methods (2). The CIS method is not accurate and does not handle doubly excited states, which are very important in a variety of molecules (e.g., polyenes). The CASPT2/CAMP2 method is in principle very accurate, but is still not sufficiently economical to apply widely and it is not a black-box method. Density functional theory (DFT) offers a few economical methods, chiefly the time-dependent DFT (TDDFT) method (3), which has become extremely popular in recent time, and is available in most MO-based software packages. However, the method does not handle doubly excited states explicitly and has other problems. Therefore despite its success, it is still questionable as a rigorous theory (4). *Ab initio* VB calculations for excited states are possible, but have so far been rather rare (5,6), and are still not routinely applicable. On the other hand, there are semiempirical VB methods, which are suitable for the calculation of excited states in some restricted classes of molecules (e.g., polyenes, aromatic molecules, polyenyl radicals, and related organic molecules) (7–10). These semiempirical methods are remarkably accurate at their computational domains, and lead to results at par with CASPT2 (10). Some VB methods are currently under development (11), but the performance and wide applicability of these methods is still not established.

One of the major problems in applying quantum chemical calculations to excited states is the restricted ability to interpret the calculations in large CI expansions, such as CIS and CASPT2. This limitation often does not exist in VB theory, which in many cases can assign a few chemical structures to describe a given excited state. As such, the major goal of this chapter is to teach a conceptual VB approach to excited states, based on the qualitative VB theory discussed throughout Chapters 1–6.

Scheme 7.1 shows a generic diagram of VB configurations for a molecule having an average of one electron per atom. This can be the simple H_2 or

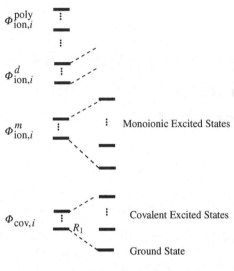

Scheme 7.1

hydrogen halide molecules, a π-system; in olefins, in higher polyenes, in cyclic systems, such as benzene and cyclobutadiene, in polycyclic systems, such as napthalene and anthracene, or in polyenyls radicals, such as allyl and higher members, H_n chains, and so on. The lowest energy configurations are the covalent ones, $\Phi_{cov,i}$, which include all the possible and non-redundant ways of pairing the electrons on the various sites (e.g., the Rumer structure set discussed in Chapters 3 and 4). Higher above in energy are the monoionic structures, $\Phi_{ion,i}^m$, in which one electron is transferred from one site to another to form a single pair of positively and negatively charged atoms. Still higher are multiply ionic configurations, ordered according to the rank of ionic pairs; the uppermost block contains the polyionic structures, where all the electrons (except for cases with an odd number) are paired in negatively charged sites. The ground state of the system will originate from the covalent configurations. As we will see later, the lowest energy excited states of the system will originate partly from the covalent structures and partly from the ionic structures.

Next we discuss examples of the two types of excited states. To focus the treatment we deal only with the excited states of the smallest spin eigenvalue for a given system, that is, singlet states for a closed-shell molecule and doublet states for radicals.

7.1 EXCITED STATES OF A SINGLE BOND

Let us take H_2 as an example, and construct the molecular states from the familiar minimal VB configuration set shown in Fig. 7.1a; these are the covalent and two ionic structures. Since the molecule has left-right symmetry we can instantly make two symmetry-adapted configurations from the two

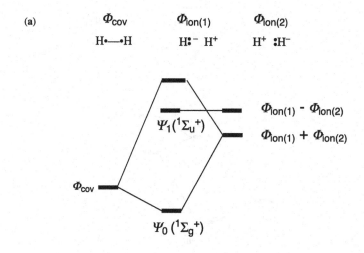

FIGURE 7.1 (a) The VB structures and their mixing diagram leading to the states of the H−H bond. (b) A Schematic MO representation of the first singlet excited state and its correspondence to the VB representation.

ionic structures. The positive combination has a symmetry match (Σ_g^+) with the covalent structure, and hence the two structures will mix leading to the ground state, as well as the somewhat high-lying excited state, which is not going to concern us any further. The negative combination of the ionic structures has Σ_u^+ symmetry and cannot therefore mix with the covalent configuration. As such, this resonating ionic state will be a pure state and become the first singlet excited state of the molecule, $\Psi_1(\Sigma_u^+)$.

One may wonder how this $\Psi_1(\Sigma_u^+)$ VB-state relates to the more familiar $\sigma_g\sigma_u$ MO-excited state of H_2, in Fig. 7.1b. This bridge can be made easily using the MO−VB mapping technique discussed in Chapter 4 or by simple multiplication of the diagonal term in the $\sigma_g\sigma_u$ MO-state after expressing the MOs in terms of the AOs. So doing will instantly demonstrate that the $\sigma_g\sigma_u$ MO-excited state is identical to the resonating ionic state, as shown in Fig. 7.1b.

Another question one may have is: Where do I see the antibonding character in the $\Psi_1(\Sigma_u^+)$ VB state? Here too, use of the semiempirical theory in Chapter 3 and the rules for taking matrix elements between VB determinants, will show that the matrix element between the two ionic determinants is $2\beta S$, where β and S are the corresponding resonance and overlap integrals between the two AOs of the H atoms. As such, the negative combination of the ionic configurations

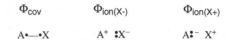

Φ_{cov} $\Phi_{ion(X-)}$ $\Phi_{ion(X+)}$

A•—•X   A$^+$:X$^-$ A:$^-$ X$^+$

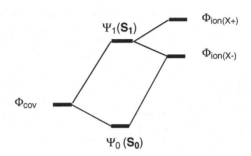

FIGURE 7.2   The VB structures and their mixing diagram leading to the states of an A–X bond (A = H, R$_3$C, etc.; X = an electronegative group, for example., Cl, OH).

suffers from an antibonding interaction, much as the corresponding $\sigma_g\sigma_u$ state in the MO formulation. Of course, with a good basis set, the orbitals of this excited state will try to minimize their antibonding interaction, and are expected to be very different than those of the ground state. The example of H$_2$ is an archetype for all homonuclear single bonds, such as Li$_2$, the π-bond in olefins, and so on. In all these cases, the singlet-excited state will be the resonating ionic state in VB theory.

Now, let us consider a heteropolar bond A–X, in Fig. 7.2, where A can be a hydrogen or an alkyl group, R$_3$C, whereas X is an electronegative group, for example, a halogen. Once again, the VB structure set involves the covalent and two oppositely ionic structures; other structures, which involve distribution of the lone-pair electrons on X are not considered. Now the ground state will be nascent from the covalent structure, while the singlet excited state, S_1, will be dominated by the lowest energy ionic structure, $\Phi_{ion(X^-)}$. Here, the antibonding interaction will involve the mixing with the covalent structure, whereas the oppositely ionized structure will mix in to minimize this antibonding interaction. These excited states are known for hydrogen halides (12), and are responsible for the heterolytic cleavage of C–X bonds in solution (via covalent–ionic crossing due to solvation, see Exercise 6.2) in the classical S$_N$1 mechanism (13).

7.2   EXCITED STATES OF MOLECULES WITH CONJUGATED BONDS

In conjugated molecules, there are a few covalent structures (see Scheme 7.1) and in most cases, the lowest lying excited state, or one of the lowest lying ones, is covalent. Among the most well-known covalent states are the so-called "dark" (or hidden) states of polyenes, having A_g symmetry (identical to the symmetry of ground state), and being dipole forbidden (14,15). These states are responsible for much of the photochemistry of polyenes (14). Similarly, the dipole forbidden excited states of aromatics, of B_{2u} symmetry (16–18), and of polynuclear aromatics, and so on, are all covalent excited states (19). There are of course also ionic states that will arise from the sea of ionic structures (Scheme 7.1), for example, the B_u^+ states of polyenes, which are dominated by monoionic structures (7,20). Here and thereafter, we will mention ionic states in passing while focusing the discussion on the covalent excited states.

7.2.1   Use of Molecular Symmetry to Generate Covalent Excited States Based on Valence Bond Theory

A convenient starting point for conceptualizing excited states is to begin with those cases where molecular symmetry can assist us to generate the VB states. Some examples are discussed below.

The Allyl Radical   The case of allyl radical in Fig. 7.3, is the simplest example of a π-system where the VB states can be generated by use of symmetry considerations (21). Allyl radical has two VB structures that are labeled in Fig. 7.3a as K_r and K_l, where K denotes a Kekulé structure (meaning a structure with maximum pairing), while the subscripts signify the location of the double bond on the right- and left-hand sides, respectively. The corresponding wave functions are given in Equation 7.1a and 7.1b, where normalization constants are dropped:

$$\Phi(K_r) = |ab\bar{c}| - |a\bar{b}c| \tag{7.1a}$$

$$\Phi(K_l) = |a\bar{b}c| - |\bar{a}bc| \tag{7.1b}$$

The a, b, and c terms in the determinants are the p_π AOs of allyl in successive order from left to right.

As shown in the VB mixing diagram (22) in the figure, the two structures mix and lead to bonding and antibonding combinations. Since the two structures are mutually transformable by the symmetry operations, σ_v and C_2 of the molecular point group, the negative combination will transform as A_2, while the positive one is B_1. Written as in Equation 7.1, it is seen that the negative combination of the two Kekulé structures will be antisymmetric with respect to the plane bisecting the CCC angle, and will lead to the $^2A_2$ state, while the positive combination is the $^2B_1$ excited state. Based on Fig. 7.3a, the $^2A_2$ is the ground state, while $^2B_1$ is the excited state (6,21).

The rationalization of the state ordering in Fig. 7.3 can be achieved by merely inspecting the wave function (one can always derive all the mixing matrix elements and verify these qualitative arguments as done for the isoelectronic problem of the $H_3^\bullet$ species in Chapter 6). Let us write the wave

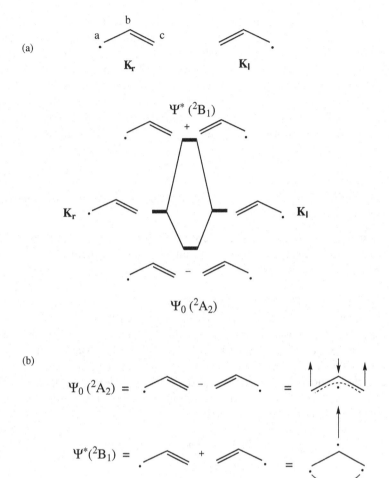

FIGURE 7.3   (a) Kekulé structures for allyl radical and their mixing diagram leading to the ground and first covalent excited states. (b) Spin density distribution in the ground and excited states.

function for the negative combination as follows:

$$\Psi(^2A_2) = \Phi(K_r) - \Phi(K_l) = -2|a\bar{b}c| + |ab\bar{c}| + |\bar{a}bc| \qquad (7.2)$$

It is seen that the coefficient of the $|a\bar{b}c|$ determinant, which is the spin-alternant determinant [or the quasiclassical (QC) determinant discussed in Chapters 3–5], is doubled compared with the other two. As such, using it once with $|ab\bar{c}|$ we account for pairing of the electrons in b–c, namely, a double bond on the right-hand side, while using it the second time with $|\bar{a}bc|$ gives rise to the double bond across a–b, on the left-hand side. This is the delocalized state of allyl radical in the ground state, as depicted in Fig. 7.3a and b. We can further see by inspecting the wave function in Equation 7.2 that the spin density distribution will be dominated by the spin-alternant determinant that has the largest weight in this state. Therefore, the ground state will involve positive spin density on the terminal atoms, a and c, and an excess of negative spin density on the central carbon, b. This spin-density distribution conforms with the experimental data based on electron spin resonance (EPR) coupling constants (23). The solution of Exercise 7.1 demonstrates these trends in a more quantitative manner.

The positive combination of the Kekulé structures, in Equation 7.1a and b, leads to the $^2B_1$ state with a wave function expressed by Equation 7.3:

$$\Psi(^2B_1) = \Phi(K_r) + \Phi(K_l) = |ab\bar{c}| - |\bar{a}bc| \tag{7.3}$$

It is seen that in this state the QC determinant $|a\bar{b}c|$ vanishes from the wave function, and we are left with a negative combination of the remaining two determinants. We recall that the QC determinant is the one that enables the pairing-up of the electrons with α and β spins in an alternating manner in the ground state. Since this determinant vanishes at the excited state, we expect that there will be changes in the bonding from the mode allowed by the QC determinant. Inspection of the wave function shows that in the $^2B_1$ state the electrons on a and c are coupled to form a bond pair, and there remains a single electron with spin α on atom b. Note that the VB structures of the $^2A_2$ and $^2B_1$ states of allyl radical are reminiscent of the resonant and antiresonant states of $H_3{}^\bullet$, described in Chapter 6, Section 6.11.1. Thus, the $^2B_1$ state is the positive combination of the two Kekulé structures, but at the same time this combination corresponds to a single structure as shown in Figure 7.3b. This bonding feature of $^2B_1$ will manifest in the photochemistry of allyl radical that is expected to give rise to cyclopropyl radical (consult the related discussion of photochemistry in Chapter 6).

As we move to the next member in the series, pentadienyl radical, the ground and excited state properties will switch, in their symmetry and spin density distribution. This example is treated in detail in Exercise 7.2, and the switch is general throughout the series. Scheme 7.2 illustrates this switching pattern for the series $C_{2n-1}H_{2n+1}$, by depicting the sites of the large positive spin densities using arrows (neglecting the negative spin densities) in the ground and excited states. It can be seen that all the polyenyl radicals with even n will possess $^2A_2$ ground states with large positive spin densities on the two carbon atoms that flank the central atom, and $^2B_1$ excited states with large spin density on the central atom. In contrast, the radicals with odd n will possess $^2B_1$ ground states with large positive spin densities on the central atom, and $^2A_2$ excited states with large spin densities on the two carbon atoms that flank the central atom. The spin densities and state ordering in allyl radical correspond to the cases with even n ($n = 2$), while the pentadienyl radical, which is an archetype of the odd n members ($n = 3$), is worked out in

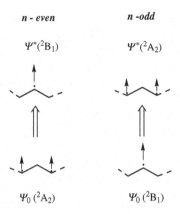

Scheme 7.2

Exercise 7.2 and later also in Exercise 8.5 (where the spin densities are derived in details). The detailed generalization of the treatment to the entire family of the polyenyl radicals has appeared, and the interested reader may wish to consult the original literature (21).

Benzene   The patterns noted above for allyl radical are general for conjugated systems with an average of one electron per site. In all these cases, we will find that one of the combinations of the Kekulé structures corresponds to a third and new structure (or to a set of different ones), and this arises from the fact that the Kekulé structures of these systems share common determinant(s), the QC ones, that are doubled in one combination and eliminated in the other, as the two structures mix (recall Section 5.4 on aromaticity). As such, the two combinations will possess different bond-pairing features, which in the case of the excited-state combination will also indicate qualitatively the photochemical products expected from the molecule.

Benzene is another molecule where molecular symmetry can be useful for construction of the ground and covalent excited states (5,6,16,17,24). The molecule possesses five covalent VB Rumer structures, two are the Kekulé structures, and the three Dewar types, shown in Fig. 7.4a. It is seen that the Kekulé structures are mutually transformable by the D_{6h} point group operations i, C_2, and σ_v. Consequently, we can make two linear combinations; the positive one is the totally symmetric A_{1g} state, which gives rise to the ground state, while the negative one is the B_{2u} state, which corresponds to the excited state. This is shown in the VB mixing diagram in Fig. 7.4b.

One can further consider the refinement of the picture by mixing in the three Dewar structures, D_{1-3}. The Dewar types remain unchanged under the i symmetry operation, and hence will have a g symmetry and form the A_{1g} and E_{2g} combinations, as shown in Fig. 7.4c; here the A_{1g} combination is the positive combination of all the Dewar structures, whereas, the E_{2g}

(a)

$K_1 \qquad K_2 \qquad D_1 \qquad D_2 \qquad D_3$

$K_1 \xrightarrow[D_{6h}]{C_2,\ \sigma_v,\ i} K_2 \Rightarrow \begin{cases} \Psi_0 - A_{1g} \\ \Psi^* - B_{2u} \end{cases} \qquad \Gamma(D_1,\ D_2,\ D_3) = A_{1g} + E_{2g}$

(b)

$\Psi^*\ (K_1 - K_2)$

$K_1 \qquad\qquad K_2$

$\Psi_0\ (K_1 + K_2)$

(c)

$D_1 - D_3 \qquad 2D_2 - D_1 - D_3$

$D_{1\text{-}3} \qquad\qquad E_{2g}$

A_{1g}

$D_2 + D_1 + D_3$

(d)

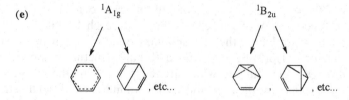

$\Omega_{QC} \qquad\qquad\qquad \tilde{\Omega}_{QC}$

(e)

$^1A_{1g} \qquad\qquad\qquad\qquad ^1B_{2u}$

, etc... , etc...

FIGURE 7.4 (a) Kekulé and Dewar structures of benzene, and their symmetry properties. (b) A VB mixing diagram for Kekulé structures. (c) The symmetry adapted combinations nascent from the Dewar structures. (d) The two QC determinants of benzene and the carbon numbering system. (e) Bonding properties of the ground and excited states of benzene.

states are two combinations, $D_1 - D_3$ and $2D_2 - (D_1 + D_3)$. These three combinations of the Dewar structures are related to one another by the C_3 symmetry operation that rotates the benzene frame by 120°, and hence one can generate the states precisely as one generates the Hückel MOs of cyclopropenium cation. Since the E_{2g} combinations have no symmetry match with either of the combinations of the Kekulé structures, they will give rise to the doubly degenerate covalent excited state, $^1E_{2g}$. Thus, altogether the covalent structures give rise to four covalent excited states; $^1B_{2u}$ (nascent from the Kekulé structures), $^1E_{2g}$ and $^1A_{1g}$ (nascent from the Dewar structures).

Before continuing with the discussion, it is instructive to relate these VB states to those that arise from the MO picture of benzene. Benzene possesses degenerate pairs of HOMOs and LUMOs, and hence the excitations from the two HOMOs to the two LUMOs give rise to four singlet excited states of the following symmetries: B_{2u}, E_{2g}, and B_{1u}. As can be seen from Fig. 7.4a–c, the B_{2u} is the covalent state made from the Kekulé structures, while E_{2g} is the covalent state made from the Dewar structures. The B_{1u} state is predominantly ionic, made from the monoionic VB structures (see those in Chapter 5) (5). The excited A_{1g} state made from the Dewar VB structures (Fig. 7.4c) corresponds to higher rank MO excitation.

Turning back to the covalent states, we note that the Dewar A_{1g} combination will mix into the ground-state combination of the two Kekulé structures; this will stabilize the $^1A_{1g}$ ground state and will generate an antibonding A_{1g} combination dominated by the Dewar structures. As already shown, however, the use of CF orbitals minimizes the mixing of the A_{1g} combination of the Dewar types into the Kekulé structures (5). Therefore, for qualitative purposes we will consider henceforth only the Kekulé structures as constituents of the covalent ground and first excited states.

Thus, as we did for the allyl radical case, here too the bonding characteristics of the two covalent states of benzene can be deduced from the respective wave functions. As discussed in Chapter 5, each Kekulé structure can be generated by a product of the corresponding bond wave functions, each having a spin factor $\alpha(1)\beta(2)-\beta(1)\alpha(2)$. Each Kekulé structure possesses the two spin alternant QC determinants, which are related to each other by a cyclic permutation of the spins over the ring, and are shown in Fig. 7.4d. To illustrate clearly the building blocks of the two Kekulé structures, we express them as a sum of all the permutations of the QC determinants, as follows:

$$\Phi(K_1) = [1 - (P_{23} + P_{45} + P_{61})] \bullet (\Omega_{QC} - \tilde{\Omega}_{QC}) \qquad (7.4a)$$

$$\Phi(K_2) = [1 - (P_{12} + P_{34} + P_{56})] \bullet (\Omega_{QC} - \tilde{\Omega}_{QC}) \qquad (7.4b)$$

Here, the first QC determinant, Ω_{QC}, is the spin-alternant determinant starting with spin α on position 1 and ending with β on atom 6, while $\tilde{\Omega}_{QC}$ is the second QC determinant; the two being related by a cyclic permutation of the spins, as indicated in Fig. 7.4d. The P_{ij} terms are spin permutations of the spins in the positions i and j. It is seen that both $\Phi(K_1)$ and $\Phi(K_2)$ involve the two QC determinants and all the determinants that arise from the QCs by successive permutations of spin along the perimeter of the molecule.

As we already argued, the ground state must conserve the QC determinants due to their lower energies, and this is going to be the positive combination, as shown in Fig. 7.4a and b and expressed in Equation 7.5:

$$
\begin{aligned}
\Psi(A_{1g}) &= \Phi(K_1) + \Phi(K_2) \\
&= [2 - (P_{12} + P_{23} + P_{34} + P_{45} + P_{56} + P_{61})] \bullet (\Omega_{QC} - \tilde{\Omega}_{QC})
\end{aligned}
\tag{7.5}
$$

One can see that all the permutations, which appear in Equation 7.5, are successive transpositions of $\alpha\beta$ spins, and hence, the mixing of the QC determinants with any one of the permuted ones will generate bonding across the C–C periphery of benzene. This picture corresponds to the delocalized benzene picture, in Fig. 7.4e, known to every chemist (note that we added in the figure the Dewar structures alongside the bonding combination of the two Kekulé structures).

In contrast, the excited state is the negative combination that eliminates the two QC determinants, and is expressed as follows:

$$
\begin{aligned}
\Psi(B_{2u}) &= \Phi(K_1) - \Phi(K_2) = [(P_{12} + P_{34} + P_{56}) \\
&\quad - (P_{23} + P_{45} + P_{61})] \bullet (\Omega_{QC} - \tilde{\Omega}_{QC})
\end{aligned}
\tag{7.6}
$$

It is seen that this wave function retains only the permutations of the QC determinants (i.e., $P_{ij}\Omega_{QC}$ terms), and consequently the actual bonding in the excited benzene will change and will not be represented anymore by the alternating double bonds as in each Kekulé structure. Thus, since all the permutations transpose the spins of two successive atoms, the permuted QC determinants (e.g., $P_{12}\Omega_{QC}$), in the B_{2u} excited state, will possess some identical neighboring spins along the benzene perimeter, thus contributing antibonding interactions that thereby weaken the bonding across the circumference of the molecule. Therefore, the bonding in the excited state will shift from being around the perimeter to bonding interactions across the ring, resulting in equal probabilities of finding a short bond between adjacent atoms or ones in a meta position (1,3). Figure 7.4e shows these changes schematically as a series of structures having spin pairing as in benzvalene (the more precise bonding changes are worked out in Exercise 7.4). These bonding patterns will of course manifest in the photochemistry of benzene (note that the spin pairing of benzvalene in a planar structure corresponds to meta-diradical pairs that will give rise to benzvalene and other products too) (25). Another property of the B_{2u} state that was discussed in Chapter 6 is the exalted frequency of the mode b_{2u} in the excited state (16,17,24). Recall that the b_{2u} mode is the one leading to crossing of the Kekulé structures in the VBSCD. *Any mode that leads to crossing of VB structures will have a low frequency in the ground state and a high one (exalted) in the excited state.*

Cyclobutadiene   The case of cyclobutadiene is analyzed in Fig. 7.5 (24). Here (Fig. 7.5a) the two Kekulé structures are mutually transformable via C_4, but are symmetric with respect to the i and σ_v symmetry operations of the D_{4h} symmetry group. Accordingly, unlike benzene, here the ground state $^1B_{1g}$ is the negative combination, while the excited state, $^1A_{1g}$, is the positive combination (one can verify that the matrix elements between the structures is positive, since it is negatively signed). The corresponding VB mixing diagram is shown in Fig. 7.5a. The resulting $^1A_{1g}/^1B_{1g}$ states correspond to the covalent diradicaloid states of cyclobutadiene, which arise from permuting the two electrons in the two nonbonding MOs (φ_1 and φ_2) of the molecule as shown in Fig. 7.5b. The third state is the negative combination of the φ_1^2 and φ_2^2 configurations and it corresponds to the combination of the diagonal ionic structures (26).

To deduce the bonding features of the covalent states, $^1A_{1g}$ and $^1B_{1g}$, we express as we did before for benzene, the Kekulé structures as follows:

$$\Phi(K_1) = [+1 - (P_{23} + P_{41})] \bullet (\Omega_{QC}) + \tilde{\Omega}_{QC} \qquad (7.7a)$$

$$\Phi(K_2) = [-1 + (P_{12} + P_{34})] \bullet (\Omega_{QC}) - \tilde{\Omega}_{QC} \qquad (7.7b)$$

where the QC determinants are shown in Fig. 7.5c. By using these expressions, the ground state, the one that conserves the QC determinants in the wave function, is now the *negative combination*, and is hence the $^1B_{1g}$ state:

$$\Psi(^1B_{1g}) = \Phi(K_1) - \Phi(K_2) = [2 - (P_{12} + P_{13} + P_{24} + P_{34})] \bullet (\Omega_{QC}) + 2\tilde{\Omega}_{QC}$$
$$(7.8)$$

Since the spin permutations are successive, the $^1B_{1g}$ state will possess bonding across the C–C circumference (Fig. 7.5c). In contrast, the excited state is the positive $^1A_{1g}$ combination, which eliminates the QC determinants:

$$\Psi(^1A_{1g}) = \Phi(K_1) + \Phi(K_2) = [(P_{12} + P_{34}) - (P_{23} + P_{41})] \bullet (\Omega_{QC}) \qquad (7.9)$$

As shown in Fig. 7.5c, now the bonding across the circumference is phased out and is replaced by diagonal 1,3 and 2,4 bonding. Exercise 7.3 works out these bonding changes. This change of bonding will manifest in the photochemistry of cyclobutadiene. As in the case of benzene here too, one mode, the b_{1g} mode, leads to crossing of the Kekulé structures in the VBSCD, and hence its frequency increases in the $^1A_{1g}$ excited state (in the ground state this frequency is imaginary, while in the $^1A_{1g}$ excited state it becomes 2089 cm^{-1}) (24).

The above trends are general for aromatic and antiaromatic systems with one electron per site.

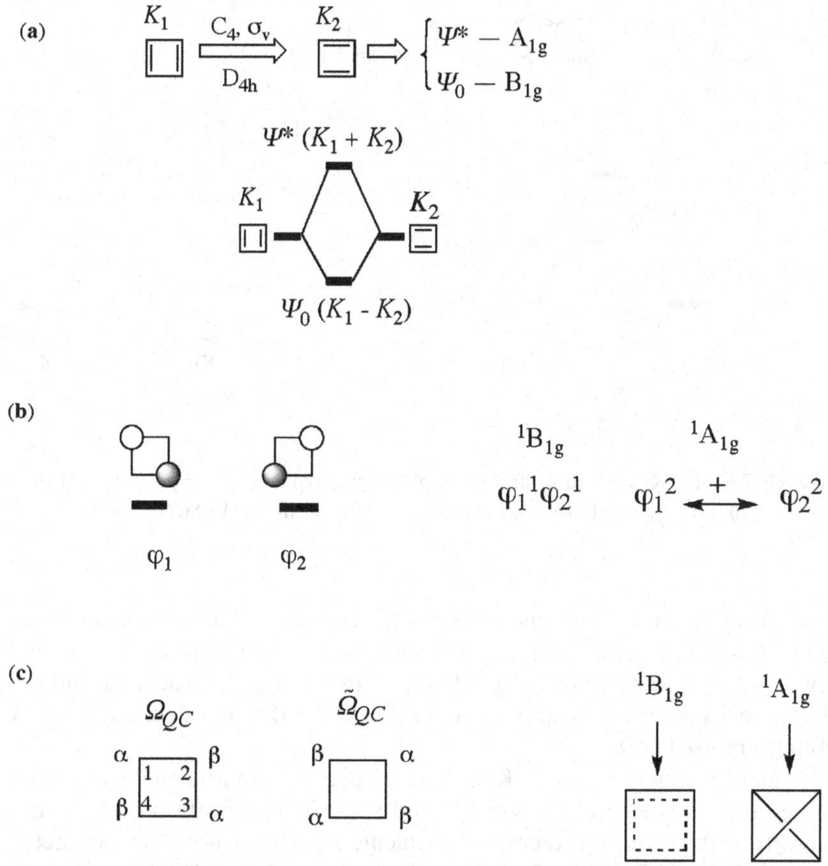
(a)
K_1
C_4, σ_v
D_{4h}
K_2
Ψ^* — A_{1g}
Ψ_0 — B_{1g}
Ψ^* (K_1 + K_2)
K_1
K_2
Ψ_0 (K_1 - K_2)
(b)
φ_1
φ_2
$^1B_{1g}$
$^1A_{1g}$
$\varphi_1{}^1\varphi_2{}^1$
$\varphi_1{}^2$ + $\varphi_2{}^2$
(c)
Ω_{QC}
$\tilde{\Omega}_{QC}$
α β β α
1 2 4 3
β α α β
$^1B_{1g}$
$^1A_{1g}$

FIGURE 7.5 (a) Kekulé structures of cyclobutadiene, their symmetry properties, and their VB mixing, yielding the covalent ground and excited states. (b) The diagonal MOs of cyclobutadiene and the corresponding ground and excited covalent states in the MO representation. (c) The two QC determinants of cyclobutadiene, the carbon numbering system, and the bonding properties of the covalent ground and excited states.

Naphthalene, Anthracene, and Linear Polyacenes Linear polyacenes possess two low-lying excited states, which in the Platt notation used in spectroscopy, are labeled as ^1L_a and ^1L_b; the former is $^1B_{1u}$ and the latter is $^1B_{2u}$ (7,19). The $^1B_{1u}$ state is a monoionic state, while $^1B_{2u}$ is a covalent excited state, which as in benzene, is dipole forbidden, but can be accessed by two photon spectroscopy. In naphthalene and anthracene, $^1B_{2u}$ is the second excited state, but in higher acenes, $^1B_{2u}$ becomes the first excited state (16,19). We will now show how to construct and conceptualize this excited state. Already at the outset, we encounter here a problem of choice, namely, that the set of covalent Rumer VB structures involves many structures (e.g., 42 in naphthalene), and one has to restrict the VB structure set in order to understand the constitution of these

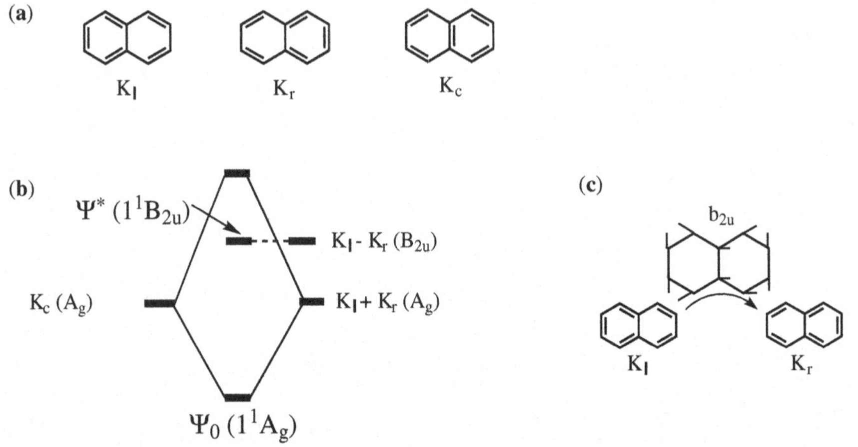

FIGURE 7.6 (a) Kekulé structures of naphthalene. (b) The corresponding VB mixing diagram. (c) The b_{2u} mode that interchanges K_l and K_r in the *VBSCD*.

states. As in the case of benzene, here too, the covalent space involves a few lower energy Kekulé structures, and many higher ones with long bonds (as in the Dewar types). Our natural choice is to focus on the Kekulé structures, and this is justified because these structures dominate the VB wave function when CF orbitals are used (27).

Figure 7.6 shows the three Kekulé structures of naphthalene. Two of them are designated as K_l and K_r (where l and r signify the location of the benzene ring with sextet), and represent the annulenic resonance along the perimeter of the naphthalene. The third Kekulé structure is labeled as K_c to signify that it has a double bond in the center. Pairing K_c with any one of the $K_{l,r}$ subset will account for resonance in the left- or right-hand benzene ring of naphthalene. The K_l and K_r structures are mutually interchangeable by the i, C_2, and σ_v symmetry operations of the D_{2h} point group of naphthalene. A positive combination therefore transforms as A_g, whereas a negative one transforms as B_{2u}. Figure 7.6b shows the mixing of these symmetry-adapted wave functions with K_c to yield the final states. Based on symmetry-match, K_c can mix only with the positive combination, leading to the ground state, $\Psi_0(^1A_g)$, and a high lying covalent excited state of the same symmetry. In contrast, the negative combination of K_l and K_r does not find a symmetry match and remains unchanged as the first-excited covalent state of naphthalene with B_{2u} symmetry.

Figure 7.6c shows that the vibrational mode b_{2u} interchanges the Kekulé structures K_l and K_r. Since the excited state is made only from these Kekulé structures, then by analogy to benzene, one will expect that the frequency of this mode will be larger in the B_{2u} excited state than in the A_g ground state, which is indeed what is observed by two-photon spectroscopy (16,24). Another analogy to benzene concerns the bonding characteristics of this excited state. Thus, since this excited state will eliminate the QC determinants in its wave function, it will no longer involve annulenic conjugation, and its spin pairing will involve remote carbons (e.g., meta carbons), which will be expressed in the photochemical products of the molecule.

Figure 7.7a shows the four classical Kekulé structures of anthracene (16,19,24). Two of the structures involve resonance in the central benzenic ring and are therefore labeled as K_{1B} and K_{2B}. The other two involve annulenic resonance along the molecule perimeter, and are labeled accordingly as K_{1A} and K_{2A}. The structures of the types A and B form two symmetry subsets, and within each subset, the two structures are mutually transformable by the D_{2h} symmetry operations (i, C_2, and σ_v). Therefore, as shown in Fig. 7.7b, within each subset there will be a positive combination that transforms as A_g and a negative one that transforms as B_{2u}.

Figure 7.7c shows how the $K_{1,2B}$ and $K_{1,2A}$ structures spread into the symmetry adapted combinations; in each subset the lowest combination is the positive combination, which lies below the corresponding negative

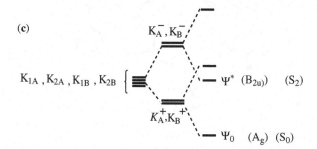

(a) K_{1B} K_{2B}

       K_{1A} K_{2A}

(b) A_g: $K_{1A} + K_{2A} \equiv K_A^+$ B_{2u}: $K_{1A} - K_{2A} \equiv K_A^-$

              $K_{1B} + K_{2B} \equiv K_B^+$ $K_{1B} - K_{2B} \equiv K_B^-$

(c) K_A^-, K_B^-

$K_{1A}, K_{2A}, K_{1B}, K_{2B}$ { Ψ^* (B_{2u}))   (S_2)

                           K_A^+, K_B^+

                                         Ψ_0 (A_g) (S_0)

FIGURE 7.7   (a) Kekulé structures of anthracene. (b) Symmetry adapted combinations of the Kekulé structures. (c) The VB mixing of the symmetry adapted VB combinations and the resulting covalent ground and excited states.

combination. This is so because the positive combination involves bonding either along the perimeter (set A) or in the central benzenic moiety (set B), while the negative combinations contain antibonding in the same corresponding regions. In the next step in Fig. 7.7c, we mix the positive A_g combinations, leading to the ground state, which will be given by Equation 7.10, dropping the normalization constant:

$$\Psi_0(1A_g) = \Phi(K_{1A}) + \Phi(K_{2A}) + \Phi(K_{1B}) + \Phi(K_{2B}) \qquad (7.10)$$

There will be a corresponding antibonding combination of the same symmetry, which will be a high lying covalent excited state. Similarly, we mix the two B_{2u} combinations, and the bonding combination becomes the *first covalent excited state* of the molecule (the S_2 state (16,19,24)), with a wave function expressed as:

$$\Psi^*(^1B_{2u}) = \Phi(K_{1A}) - \Phi(K_{2A}) + \Phi(K_{1B}) - \Phi(K_{2B}) \qquad (7.11)$$

Since the ground state and the $^1B_{2u}$ excited state are made from two sets of Kekulé structures that are interchanged by two b_{2u} modes, one will expect to find in the spectrum of the B_{2u} state two b_{2u} modes with exalted frequencies as shown in Fig. 7.8. As can be seen in the figure, one of these modes is benzenic, the other is annulenic. These two exalted modes are indeed observed in the B_{2u} spectrum of anthracene (16,19). The bonding features of the B_{2u} state are expected to involve spin pairing between nonconsecutive carbons

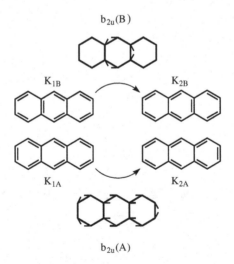

$b_{2u}(B)$

K_{1B} K_{2B}

K_{1A} K_{2A}

$b_{2u}(A)$

FIGURE 7.8 The b_{2u} mode that interchanges K_{1B} and K_{2B} and K_{1A} and K_{2A} for anthracene in the corresponding VBSCDs.

(e.g., meta) which will manifest in the product of the photochemistry of the molecule.

Naphthalene and anthracene are archetypes of the even and odd members of the polyacene series. In each subseries, one can start by classifying the classical Kekulé structures by using the symmetry operations i, C_2, and σ_v of the D_{2h} point group. Then one can form symmetry-adapted linear combinations of the mutually transformable Kekulé structures and deduce their bonding characteristics. Finally, these ^1A_g and $^1B_{2u}$ symmetry-adapted combinations are allowed to mix and form the states of interest, the ground and first covalent excited states (16).

7.2.2 Covalent Excited States of Polyenes

As already mentioned, polyenes possess covalent excited state of A_g symmetry, which are dipole forbidden, and hence are called the "dark" (hidden) states. These states play a major role in the photochemistry of polyenes (25), and hence, one would like to derive these states and understand their features. In MO theory, these states involve double excitation from the HOMO-1 to the LUMO and simultaneously from the HOMO to the LUMO + 1 (14,20). In VB theory, the same excitations can be described simply, as done throughout this chapter, in terms of the covalent VB structures. The following treatment is based on a recent quantitative semiempirical calculations of a long series of polyenes (10). At the outset, we recognize that one basic difference compared with the cyclic systems is that polyenes possess a single Kekulé structure (the classical perfectly paired structure), while all the other structures in the covalent structure set involve long bonds. One of these long-bond structures, where the long bond is across the two termini of the polyene, corresponds to the Kekulé structure of the corresponding ring. This Kekulé-type long-bond structure is an important constituent of the "dark" state. Therefore, if we wish to understand the "dark" state, now we must consider also the long-bond structures along with the perfectly paired structure, as is done in the following two examples.

Butadiene and Hexatriene   The two covalent VB structures for butadiene are depicted in Fig. 7.9a as R_1 and R_2. Under the symmetry operations of the point group, the two structures transform as A_g (since both are unchanged by i), and will therefore mix to give rise to two states of the same symmetry. The corresponding wave functions are written in Equations 7.12a and b, with normalization constants that neglect all overlap terms between orbitals on atoms a–d.

$$\Phi(R_1) = 0.5(|a\bar{b}c\bar{d}| + |\bar{a}bc\bar{d}| - |\bar{a}bc\bar{d}| - |a\bar{b}\bar{c}d|) \qquad (7.12a)$$

$$\Phi(R_2) = 0.5(|ab\bar{c}\bar{d}| + |\bar{a}\bar{b}cd| - |a\bar{b}\bar{c}d| - |\bar{a}b\bar{c}d|) \qquad (7.12b)$$

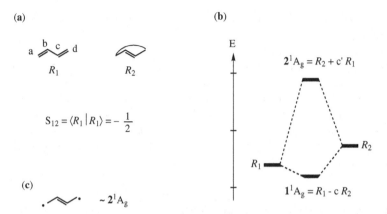

FIGURE 7.9 (a) Covalent resonance structures for butadiene and their overlap integral. (b) A VB mixing diagram of the resonance structures yielding the covalent ground and excited states. (c) The major 1,4-biradical character of the 2^1A_g state.

Since R_1 and R_2 possess two common determinants, in their corresponding wave functions, they overlap and their overlap integral is $-1/2$, as shown in Fig. 7.9a. The negative sign of the overlap means that the two resonance structures will possess an effective mixing matrix element that will be negatively signed, and therefore, the lowest state due to their mixing will be a negative combination of the two structures, while the antibonding state will be the positive combination; both transforming as A_g.

The corresponding VB mixing diagram in Fig. 7.9b expresses this information in a perturbative manner (work out Exercise 7.6 for a full quantitative solution). Thus the 1^1A_g state is the negative combination, where R_1 mixes some of R_2. This mixing accounts for some resonance stabilization of the ground state of butadiene. In contrast, the 2^1A_g excited state is generated from R_2 mixed with some of R_1 in an antibonding fashion. As such, the hidden excited state has a large character of the 1,4-diradical with a double bond in the middle C–C linkage, as shown in Fig. 7.9c. This description already gives us some idea that upon excitation of butadiene, we expect to obtain cyclobutene as one of the products. In addition, if the terminal double bonds are substituted so they have cis and trans isomers, the isomeric identity will be scrambled in the 2^1A_g state, and upon decay to the ground state will give rise to a mixture of *cis* and *trans* dienes. However, as discussed in Chapter 6, and especially in Exercise 6.14, the expected set of photoproducts can be deduced from the electronic structure of the conical intersection between R_1 and R_2. In the conical intersection (25), the "excited state" is an equal mixture of the two structures (i.e., $R_1 + R_2$) and the corresponding wave function is dominated by diagonal bonding as discussed above for cyclobutadiene.

Hexatriene, like benzene, has five covalent structures, which are organized in order of increasing energy in Fig. 7.10a. The lowest one, R_1, is the perfectly paired structure; the others R_2-R_5 have one and two long bonds, and are therefore higher in energy than R_1. The VB structure R_4 is analogous to the second Kekulé structure in benzene, and hence is also labeled as R_k. Note that R_1 and R_k have opposite bond alternation propensities due to the location of the double bonds. Therefore, the energy of R_k will be considerably lowered when the geometric bond alternation that stabilizes R_1 reverses its sense.

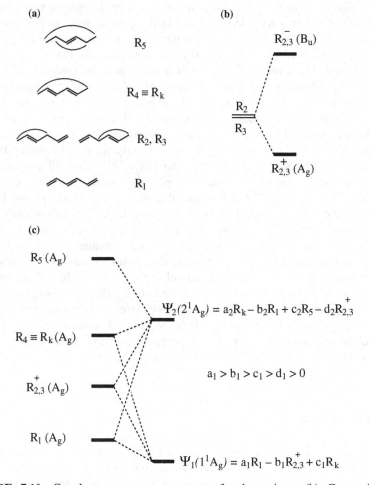

FIGURE 7.10 Covalent resonance structures for hexatriene. (b) Generation of symmetry adapted combinations of R_2 and R_3. (c) A VB mixing diagram of the symmetry adapted VB combinations yielding the covalent ground and excited states.

Structures R_2 and R_3 are mutually transformable by the symmetry operations i, C_2, and σ_v, and can therefore be symmetry adapted, as shown in Fig. 7.10b, into the positive combination of symmetry A_g and a negative one of symmetry B_u. Although the B_u combinations gives rise to one of the excited states of the hexatriene ($1B_u^-$) (7,8,10), we will not deal with it anymore and focus on the dark state, 2^1A_g. To this end, we show in Fig. 7.10c the VB mixing diagram of all the VB structures that give rise to the A_g states. The ground state is nascent from R_1 that mixes the R_k structure and $R_{2,3}^+$ combination in a bonding fashion. The negative sign of $R_{2,3}^+$ mixing can be deduced from the overlap with R_1, as well as by analogy to butadiene, since both R_2 and R_3 are related to the dienic moieties of R_1 as the corresponding long-bond structure that mixes in a negative sign (Fig. 7.9b). Similarly, the positive sign of R_k mixing in the ground state can be thought of by analogy to the mixing of the two Kekulé structures in the ground state of benzene (Fig. 7.4). The 2^1A_g excited state, on the other hand, is nascent from R_k with a negative combination with R_1 and smaller contributions from the other A_g combinations. Thus, the hidden state of hexatriene is an analogue of the B_{2u} excited state of benzene. This analogy carries over to vibrational modes of the two states, and as in benzene, one also finds here a C=C stretching mode (this time a_g) that has a low frequency in the ground state and a high frequency in the hidden state. However, one must remember not to overstretch this analogy, since the 2^1A_g state in the linear polyene is predominantly the 1,6-diradicaloid R_k structure. As such the $1^1A_g - 2^1A_g$ transition is simply a shift of all the double bonds from their positions in the ground state to the erstwhile single bonds. This also means that the ground state and the hidden excited state exhibit opposing bond alternation; the long bonds in the ground state become short in the excited state and vice versa (8,10). In fact, as shown by the VB calculations (10), the picture obtained for hexatriene can be generalized for longer polyenes; the ground state of a polyene will be dominated by R_1, whereas the covalent excited state by R_k. Therefore, the geometries of the ground state and excited state will exhibit opposite bond alternation, and the frequency of the C=C stretching mode will undergo exaltation in the excited state. This topic and others related to the photochemistry and photophysics of polyenes are discussed in more detail in the original lliterature (10).

7.3 A SUMMARY

This chapter focused on a special class of excited states of organic molecules hoping to provide a conceptual qualitative frame that would allow thinking about these excited states and discussing some of their features from the bonding characteristics of the VB wave function. One general rule is already apparent from the treatments of the various molecules: In all annulenes, the excited covalent state changes the bonding from a delocalized-perimeter mode to a 1,3-cross bonding form whereas in n-polyenes, the bonding mode will become 1,n-type. This can be called *"the 1,3-bonding and 1,n-bonding rules in covalent excited states"*. While the topic of excited states is much broader than we can cover in a chapter of a textbook, still together with the topic of the twin states discussed in Chapter 6 this chapter demonstrates that VB theory can be a powerful conceptual tool in this area, as has been recognized since the pioneering studies in photochemical reactivity and spectroscopy (7,28,29). Chapter 8 discusses some of the same problems and related ones, using an approach that originates from Physics (spin-Hamiltonian theory). It will be shown that while the approaches in Chapters 7 and 8 use different languages, they are in fact very similar.

REFERENCES

1. J. B. Foresman, M. Head-Gordon, J. A. Pople, M. J. Frisch, *J. Phys. Chem.* **96**, 135 (1992). Toward a Systematic Molecular Orbital Theory for Excited States.

2. L. Serrano-Andrés, M. Merchán, I. Nebot-Gil, R. Lindh, B. O. Roos, *J. Chem. Phys.* **98**, 3151 (1993). Toward an Accurate Molecular Orbital Theory for Excited States: Ethene, Butadiene, and Hexatriene.

3. E. Runge, E. K.U. Gross, *Phys. Rev. Lett.* **52**, 997 (1984). Density Functional Theory for Time-Dependent Systems.

4. L. Serrano-Andrés, M. Merchán, *J. Mol. Struct. (Theochem)* **729**, 99 (2005). Quantum Chemistry of the Excited State: 2005 Overview.

5. E. C. da Silva, J. Gerratt, D. L. Cooper, M. Raimondi, *J. Chem. Phys.* **101**, 3866 (1994). Study of the Electronic States of the Benzene Molecule Using Spin-Coupled Valence Bond Theory.

6. J. M. Oliva, J. Gerratt, P. B. Karadakov, M. Raimondi, *J. Chem. Phys.* **106**, 3663 (1997). Study of the Electronic States of the Allyl Radical Using Spin-Coupled Valence Bond Theory.

7. C. Sandorfy, *Electronic Spectra and Quantum Chemistry*. Prentic-Hall, Englewood Cliffs, NJ, 1964.

8. M. Said, D. Maynau, J. P. Malrieu, *J. Am. Chem. Soc.* **106**, 580 (1984). Excited-State Properties of Linear Polyenes Studied Through a Nonempirical Heisenberg Hamiltonian.

9. W. Wu, S-.J. Zhong, S. Shaik, *Chem. Phys. Lett.* **292**, 7 (1998). VBDFT(s): A Hückel-type Semiempirical Valence Bond Method Scaled to Density Functional Energies. Application to Linear Polyenes.

10. W. Wu, D. Danovich, A. Shurki, S. Shaik, *J. Phys. Chem. A* **104**, 8744 (2000). Using Valence Bond Theory to Understand Electronic Excited States: Application to the Hidden Excited State (2^1A_g) of $C_{2n}H_{2n+2}$ ($n = 2$–14) Polyenes.

11. W. Wu et al., *Ab Initio* Valence Bond Methods for Excited States. In preparation.

12. S. M. Hurley, T. E. Dermota, P. D. Hydutzky, A. W. Castleman, Jr., *Science* **298**, 202 (2002). Dynamics of Hydrogen Bromide Dissolution in the Ground and Excited States.

13. K. S. Peters, *Acc. Chem. Res.* **40**, 1 (2007). Dynamic Processes Leading to Covalent Bond Formation for S_N1 Reactions.

14. B. E. Kohler, *Chem. Rev.* **93**, 41 (1993). Octatetraene Photoisomerization.

15. G. Orlandi, F. Zerbetto, M. Z. Zgierski, *Chem. Rev.* **91**, 867 (1991). Theoretical Analysis of Spectra of Short Polyenes.

16. S. Shaik, S. Zilberg, Y. Haas, *Acc. Chem. Res.* **29**, 211 (1996). A Kekule-Crossing Model for the "Anomalous" Behavior of the b_{2u} Mode of Aromatic Hydrocarbons in the Lowest Excited $^1B_{2u}$ State.

17. S. Shaik, A. Shurki, D. Danovich, P. C. Hiberty, *J. Am. Chem. Soc.* **118**, 666 (1996). Origins of the Exalted b_{2u} Frequency in the First Excited State of Benzene.

18. L. Goodman, M. J. Berman, A. G. Ozkabak, *J. Chem. Phys.* **90**, 2544 (1989). The Benzene Ground State Potential Surface. III. Analysis of b_{2u} Vibrational Mode Anharmonicity Through Two-Photon Intensity.

19. S. Zilberg, Y. Haas, S. Shaik, *J. Phys. Chem.* **99**, 16558 (1995). Electronic Spectrum of Anthracene: An *ab-Initio* Molecular Orbital Calculations Combined with a Valence Bond Interpretation.

20. K. Schulten, M. Karplus, *Chem. Phys. Lett.* **14**, 305 (1972). On the Origin of a Low-Lying Forbidden Transition in Polyenes and Related Molecules.

21. Y. Luo, L. Song, W. Wu, D. Danovich, S. Shaik, *ChemPhysChem* **5**, 515 (2004). The Ground and Excited States of Polyenyl Radicals $C_{2n-1}H_{2n+1}$ ($n = 2$–13): A Valence Bond Study.

22. S. S. Shaik, in *New Theoretical Concepts for Understanding Organic Reactions*, J. Bertran, I. G. Csizmadia, Eds., NATO ASI Series, **C267**, Kluwer Academic Publ., 1989, pp. 165–217. A Qualitative Valence Bond Model for Organic Reactions.

23. R. W. Fessenden, R. H. Schuler, *J. Chem. Phys.* **39**, 2147 (1963). Electron Spin Resonance Studies of Transient Allyl Radicals.

24. S. Shaik, A. Shurki, D. Danovich, P. C. Hiberty, *Chem. Rev.* **101**, 1501 (2001). A Different Story of π-Delocalization—The Distortivity of π-Electrons and Its Chemical Manifestations.

25. M. A. Robb, M. Garavelli, M. Olivucci, F. Bernardi, *Rev. Comput. Chem.* **15**, 87 (2000). A Computational Strategy for Organic Photochemistry.

26. A. Shurki, P. C. Hiberty, F. Dijkstra, S. Shaik, *J. Phys. Org. Chem.* **16**, 731 (2003). Aromaticity and Antiaromaticity: What Role Do Ionic Configurations Play in Delocalization and Induction of Magnetic Properties?

27. M. Sironi, D. L. Copper, J. Gerratt, M. Raimondi, *J. Chem. Soc. Chem. Commun.* 675 (1989). The Modern Valence Bond Description of Naphthalene.

28. W. Th. A. M. Van der Lugt, L. J. Oosterhoff, *J. Am. Chem. Soc.* **91**, 6042 (1969). Symmetry Control and Photoinduced Reactions.

29. J. Michl, *Topics Curr. Chem.* **46**, 1 (1974). Physical Basis of Qualitative MO Arguments in Organic Photochemistry.

EXERCISES

7.1. Normalize the wave functions of the $^2A_2$ and $^2B_1$ states of allyl radical (Eqs. 7.2 and 7.3 in the main text) by assuming zero-overlap between AOs. Then calculate the weights of each determinant by squaring their coefficients. From these reproduce quantitatively the spin density distribution for both states, given qualitatively in Fig. 7.3b.

7.2. Express the VB wave functions for the ground state ($^2B_1$) and first covalent excited state ($^2A_2$) of the pentadienyl radical in terms of Kekulé structures. Deduce the qualitative spin distribution change upon excitation. *Hints*: For the excited-state case, you will need to express the Kekulé structures in terms of determinants. For the ground state, you may express the wave function as the spin alternant determinant, plus some minor contributions of the two determinants that exhibit a single spin frustration (two identical neighboring spins). You may consult *ChemPhysChem.* **5**, 515 (2004).

7.3. Write the wave function for the $^1A_{1g}$ excited state of cyclobutadiene in terms of AO-based determinants. Show the transformation in bonding features as cyclobutadiene is excited to its $^1A_{1g}$ state (see Fig. 7.5c).

7.4. The aim of this exercise is to compare the bonding features of the ground versus first excited states of benzene, with a method different from that of the preceding exercise. The bond index for an $r-s$ bond will be taken as the probability of finding a spin alternation from r to s in the wave function. Thus, the $r-s$ bond index can be estimated by summing up the squares of the coefficients of the determinants displaying an r-s spin alternation, in the wave function of the state under consideration.

    a.   Write the wave function Ψ_0 for the ground state of benzene ($^1A_{1g}$), in terms of AO-based determinants. Calculate the bond index for short bonds (1,2 bonds), *meta* bonds (1,3) and *para* bonds (1,4).

    b.   Do the same for the $^1B_{2u}$ excited state Ψ^*.

7.5. Use the Kekulé structure set to construct the ground and covalent excited states for pentalene. You may consult *Chem. Rev.* **101**, 1501 (2001).

7.6. Use the semiempirical theory in Chapter 3 to obtain quantitative expressions for the energies and wave functions of the 1^1A_g and 2^1A_g states of butadiene. *Hint*: Express the energies of the two Rumer structures relative to the QC determinant (the spin alternant determinant). Deduce the matrix element between the structures keeping only the close neighbor $2\beta S$ term (for simplicity define $\lambda = -2\beta S$). Neglect overlap in the normalization constant.

Answers

Exercise 7.1 The spin density distribution in a state Ψ is given by the expectation value, $\langle\Psi|\Sigma_r\rho_r|\Psi\rangle$, where ρ_r is a local excess spin density operator on site r, with local expectation value of $+1$ for spin α and -1 for spin β. Thus, for a VB wave function given as a linear combination of determinants with coefficients C_i, the spin density in site r will be

$$< \rho_r > = N^2 \sum_i C_i^2 \delta_{ir}; \delta_{ir} = +1 \text{ or } -1 \qquad (7.\text{Ans}.1)$$

Normalizing the wave function for the $^2A_2$ ground state of allyl, we get

$$\Psi(^2A_2) = 6^{-1/2}(-2|a\bar{b}c| + |ab\bar{c}| + |\bar{a}bc|) \qquad (7.\text{Ans}.2)$$

Therefore, the spin densities on atoms a, b, and c of allyl will be the following:

$$< \rho_a >= 1/6[4+1-1] = 2/3 \qquad (7.\text{Ans}.3\text{a})$$
$$< \rho_b >= 1/6[-4+1+1] = -1/3 \qquad (7.\text{Ans}.3\text{b})$$
$$< \rho_c >= 1/6[4+1-1] = 2/3 \qquad (7.\text{Ans}.3\text{c})$$

It is seen that in the ground state the terminal carbon atoms a and c will involve a positive excess spin density of $+2/3$, while the central carbon b a negative spin density of $-1/3$. This spin polarization is close to the experimentally deduced value. Of course, one cannot expect full accuracy using covalent only structures.

The wave function for the excited state is

$$\Psi(^2B_1) = 1/\sqrt{2}(|\bar{a}bc| - |ab\bar{c}|) \qquad (7.\text{Ans}.4)$$

The corresponding spin densities are

$$< \rho_a >= 1/2[1-1] = 0 \qquad (7.\text{Ans}.5\text{a})$$
$$< \rho_b >= 1/2[1+1] = +1 \qquad (7.\text{Ans}.5\text{b})$$
$$< \rho_c >= 1/2[-1+1] = 0 \qquad (7.\text{Ans}.5\text{c})$$

Exercise 7.2 The pentadienyl radical is an odd member of the $C_{2n-1}H_{2n+1}$ polyenyl radicals ($n = 3$). It possesses three Kekulé-type resonance structures, shown in Fig. 7.Ans.1a. One resonance structure has a radical at the central carbon and is labeled R_c, the other two place the radical either on the right- or left-hand carbon atoms and are labeled accordingly as R_r and R_l, respectively. There are other structures with long bonds, but we are going to neglect them in this treatment. This will affect the calculated values of the spin density, and

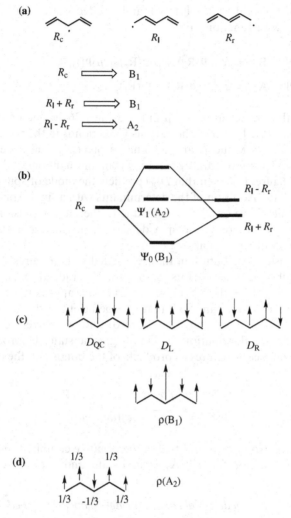

FIGURE 7.Ans.1   (a) Resonance structures of pentadienyl radical and their symmetry properties. (b) The VB mixing of the resonance structures. (c) The quasiclassical (spin alternant) determinant that dominates the ground-state wave function, and the corresponding secondary determinants, and the resulting spin density distribution (ρ) in the ground state. (d) The spin distribution in the covalent excited state.

therefore, the advanced reader is advised to consult Exercise 8.5 for a full quantitative treatment.

The R_c structure transforms as B_1 (in C_{2v}), while R_r and R_l are mutually transformable via the σ_v and C_2 symmetry elements, and therefore their linear combinations transform as B_1 and A_2. Part (b) shows the VB mixing diagram leading to the ground state $\Psi_0(^2B_1)$ and excited state $\Psi_1(^2A_2)$, and the corresponding wave functions are:

$$\Psi_0(^2B_1) = N[\lambda\Phi(R_c) + \mu\Phi(R_r) + \mu\Phi(R_l)] \tag{7.Ans.6}$$

$$\Psi_1(^2A_2) = 2^{-1/2}[\Phi(R_r) - \Phi(R_l)] \tag{7.Ans.7}$$

The values of the coefficients λ and μ in Equation 7.26 depend on the geometry of the hexatriene (the length of the various C–C bonds in the molecule). Thus, since the coefficients λ and μ are not known precisely, the easiest way is to deduce the spin density pictorially from the spin alternant determinant, D_{QC} in part (c) of the figure. It is seen that D_{QC} predicts the mode of spin polarization in the ground state. Addition of the determinants with a single spin frustration, D_R and D_L, which have smaller weights due to the Pauli repulsion in the spin frustrated sites, will increase the spin density in the middle carbon and will decrease it on all other carbons.

The spin density distribution in the $^2A_2$ excited state requires the derivation of all the contributing determinants as done for allyl radical. A full treatment is given in Exercise 8.5, while here we provide an approximate description. Already at the outset one can recall that the coefficient of the QC determinant in the excited state's wave function is zero, and we therefore expect very different spin density distribution than in the ground state. To proceed, we first express the resonance structures as products of the bonds and the odd electron. Thus

$$\Phi(R_r) = |(a\bar{b} - \bar{a}b)(c\bar{d} - \bar{c}d)e| \tag{7.Ans.8}$$

$$\Phi(R_l) = |a(b\bar{c} - \bar{b}c)(d\bar{e} - \bar{d}e)| \tag{7.Ans.9}$$

Expanding these expressions and taking the negative combination, as required by the above expression for the $^2A_2$ excited state, and normalizing it leads to

$$\Psi_1(^2A_2) = 6^{-1/2}(-|\bar{a}bcd\bar{e}| - |a\bar{b}cd\bar{e}| + |\bar{a}bc\bar{d}e| - |a\bar{b}c\bar{d}e| + |ab\bar{c}d\bar{e}| + |ab\bar{c}\bar{d}e|) \tag{7.Ans.10}$$

By using this expression, the expectation value of the spin density can be determined immediately and is shown in part (d) of the figure. The distribution is very different than in the ground state. Note that absolute values of the spin are identical on all carbons since we neglected the long-bond structures. When these are added in the computations, the spin distribution on the terminal carbons decreases and on the internal carbons it increases. The middle carbon retains a spin density of $-1/3$. More details are given in Exercise 8.5.

Exercise 7.3   Labeling as a, b, c, and d the π AOs along the ring, the respective unnormalized wave functions for the Kekulé structures K_1 and K_2 of cyclobutadiene can be obtained by the product of the bond wave functions along the perimeter (e.g., $\Phi_{a-b} = |a\bar{b}| - |\bar{a}b|$):

$$\Phi(K_1) = |(b\bar{c} - \bar{b}c)(d\bar{a} - \bar{d}a)| = -|a\bar{b}\bar{c}d| + |\bar{a}\bar{b}cd| + |abc\bar{d}| - |a\bar{b}c\bar{d}|$$

$$(7.\text{Ans}.11)$$

$$\Phi(K_2) = |(a\bar{b} - \bar{a}b)(c\bar{d} - \bar{c}d)| = |a\bar{b}c\bar{d}| - |\bar{a}bc\bar{d}| - |a\bar{b}\bar{c}d| + |\bar{a}b\bar{c}d| \quad (7.\text{Ans}.12)$$

The wave function of the $^1A_{1g}$ excited state is the sum of $\Phi(K_1)$ and $\Phi(K_2)$ and is given as:

$$\Psi(^1A_{1g}) = \Phi(K_1) + \Phi(K_2) = +|a\bar{b}c\bar{d}| - |\bar{a}bc\bar{d}| - |a\bar{b}\bar{c}d| + |\bar{a}b\bar{c}d| \quad (7.\text{Ans}.13)$$

It is easy to verify that this wave function corresponds to coupling of the diagonal carbon atoms, that is a-c and b-d, by writing this wave function and showing its identity to $\Phi(K_1) + \Phi(K_2)$:

$$\Phi(K_{a-c,b-d}) = |(a\bar{c} - \bar{a}c)(b\bar{d} - \bar{b}d)| = |a\bar{b}\bar{c}d| - |\bar{a}bc\bar{d}| - |\bar{a}bc\bar{d}|$$
$$+ |\bar{a}\bar{b}cd| = \Phi(K_1) + \Phi(K_2)$$

$$(7.\text{Ans}.14)$$

Exercise 7.4 By labeling the π AOs along the ring as a, b, c, d, e, and f, the respective unnormalized wave functions for the ground and excited states of benzene can be obtained as in the preceding exercise by the product of the bond wave functions along the perimeter. After expanding the product, the wave function for the ground and the excited state of benzene will be described by the following collection of VB determinants:

$$\Psi_0 = 2|abc\bar{d}e\bar{f}| - |abc\bar{d}\bar{e}f| - |ab\bar{c}d\bar{e}f| + |a\bar{b}\bar{c}de\bar{f}| - |\bar{a}bc\bar{d}e\bar{f}| + |\bar{a}bc\bar{d}\bar{e}f|$$
$$+ |\bar{a}\bar{b}cde\bar{f}| - 2|\bar{a}\bar{b}c\bar{d}ef| + |ab\bar{c}\bar{d}ef| + |a\bar{b}c\bar{d}\bar{e}f| - |abc\bar{d}\bar{e}f|$$
$$+ |\bar{a}b\bar{c}de\bar{f}| - |a\bar{b}\bar{c}d\bar{e}f| - |\bar{a}b\bar{c}d\bar{e}f|$$

$$(7.\text{Ans}.15)$$

$$\Psi^* = -|ab\bar{c}\bar{d}ef| - |abc\bar{d}\bar{e}f| + |ab\bar{c}d\bar{e}f| - |a\bar{b}c\bar{d}ef| + |abc\bar{d}\bar{e}f|$$
$$+ |\bar{a}b\bar{c}d\bar{e}f| - |\bar{a}bc\bar{d}\bar{e}f| - |\bar{a}b\bar{c}d\bar{e}f| + |a\bar{b}c\bar{d}\bar{e}f| - |\bar{a}bc\bar{d}e\bar{f}|$$
$$+ |a\bar{b}c\bar{d}e\bar{f}| + |\bar{a}\bar{b}cde\bar{f}|$$

$$(7.\text{Ans}.16)$$

The normalized $r-s$ bond indices giving probabilities of having an $r-s$ bond are summarized in Table 7.Ans.1.

It is seen that, in the ground state Ψ_0, the indices are 0.8 for short bonds, 0.6 for bonds between carbons in para positions, and 0.4 for bonds between meta carbon atoms. On the other hand, in the excited state the indices for all short and meta bonds are the same, while the para bonds have a lower index.

TABLE 7.Ans.1   Indices for the $r-s$ Bond in the Ground and Excited States of Benzene

	a–b	a–c	a–d	a–e	a–f	b–c	b–d	b–e	b–f	c–d	c–e	c–f	d–e	d–f	e–f
Ψ_0	0.8	0.4	0.6	0.4	0.8	0.8	0.4	0.6	0.4	0.8	0.4	0.6	0.8	0.4	0.8
Ψ_*	0.4	0.4	0.2	0.4	0.4	0.4	0.4	0.2	0.4	0.4	0.4	0.2	0.4	0.4	0.4

Exercise 7.5 Pentalene has two Kekulé structures, which are depicted in Fig. 7.Ans.2. As can be seen, the two structures remain unchanged by the i symmetry element, but are mutually transformable via C_2 and σ_v. As such, the positive linear combination will be A_{1g}, while the negative one will be B_{1g}. As we argued in Chapter 5, the ground state of antiaromatic molecules is the negative combination, because it is the one that conserves the lowest energy QC determinants. Based on that we can construct a VB mixing diagram and assign the ground state as $^1B_{1g}$ and the covalent excited state as $^1A_{1g}$. Note that the mode that interchanges the two Kekulé structures in the VBSCD is b_{1g} and therefore, this mode has an imaginary frequency in the ground state and a real one with a high value (2506 cm^{-1}) in the covalent excited state.

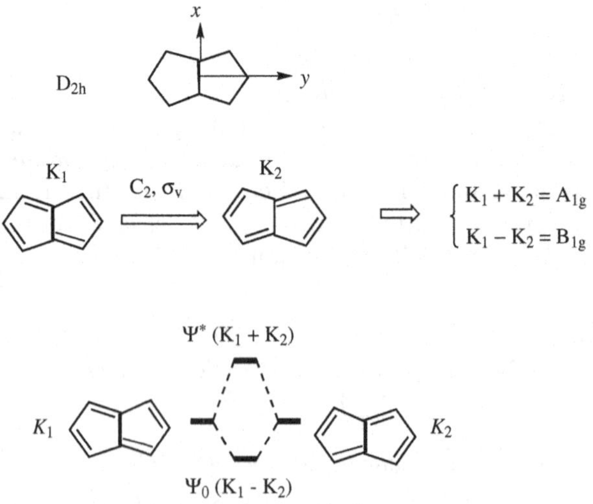

FIGURE 7.Ans.2 Pentalene, its Kekulé structures, and their VB mixing to produce the ground state and covalent excited state.

Exercise 7.6 The two covalent VB structures for butadiene are depicted in Fig. 7.9a as R_1 and R_2. The corresponding normalized wave functions are written below:

$$\Phi_1 = 0.5(|a\bar{b}c\bar{d}| + |\bar{a}bc\bar{d}| - |\bar{a}bc\bar{d}| - |a\bar{b}\bar{c}d|) \tag{7.Ans.17}$$

$$\Phi_2 = 0.5(|ab\bar{c}\bar{d}| + |\bar{a}\bar{b}cd| - |a\bar{b}c\bar{d}| - |\bar{a}b\bar{c}d|) \tag{7.Ans.18}$$

Since R_1 possesses two bonds and a nonbonding interaction between the bonds, its energy relative to the energy of the spin alternant determinant, taken as the zero:

$$E(\Phi_1) = -2\lambda + 0.5\lambda = -1.5\lambda \qquad (7.\text{Ans}.19)$$

Note that λ is $-2\beta S$.

The VB structure R_2 has only one bond and two nonbonding interactions, leading to the same energy as the spin-alternant determinant:

$$E(\Phi_2) = -\lambda + 2 \times 0.5\lambda = 0 \qquad (7.\text{Ans}.20)$$

Φ_1 and Φ_2 have two determinants in common, and their overlap is $S_{12} = -0.5$.

Remembering that the energy of the spin-alternant is taken as zero, and that determinants interact only if they differ by only one spin permutation between adjacent atoms (in which case their matrix element is λ), the Hamiltonian matrix element between the two Rumer structure is

$$\langle \Phi_1|H|\Phi_2\rangle = 1.5\lambda \qquad (7.\text{Ans}.21)$$

Given these matrix elements one can diagonalize the Hamiltonian and find the energies (E_0 and E^*) and the wave functions (Ψ_0 and Ψ^*) of the ground and excited states, respectively

$$E_0 = -1.732\lambda \qquad E^* = 1.732\lambda \qquad (7.\text{Ans}.22)$$
$$\Psi_0 = 0.82\Phi_1 - 0.30\Phi_2 \qquad \Psi^* = 0.82\Phi_1 + 1.12\Phi_2 \qquad (7.\text{Ans}.23)$$

8 Spin Hamiltonian Valence Bond Theory and its Applications to Organic Radicals, Diradicals, and Polyradicals

There are additional brands of VB theory that originate in physics and are based on definitions of phenomenological model Hamiltonians (1) designed to treat special classes of materials. For example, the Hubbard Hamiltonian (2), often used to treat conducting materials, is based on two effective parameters, the transfer integral, which is the β resonance integral, and the on-site repulsion integral. Matsen has shown how the Hubbard Hamiltonian can be used to formulate a VB theory for molecules and to bridge between VB theory and Hückel MO theory (3). Another class of phenomenological Hamiltonians are called magnetic- or spin-Hamiltonians or simply Heisenberg Hamiltonians (4), and are designed to treat the spin states of species having an average of one electron per site or orbital; such species are, for example, neutral π-systems or the spin states of transition metal complexes. There are other model Hamiltonians, which are related to one another and these relationships are reviewed and explained by Klein (5). Because of the clear insight provided by these model Hamiltonians, and because of their relationships to the qualitative VB theory, introduced in Chapter 3, we focus this chapter on the Heisenberg VB Hamiltonian and its application to radicals, diradicals, and polyradicals. The Heisenberg spin-Hamiltonian approach, first presented as a phenomenological tool, has been given a firm theoretical basis by Anderson (6), and later extented and extensively applied by Malrieu and co-workers (7–10), and Klein and co-workers (5,11). Wu et al. demonstrated that the same problem can be reformulated using the approach discussed and employed in Chapter 3, in which case one can obtain the states of a molecule by solving a Hückel matrix of a molecule having the same connectivity as the VB mixing problem (12).

Unlike the theory discussed in Chapter 3, which relies on VB structures that are eigenfunctions of both the S_z and S^2 operators, Heisenberg–Hamiltonian theory uses, as basis functions, VB determinants that are eigenfunctions of the S_z operator only. The reader has by now some background on the VB

---

determinants of a bond, and on spin-alternant determinants that can be used to define the nonbonding reference energy for VB structures, and so on. In fact, Chapters 5 and 7 showed that using the spin-alternant VB determinants provides profound insight into aromaticity and antiaromaticity in benzene and cyclobutadiene and their analogues, as well as into properties of polyenyl radicals, in the ground and excited states. As such, the spin-Hamiltonian VB theory is best suited for homonuclear assemblies with one electron per site, that is, typically conjugated molecules, and essentially deals with neutral states. In its qualitative version (topological Hamiltonian), it is as simple to use as Hückel MO theory. Based on the molecular graph alone, this theory leads to deduction of qualitative rules, related to ground-state properties, such as spin multiplicity of diradicals or polyradicals, spin distribution in free radicals or high-spin states, and isomerization energies. This theory also possesses a quantitative nonempirical version (10), which has proven itself to be accurate for predicting ground state equilibrium geometries, rotational barriers, geometry relaxation in some excited states, and vertical as well as adiabatic transition energies.

The spin-Hamiltonian VB theory rests on the same principles as the qualitative theory presented in Chapter 3, with some further simplifying assumptions. This chapter describes the method and focuses on its qualitative applications.

8.1 A TOPOLOGICAL SEMIEMPIRICAL HAMILTONIAN

A difference between the qualitative VB theory, discussed in Chapter 3, and the spin-Hamiltonian VB theory is that the basic constituent of the latter theory is the AO-based determinant, without any *a priori* bias for a given electronic coupling into bond pairs like those used in the Rumer basis set of VB structures. The bond coupling results from the diagonalization of the Hamiltonian matrix in the space of the determinant basis set. The theory is restricted to determinants having one electron per AO. This restriction does not mean, however, that the ionic structures are neglected since their effect is effectively included in the parameters of the theory. Nevertheless, since ionicity is introduced only in an effective manner, the treatment does not yield electronic states that are ionic in nature, and excludes molecules bearing lone pairs. Another simplification is the zero-differential overlap approximation, between the AOs.

Let us now briefly describe the principles of the method and the rules for the construction of the Hamiltonian matrix. For the sake of consistency, rather than the original formulation (7–10), here we use a formulation that is in harmony with the VB theory in Chapter 3. The method can be summarized by the following principles: (a) All overlaps between AOs are set to zero. (b) The energy of a VB determinant Ω_i is proportional to the number of Pauli repulsions that exist between electrons with identical spins, which occupy adjacent AOs:

$$E(\Omega_i) = \sum_{r\uparrow, s\uparrow} g_{rs} \qquad (8.1)$$

Here, g_{rs} is a parameter that is quantified either from experimental data, or is calculated by an *ab initio* method as one-half of the singlet–triplet excitation energy gap of the r—s bond. In terms of the qualitative theory in Chapter 3, g_{rs} is therefore identical to the key quantity $-2\beta_{rs}S_{rs}$. This empirical quantity incorporates the effect of the ionic components of the bond, albeit in an implicit way. (c) The Hamiltonian matrix element between two determinants differing by one spin permutation between orbitals r and s is equal to g_{rs}. Only close neighbor g_{rs} elements are taken into account; all other off-diagonal matrix elements are set to zero. An example of a Hamiltonian matrix is illustrated in Scheme 8.1 for 1,3-butadiene.

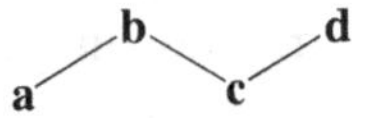

$$
\begin{array}{cccccc}
|a\bar{b}c\bar{d}| & |a\bar{b}c\bar{d}| & |a\bar{b}c\bar{d}| & |a\bar{b}c\bar{d}| & |ab\bar{c}\bar{d}| & |\bar{a}bc\bar{d}| \\[4pt]
0 & 0 & g_{ab} & g_{cd} & 0 & g_{bc} \\
 & 0 & g_{cd} & g_{ab} & g_{bc} & 0 \\
 & & g_{bc} & 0 & 0 & 0 \\
 & & & g_{bc} & 0 & 0 \\
 & & & & g_{ab}+g_{cd} & 0 \\
 & & & & & g_{ab}+g_{cd}
\end{array}
$$

Scheme 8.1

As a rule, diagonalization of the Hamiltonian matrix provides the energy of the ground state relative to the nonbonding state (the spin-alternant determinant), and in addition leads to the entire spectrum of the lowest neutral excited states. Figure 8.1 shows these states for the simple case of ethylene. It can be seen that the spin-Hamiltonian model generates a ground state with a π-bond and a singlet spin quantum number with energy $-g$, and an excited state with a triplet spin quantum number with energy $+g$. This figure also shows its relationship to the qualitative VB theory in Chapter 3, in which the singlet state of the π-bond is stabilized by $2\beta S$, while the triplet state is destablized by the same quantity with a negative sign, $-2\beta S$, the corresponding energy gap being $2g$ and $-4\beta S$ in the respective approaches. Based on Fig. 8.1 and Scheme 8.1, it is clear that all the information needed for a calculation of the neutral electronic states of polyenes is contained in the ethylene molecule. This property can also be exploited below to build a nonempirical geometry-dependent spin-Hamiltonian (10), or the equivalent qualitative theory of Chapter 3 (12).

As already reasoned throughout this book, the lowest energy determinants are those possessing maximum spin-alternation. In a linear polyene or any

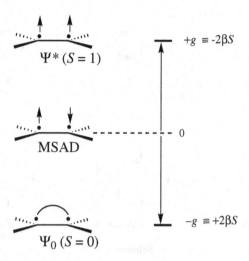

FIGURE 8.1 Energies of the singlet (Ψ_0) and triplet excited (Ψ^*) states of ethylene relative to the most spin-alternant determinant (MSAD).

alternant hydrocarbon, this determinant is the quasiclassical state (see definition in Chapters 3, 5 and 7), but in nonalternant hydrocarbons there will always be at least one interaction with identical spins (so-called "spin frustration"). Hence, the more general name for this determinant is the "most spin-alternant determinant" (MSAD). As a consequence of its low energy, the MSAD will have the largest coefficient in the wave function and will play the major role in the electronic structure. Thus, using determinants, a molecule can be viewed as a collective spin-ordering: The electrons tend to occupy the molecular space (i.e., the various atomic centers) in such a way that *an electron of α spin will be surrounded by as many β spin electrons as possible, and vice versa*. This property leads to some qualitative rules that are outlined below.

8.2 APPLICATIONS

8.2.1 Ground States of Polyenes and Hund's Rule Violations

A simple principle of the spin-Hamiltonian VB theory, first formulated by Ovchinnikov (13), applies to alternant conjugated molecules, that is, those molecules that possess fully spin-alternant determinants. The rule is stated as follows:

Ovchinnikov's Rule   *The lowest state of an alternant conjugated molecule will have the multiplicity associated with the S_z value of its MSAD, that is, it will be a singlet if $n_\alpha = n_\beta$ in the MSAD, a doublet if $n_\alpha = n_\beta + 1$, a triplet if $n_\alpha = n_\beta + 2$, and so on.*

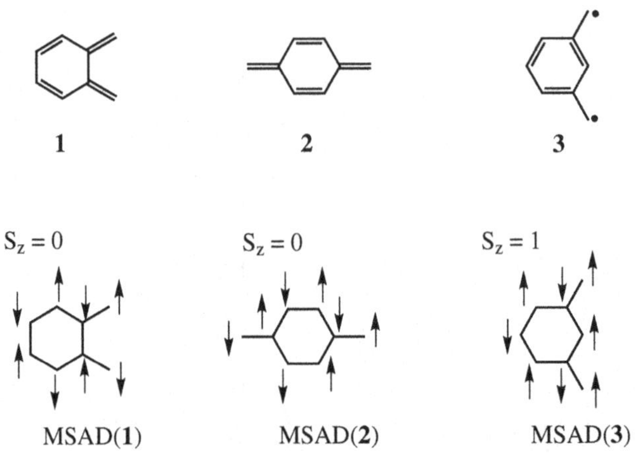

Scheme 8.2

To apply the rule, take, for example, the MSAD of *o*-xylylene, **1**, *p*-xylylene, **2** and *m*-xylylene, **3**, in Scheme 8.2. Both **1** and **2** have $S_z = 0$, while **3** has $S_z = 1$. It follows that *o*- and *p*-xylylenes will have singlet ground states, while *m*-xylylene will possess a triplet ground state. The prediction is correct, but not particularly surprising, since **1** and **2** can each be described by a perfectly paired Kekulé structure, while **3**, which cannot, is a diradical species that can be reasoned to have a triplet spin based on Hund's rule.

More intriguing are the predictions of Hund's rule violations in Scheme 8.3. Let us consider, for example, 2,3-dimethylene-butadiene and 1,3-dimethylene-butadiene, **4** and **5**. These are two species that do not have a perfectly paired Kekulé structure, and that are therefore diradicaloids. Now, according to Ovchinnikov's rule, the MSADs of these two species have different S_z values, $S_z = 0$ for **4** versus $S_z = 1$ for **5**. In this case, the spin-Hamiltonian theory predicts **4** to be a singlet diradicaloid (in violation of Hund's rule) and **5** to be a triplet, in accordance with Hund's rule; both predictions being in agreement with experiment. Another famous Hund's rule violation, the singlet state of square cyclobutadiene, is also immediately predicted by the model. Note that monodeterminantal MO calculations predict all these diradicaloids to have triplet ground states, and only after CI is the correct assignment achieved. Within MO–CI theory, violations of Hund's rule can be explained by a phenomenon called dynamic spin-polarization. The violations are predicted to take place when the degenerate singly occupied MOs form a disjointed set (14,15). In such a case, the exchange integral between the two disjointed orbitals is very small and the advantage of the triplet over the singlet is very weak at the monodeterminantal level. Therefore, the application of CI, which generally stabilizes the singlet more than the triplet, reverses the singlet–triplet ordering. In comparison, the Ovchinnikov rule is a simpler predictor, which is physically intuitive, and independent of any numerical calculation.

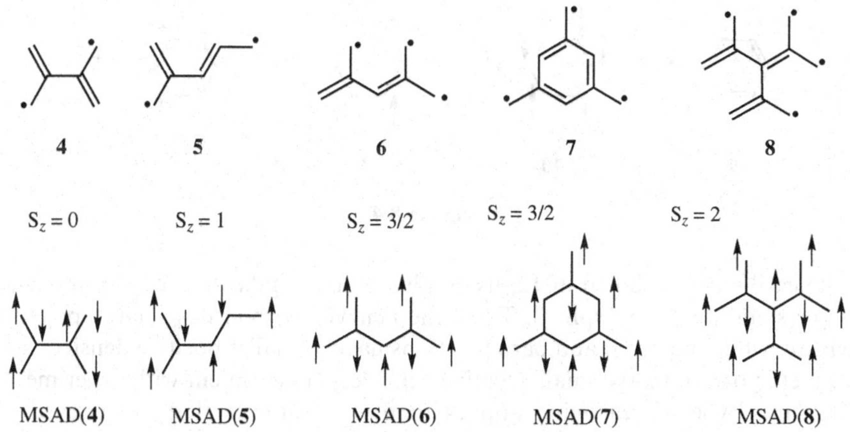

4   5   6   7   8

$S_z = 0$ $S_z = 1$ $S_z = 3/2$ $S_z = 3/2$ $S_z = 2$

MSAD(4)   MSAD(5)   MSAD(6)   MSAD(7)   MSAD(8)

Scheme 8.3

Polyradicals **6–8** in Scheme 8.3 further show the utility of the Ovchinnikov rule. For example, 2,4-dimethylenepentadienyl, **6**, is immediately seen to have a quadruplet ground state, as well as 1,3,5-trimethylenebenzene, **7**, while tripropadienemethyl, **8**, is a quintuplet (Scheme 8.3). Many other examples are offered in Exercise 8.1.

While the singlet–triplet ordering of diradicaloids can be found without any numerical calculations, quantitative singlet–triplet energy differences can be obtained through diagonalization of the spin-Hamiltonian matrix. Doing so (16), and comparing the results with those of full CI calculations in the π space in the framework of PPP theory (17), led to agreement that is not only qualitative, but also quantitative with a linear relationship between the two sets of calculated transition energies (notable exceptions were benzene and cyclobutadiene, for which higher order cyclic terms had to be included in the spin-Hamiltonian theory).

8.2.2 Spin Distribution in Alternant Radicals

As the MSADs have the largest coefficients in the ground-state wave function, one may expect that for alternant free radicals or polyradicals, *the dominant positive spin density will be largest on the atoms that bear an alpha spin in the MSAD*. An example that was analyzed in Chapter 7 is the allyl radical, where the MSAD predicts positive spin densities at positions 1, 3 and a negative density at position 2. Similarly, the MSAD of benzyl radical predicts positive spin densities on the benzylic carbon and on the ortho and para positions.

More subtle predictions can be made by considering not only the MSAD, but also the second most alternant determinant(s). Let us illustrate the reasoning by considering again the benzyl radical (**9**) in Scheme 8.4, using its MSAD (**10**) and the second most alternant determinant (**11**). Knowing that **10** has a larger coefficient than **11**, a qualitative spin distribution can be deduced

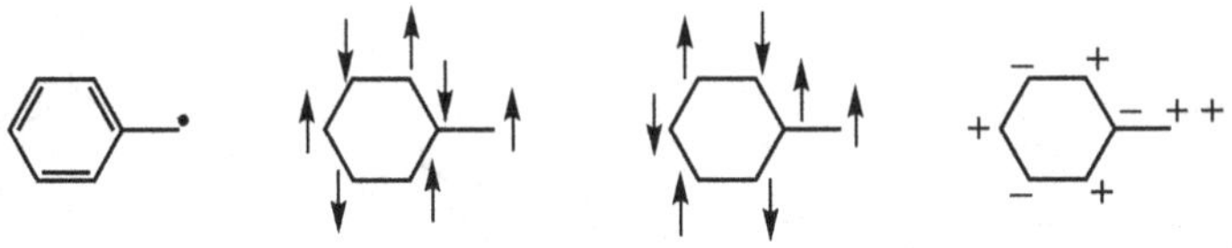

9 10 11 12

Scheme 8.4

quite easily and is shown in **12**, using plus signs to indicate excess spin α and minus signs for excess spin β. Thus, the benzylic carbon has a large positive density, while the ortho and para positions have a smaller positive density, and the meta positions have small negative densities, in agreement with experiment. The same type of reasoning applies to other examples that are treated in Exercise 8.3.

To get accurate numerical results, the spin densities can be deduced from the diagonalization of the spin-Hamiltonian matrix (9). To calculate the spin density of a given atom, one only has to sum up the weights (in this case, the squared coefficients) of all determinants, in which this atom is occupied by an α spin, while subtracting the weights of determinants in which this atom has a β spin (consult the case of allyl radical in Exercise 7.1).

8.2.3 Relative Stabilities of Polyenes

Subtle predictions can be made regarding the relative energies of two isomers having comparable Kekulé structures, like the linear *s-trans* and branched isomers of hexatriene, **13** and **14**, in Scheme 8.5. The total π energy for each of

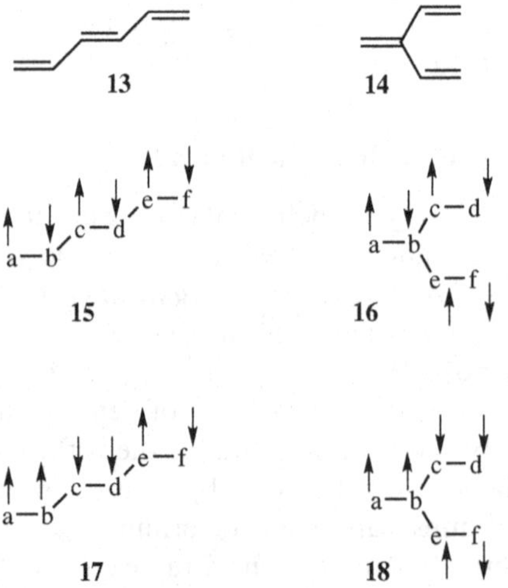

13 14 15 16 17 18

Scheme 8.5

these isomers can be evaluated as a sum of perturbations on the energy of the corresponding MSADs, **15** for the linear isomer and **16** for the branched one, by less ordered determinants, for example, **17** and **18** (Scheme 8.5). Each of the latter determinants is generated from the corresponding MSAD by the transposition of two spins along a given linkage (e.g., linkage b–c in **17** vs. **15** in Scheme 8.5), while keeping the total S_z constant. According to the above rules, the Hamiltonian matrix element between **15** and **17** is the integral g_{bc}, and the energy of **15** relative to **17** is $g_{ab} + g_{cd}$, since the spin reorganization introduces two Pauli repulsions along the $a–b$ and $c–d$ linkages.

More generally, the number of Pauli repulsions that one introduces, relative to the MSAD, by inverting the spins in a linkage $r–s$ is equal to the number of linkages that are adjacent to $r–s$. Thus, assuming that all the g integrals are the same for the sake of simplicity, the energy of a determinant Ω_{rs} that is generated by spin inversion relative to a MSAD Ω, is given by Equation 8.2:

$$E(\Omega_{rs}) - E(\Omega) = g \times n_a(rs) \tag{8.2}$$

where n_a is the number of linkages that are adjacent to $r–s$. Now, if we consider all determinants, Ω_{rs}, displaying a spin transposition between two adjacent atoms with respect to the MSAD (labeled as Ω), all the $\Omega_{rs} - \Omega$ matrix elements will be the same and all equal to the g integral, so that the total π energy that arises from a second-order perturbation correction (PT2) will be given by Equation 8.3:

$$E(PT2) = \sum_{rs} \frac{g^2}{g \times n_a(rs)} = g \sum_{rs} \frac{1}{n_a(rs)} \tag{8.3}$$

where the $E(PT2)$ energy is calculated relative to the MSAD. Therefore, it appears that the energy of a polyene is a simple topological function, related to the shape of the molecule and to the way the various linkages are connected to each other (7). Calculating the energies of the two isomers of hexatriene, **13** and **14**, is thus a simple matter. In Scheme 8.6, each linkage in **19** and **20** is labeled by the number of bonds that are adjacent to this linkage. From these numbers, the expressions for the total energies of each isomer are immediately calculated. The result in Scheme 8.6 clearly shows that the linear isomer is more stable than the branched one, in agreement with experimental facts.

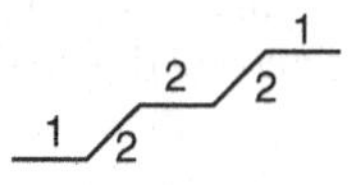

19

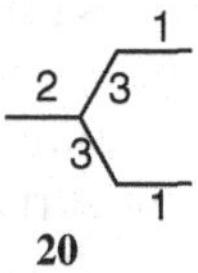

20

$E(PT2) = - (2/1 + 3/2)g$
$\quad\quad = - 3.5g$

$E(PT2) = - (2/1 + 1/2 + 2/3)g$
$\quad\quad\quad = - 3.17g$

Scheme 8.6

8.2.4   Extending Ovchinnikov's Rule to Search for Bistable Hydrocarbons

Extension of the Ovchinnikov rule to nonalternant systems (e.g., systems displaying odd-membered rings) leads to an interesting category of molecules, which display at least one Kekulé structure, but for which the MSAD has an S_z value >0, for example, **21–23** in Scheme 8.7, where the corresponding MSADs have $S_z = 1$. In such a case, the prediction of the ground-state multiplicity faces a dilemma of choice. On the one hand, the existence of a Kekulé structure is generally taken as an indication that the ground state would have the singlet state since Kekulé structure possess strong local pairings of electrons of opposite spin, as one indeed finds in alternant hydrocarbons. On the other hand, an S_z value of, for example, 1 for the MSAD is an argument in favor of a triplet ground state according to the spin-Hamiltonian VB theory. The possibility of these two conflicting tendencies, the most stable MSAD with a triplet spin vis-à-vis spin-paired Kekulé structure, might be the sign of the existence of two local minima of different multiplicities, which are sufficiently separated on the potential surface so that each state is the ground state in its own equilibrium geometry (18).

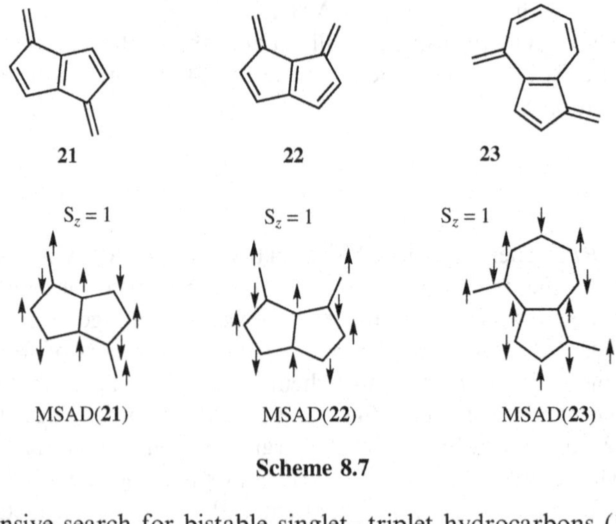

Scheme 8.7

An extensive search for bistable singlet–triplet hydrocarbons (18) revealed that the singlet ground state is generally the unique minimum in most cases (e.g., of **21–23**). However, interesting results were obtained by devising hydrocarbons in which the diradical structure can assume aromaticity in some parts of the molecule, while the Kekulé structures cannot. This extra stabilization of the triplet state may indeed lead to a stable triplet ground state, separated from an alternative singlet ground state by a nonnegligible barrier. Compound **24** in Scheme 8.8 is an example of such a bistable hydrocarbon, with a barrier of 0.14 eV between the two minima. Other candidates to bistability have been proposed on the basis of similar qualitative reasonings (18).

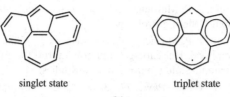

singlet state   triplet state

24

Scheme 8.8

8.3   A SUMMARY

The spin-Hamiltonian VB theory is a very simple and easy-to-use semiempirical tool that is based on the molecular graph. It is consistent with the VB theory described in Chapter 3, albeit with some simplifying assumptions and a more limited domain of application. Typically, this theory deals with the neutral ground or excited states of conjugated molecules or other homonuclear assemblies with one electron per site. For large systems, it reproduces the results of PPP full CI, while dealing with a much smaller Hamiltonian matrix.

This theory also possesses an *ab initio* based quantitative version (10,19), in which the parameters are geometry dependent and fitted on accurately calculated potential surfaces of ethylene. Despite its simplicity, the spin-Hamiltonian theory has proven itself to be accurate for predicting ground state, as well as excited state, properties and transition energies.

REFERENCES

1. D. L. Cooper, Ed., *Valence Bond Theory*, Chapters 15–23, Elsevier, New York, 2002.
2. J. Hubbard, *Proc. Roy. Soc. London, Ser. A* **276**, 238 (1963). Electron Correlations in Narrow Energy Bands.
3. (a) F. A. Matsen, *Acc. Chem. Res.* **11**, 387 (1978). Correlation of Molecular Orbital and Valence Bond States in π Systems. (b) M. A. Fox, F. A. Matsen, *J. Chem. Educ.* **62**, 477 (1985). Electronic Structure of π Systems. II. The Unification of Hückel and Valence Bond Theory.
4. W. Heisenberg, *Z. Phys.* **411**, 38, (1926). Mehrkorperproblem und Resonanz in der Quantenmechanik.
5. D. J. Klein, in *Valence Bond Theory*, D. L. Cooper Ed., Elsevier, New York, 2002, pp. 447–502. Resonating Valence-Bond Theories for Carbon π-Networks and Classical/Quantum Connections.
6. P. W. Anderson, *Phys. Rev.* **115**, 2 (1959). New Approach to the Theory of Superexchange Interactions.
7. J. P. Malrieu, in *Models of Theoretical Bonding*, Z. B. Maksic, Ed., Springer-Verlag, 1990, pp. 108–136. The Magnetic Description of Conjugated Hydrocarbons.
8. J. P. Malrieu, D. Maynau, *J. Am. Chem. Soc.* **104**, 3021 (1982). A Valence Bond Effective Hamiltonian for Neutral States of π Systems. 1. Method.

9. D. Maynau, M. Said, J. P. Malrieu, *J. Am. Chem. Soc.* **105**, 5244 (1983). Looking at Chemistry as a Spin Ordering Problem.

10. M. Said, D. Maynau, J. P. Malrieu, M. A. Garcia Bach, *J. Am. Chem. Soc.* **106**, 571 (1984). A Nonempirical Heisenberg Hamiltonian for the Study of Conjugated Hydrocarbons. Ground-State Conformational Studies.

11. (a) D. J. Klein, *J. Chem. Phys.* **77**, 3098 (1982). Ground-State Features of Heisenberg Models. (b) D. J. Klein, *Pure Appl. Chem.* **55**, 299 (1982). Valence Bond Theory for Conjugated Hydrocarbons. (c) J. Wu, T. G. Schmalz, D. J. Klein, *J. Chem. Phys.* **117**, 9977 (2002). An Extended Heisenberg Model for Conjugated Hydrocarbons.

12. (a) W. Wu, S.-j. Zhong, S. Shaik, *Chem. Phys. Lett.* **292**, 7 (1998). VBDFT(s): A Hückel-type Semiempirical Valence Bond Method Scaled to Density Functional Energies. Applications to Linear Polyenes. (b) Y. Luo, L. Song, D. Danovich, S. Shaik, *ChemPhysChem* **5**, 515 (2004). The Ground and Excited States of Polyenyl Radicals $C_{2n-1}H_{2n+1}$ ($n = 2$–13): A Valence Bond Study.

13. A. A. Ovchinnikov, *Theor. Chim. Acta (Berlin)*, **47**, 297 (1978). Multiplicity of the Ground State of Large Alternant Organic Molecules with Conjugated Bonds.

14. W. T. Borden, H. Iwamura, J. A. Berson, *Acc. Chem. Res.* **27**, 109 (1994). Violations of Hund's Rule in Non-Kekulé Hydrocarbons: Theoretical Prediction and Experimental Verification.

15. W. T. Borden, in *Encyclopedia of Computational Chemistry*, Vol. 1, P. v. R. Schleyer, N. L. Allinger, T. Clark, J. Gasteiger, P. A. Kollman, H. F. Schaefer, III, P. R. Schreiner, Eds. John Wiley & Sons, Inc., Chichester, 1998, p. 708.

16. D. Maynau, J. P. Malrieu, *J. Am. Chem. Soc.* **104**, 3029 (1982). A Valence Bond Effective Hamiltonian for Neutral States of π-Systems. 2. Results.

17. D. Döhnert, J. Koutecky, *J. Am. Chem. Soc.* **102**, 1789 (1980). Occupation Numbers of Natural Orbitals as a Criterion for Biradical Character. Different Kinds of Biradicals.

18. N. Guihery, D. Maynau, J. P. Malrieu, *New. J. Chem.* **22**, 281 (1998). Search for Singlet–Triplet Bistabilities in Conjugated Hydrocarbons.

19. M. Said, D. Maynau, J. P. Malrieu, *J. Am. Chem. Soc.* **106**, 580 (1984). Excited-State Properties of Linear Polyenes Studied through a Nonempirical Heisenberg Hamiltonian.

EXERCISES

8.1. Use Ovchinnikov's rule to predict the preferred multiplicities for the ground states of the hydrocarbon diradicaloids displayed in Scheme 8.Ex.1

8.2. Write the Heisenberg Hamiltonian of allyl radical and diagonalize it. Then write the wave functions for the ground and first neutral excited state. Show that the excited state has a positive α spin density on the central atom, as discussed in Chapter 1.

8.3. Predict the qualitative spin densities distribution in the free radicals **18** and **19** displayed in Scheme 8.Ex.2

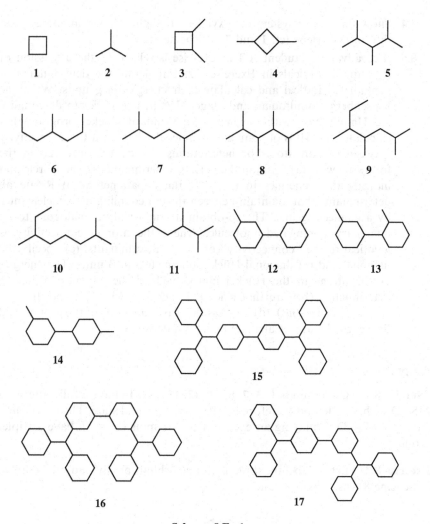

Scheme 8.Ex.1

Scheme 8.Ex.2

8.4. Show that the π system of *o*-xylylene is slightly lower in energy than that of *p*-xylylene (cf. **1** and **2** in Scheme 8.2).

8.5. (For advanced students.) This exercise is related to the discussion of pentadienyl radicals in Exercise 7.2. Enumerate the determinants of pentadienyl radical and calculate their energies in *g* units. Write the Heisenberg Hamiltonian and diagonalize it. It is easier to diagonalize the Hamiltonian with the help of a standard Hückel program that calculates the MOs of conjugated molecules based on the connectivity between carbon atoms or heteroatoms. To do this, proceed in the following way: (a) Position the various determinants Ω_i on a circle and indicate their energies in terms of the *g* parameter. (b) Relate all determinants that maintain between them a coupling matrix element *g* by a connecting line. The so-obtained graph will be analogous to an imaginary conjugated molecule, where the atomic connectivity is identical to the "connectivity" of the Ω_i determinants. (c) Specify the diagonal and off-diagonal Hückel parameters in β units. The energies of the atoms in the Hückel matrix will be the energies of the Ω_i determinants (beware that since β is negative while *g* is positive, *ng* corresponds to $-n\beta$). (d) Express the wave functions of the ground and first excited states. Calculate their spin densities.

Answers

Exercise 8.1 Compounds **1, 3, 7, 8, 11, 12, 15,** and **16** have a fully alternant MSAD with $S_z = 0$, and therefore have a singlet ground state. The other ones, having a MSAD displaying an excess of two α spins ($S_z = 1$), have a triplet ground state.

Exercise 8.2 Let us label the three π atomic orbitals of allyl radical as shown in Scheme 8.Ans.1. by *a*, *b*, and *c*.

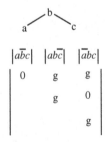

Scheme 8.Ans.1

The corresponding Heisenberg Hamiltonian derives from the connectivity of the allyl structure and is shown in the scheme below the structure. This matrix can be diagonalized with a simple Hückel program. The two lowest solutions are Ψ and Ψ^* below (as indeed shown in Exercise 7.1):

$$\Psi = 6^{-\frac{1}{2}}(-2|a\bar{b}c| + |\bar{a}bc| + |ab\bar{c}|) \qquad (8.\text{Ans}.1)$$

$$\Psi^* = 2^{-\frac{1}{2}}(|\bar{a}bc| - |ab\bar{c}|) \qquad (8.\text{Ans}.2)$$

The weights of the three determinants of Ψ^* are 0.5 each. It follows that the α spin density of the central atom, b, is $+1$.

Exercise 8.3   Combining the MSAD with a minor second most alternant determinant leads to the qualitative spin multiplicities displayed in Scheme 8.Ans.2

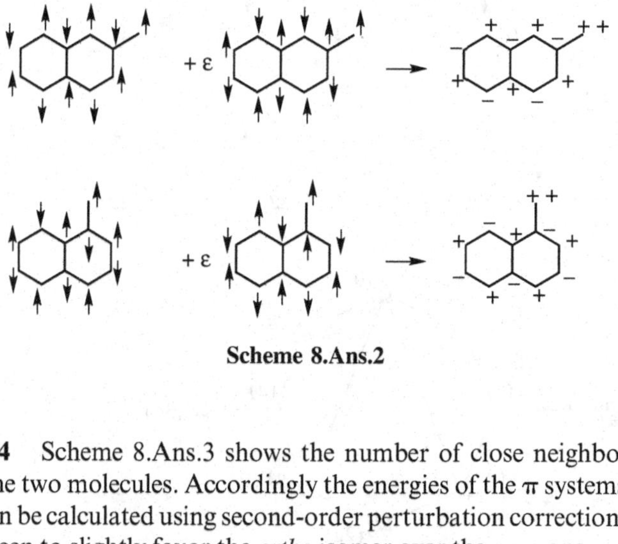

Scheme 8.Ans.2

Exercise 8.4   Scheme 8.Ans.3 shows the number of close neighbors for each linkage in the two molecules. Accordingly the energies of the π systems of the two xylylenes can be calculated using second-order perturbation correction (PT2). The results are seen to slightly favor the *ortho* isomer over the *para* one.

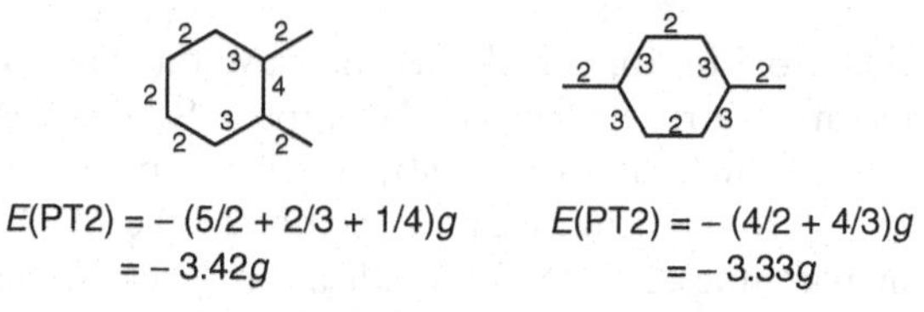

$$E(\text{PT2}) = -(5/2 + 2/3 + 1/4)g$$
$$= -3.42g$$

$$E(\text{PT2}) = -(4/2 + 4/3)g$$
$$= -3.33g$$

Scheme 8.Ans.3

Exercise 8.5   The 10 determinants, $\Omega_1 - \Omega_{10}$, are represented in Scheme 8.Ans.4.

Their energies are obtained by counting the number of spin frustrations, giving

$$E(\Omega_1) = 0 \tag{8.Ans.3a}$$

$$E(\Omega_3) = E(\Omega_5) = E(\Omega_7) = E(\Omega_8) = g \tag{8.Ans.3b}$$

$$E(\Omega_2) = E(\Omega_6) = E(\Omega_{10}) = 2g \tag{8.Ans.3c}$$

$$E(\Omega_4) = E(\Omega_9) = 3g \tag{8.Ans.3d}$$

The connecting lines between the Ω_i determinants, each representing a matrix element g, are depicted in the drawing along with the self-energies of the determinants (in g units in parentheses).

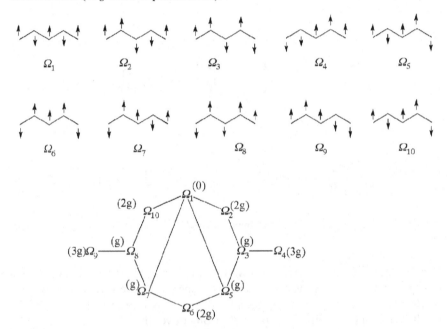

Scheme 8.Ans.4

To use a Hückel program for calculating the wave functions, one only has to replace all off-diagonal matrix elements in the graph in Scheme 8.Ans.4 by $-\beta$ and the energy of each fictitious atom by $-n\beta$, which is simply the energy of the corresponding Ω_i determinant in ng units, relative to the MSAD Ω_1, given in the graph within parentheses. The diagonalization leads to the ground state $\Psi(^2B_1)$:

$$\Psi(^2B_1) = +0.67\,\Omega_1 - 0.23\,(\Omega_1 + \Omega_{10}) + 0.23\,(\Omega_3 + \Omega_8)$$
$$- 0.05\,(\Omega_4 + \Omega_9) - 0.38(\Omega_5 + \Omega_7) + 0.20\,\Omega_6 \tag{8.Ans.4}$$

and to the first excited state $\Psi^*(^2A_2)$:

$$\Psi^*(^2A_2) = 0.22(\Omega_{10} - \Omega_2) + 0.53(\Omega_3 - \Omega_8)$$
$$+ 0.16(\Omega_9 - \Omega_4) + 0.38(\Omega_7 - \Omega_5)$$

(8.Ans.5)

The spin densities are deduced from the weights (here the squared coefficients) of the determinants for the ground and excited states. They are indicated in Fig. 8.Ans.1 (bold numbers). The spin densities arising from a VBDFT calculation (Ref. 21 of Chapter 7) are depicted in italics on the same figure. The agreement between both sets of calculated values is seen to be very good.

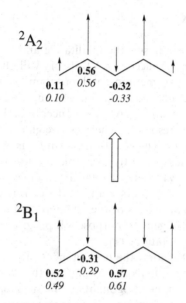

FIGURE 8.Ans.1 Calculated spin densities for the ground state ($^2B_1$) and the first excited state ($^2A_2$) of pentadienyl radical (bold numbers). The numbers in italics are those arising from the VBDFT(s) calculations.

9 Currently Available *Ab Initio* Valence Bond Computational Methods and Their Principles

9.1 INTRODUCTION

The previous chapters emphasize the qualitative and semiquantitative aspects of VB theory. However, the success of VB theory will ultimately depend on the availability of quantitative methods and program packages that are widely accessible and user-friendly, and that enable one to carry out *ab initio* VB computations to test the qualitative VB concepts and assess the native VB quantities, for example, resonance energies, weights of resonance structures, and diabatic energy curves. The goal of this chapter is to provide a snapshot of the *ab initio* VB methods currently available for use by chemists. There are plenty of semiempirical VB methods, which we do not review here. Some of these were mentioned in Chapters 7 and 8. Thus, this chapter will provide a brief description of the principles of the *ab initio* methods, while the appendix will provide a short list of general-purpose VB programs that chemists can use for conducting VB computations (1).

Compared with MO theory, the development of *ab initio* VB methods has been delayed by some 20 years, owing to the computational difficulty associated with the nonorthogonality of orbitals, often called the "$N!$ problem" (N being the number of electrons) that in a strict sense means that one has to compute all the $N!$ permutations in the VB determinant. However, since the 1980s, much methodological progress has been made by a few groups (2–10) in such a way that the so-called "$N!$ problem" has become a misnomer, since the difficulty due to nonorthogonality *does not* imply any more that the computational effort required to perform a nonorthogonal configuration interaction scales as $N!$. As a matter of fact, in actual modern implementations (11), the calculation of Hamiltonian matrix elements between nonorthogonal determinants scales as N^4. These modern methods, together with the exponential increase of the computer power, allow us to perform rigorous *ab initio* VB calculations, in which overlap integrals are explicitly considered and all one- and two-electron integrals are precisely evaluated. An additional reason for the blooming of *ab initio* VB calculations is the appearance of new VB methods (see below), which allow one to perform a variety of different calculations; of diabatic states, weak interactions free of basis-set superposition errors, assessment of hybridization, quantification of resonance energies, and so on. On one hand, the growing number of such quantitative applications shows that VB theory is slowly recovering its role as a computational method tailor-made for chemistry; one that is able to offer clear interpretation of the wave function. On the other hand, as the following survey will illustrate, there are many ways to do VB computations, unlike the situation in MO theory where one can follow a canonical route, starting with the monodeterminant MO wave function and gradually improving it by means of CI.

Historically, the first *ab initio* VB calculations were performed by using the orbitals of the free atoms. This crude approximation, which ignored the considerable rearrangements in size and shape that an AO undergoes when fragments assemble to a molecule, resulted in poor accuracy, and is no longer in use. Accordingly, the variational optimization of the orbitals that are used for the construction of the AO or FO determinants is an important feature of all modern *ab initio* methods. The orbitals that result from such an optimization process have, as a rule, a strongly local character and remain nonorthogonal in all *ab initio* VB methods. However, the degree of delocalization that these orbitals are allowed to have varies from one method to another. As has been shown in Chapter 3 with the H_2 example, both localized and semilocalized approaches have their specific advantages. Localized AOs provide a very clear understanding of the nature of chemical bonding, but the number of VB structures that are necessary to take into account will become large if many bonds have to be described in a VB manner. On the other hand, semilocalized orbitals provide extremely compact wave functions, but this compact nature obscures the classical interpretation of covalent and ionic structures. This fundamental difference between methods using localized orbitals and those using semidelocalized ones will define two distinct categories of VB methods, based on semilocalized and localized orbitals. These two classes are described in the following sections.

9.2   VALENCE BOND METHODS BASED ON SEMILOCALIZED ORBITALS

All the VB methods that deal with semilocalized orbitals use a generalization of the Coulson–Fischer idea (12), whereby a bond is described as a singlet coupling between two electrons in nonorthogonal orbitals that possess small delocalization tails resulting from the variational orbital optimization. Albeit formally covalent, this description implicitly involves some optimal contributions of ionic terms, as a decomposition of the wave function in terms of pure AO determinants would show (see Eqs. 3.5 and 3.6). For a polyatomic molecule, the Coulson–Fischer orbitals can be generalized in the following two ways:

1. One may allow the orbitals to delocalize freely over the entire molecule, and write the wave function either as a single, and formally covalent, VB structure (known as the perfect-pairing approximation, to be discussed later), or as a linear combination of several VB structures that represent all the possible pairing schemes between a given number of electrons and orbitals (e.g., 14 possible pairing schemes for a single configuration of methane with eight electrons in eight orbitals). The orbitals that emerge from such a procedure are called "overlap-enhanced orbitals" (OEOs). Most of the time, the OEOs appear to be fairly localized, but nevertheless, the shape of these orbitals, or their degree of delocalization, is an important feature to be taken into account when one interprets the wave function in terms of Lewis structures.
2. In a more restrictive definition of the CF orbitals, each orbital is allowed to delocalize only onto the atom to which it is bonded in the VB structure under consideration; such orbitals are called "bond-distorted orbitals" (BDOs).

The BDOs have the advantage of allowing an unambiguous correspondence between the mathematical expression of a VB structure and the associated Lewis structure. *On the other hand, owing to the delocalization tails in these orbitals, neither OEOs nor BDOs allow the distinction to be made between the covalent and ionic components of the bonds.*

9.2.1 The Generalized Valence Bond Method

The generalized valence bond (GVB) method was the earliest important generalization of the Coulson–Fischer idea to polyatomic molecules (13,14). The method uses OEOs that are free to delocalize over the whole molecule during orbital optimization. Despite its general formulation, the GVB method is usually used in its restricted form, referred to as GVB–SOPP, which introduces two simplifications. The first one is the perfect-pairing (PP) approximation, in which only one VB structure is generated in the calculation. The wave function may then be expressed in the simple form of Equation 9.1, as a product of so-called "geminal" two-electron functions:

$$\Psi_{GVB} = |(\varphi_{1a}\bar{\varphi}_{1b} - \bar{\varphi}_{1a}\varphi_{1b})(\varphi_{2a}\bar{\varphi}_{2b} - \bar{\varphi}_{2a}\varphi_{2b})\cdots(\varphi_{na}\bar{\varphi}_{nb} - \bar{\varphi}_{na}\varphi_{nb})| \qquad (9.1)$$

Each geminal function is a singlet-coupled GVB pair (φ_{ia}, φ_{ib}) that is associated with a particular bond or lone pair in the molecule. For example, CH_4 will have the familiar Lewis structure and its wave function will involve a product of four geminal functions, each corresponding to a C–H bond.

   The second simplification, which is introduced for computational convenience, is the *strong orthogonality* (SO) constraint, by which all the orbitals in

Equation 9.1 are required to be orthogonal to each other unless they belong to the geminal pair, that is,

$$\langle \varphi_{ia} | \varphi_{ib} \neq 0 \rangle \quad (a\text{--}b \text{ paired}) \tag{9.2a}$$

$$\langle \varphi_i | \varphi_j = 0 \rangle \quad \text{otherwise} \tag{9.2b}$$

This strong orthogonality constraint, while seemingly a restriction, is usually not a serious one, since it applies to orbitals that are not expected to overlap significantly. On the other hand, the orbitals (φ_{ia}, φ_{ib}) that are coupled together in the same GVB pair, of course, display a strong overlap. The term SOPP–GVB then describes a perfectly paired GVB wave function generated under the constraint of zero-overlap between the orbitals of different pairs.

For further mathematical convenience, each geminal function in Equation 9.1 can be rewritten, by simple orbital rotation, as an expansion in terms of natural orbitals, in Equation 9.3:

$$|\varphi_{ia}\bar{\varphi}_{ib}| - |\bar{\varphi}_{ia}\varphi_{ib}| = C_i|\phi_i\bar{\phi}_i| + C_i^*|\phi_i^*\bar{\phi}_i^*| \tag{9.3}$$

This alternative form of the geminal contains two closed-shell terms. The natural orbitals ϕ_i and ϕ_i^*, in Equation 9.3, have the shapes of localized MOs, respectively bonding and antibonding, which are orthogonal to each other. The natural orbitals are connected to the GVB pairs by the simple transformation below:

$$\varphi_{ia} = \frac{\phi_i + \lambda\phi_i^*}{\left(1 + \lambda^2\right)^{1/2}} \quad \varphi_{ib} = \frac{\phi_i - \lambda\phi_i^*}{\left(1 + \lambda^2\right)^{1/2}} \quad \lambda^2 = -\frac{C_i^*}{C_i} \tag{9.4}$$

The great computational advantage of using natural orbitals, rather than GVB pairs, in the effective equations that are solved for self-consistency, is that *all* the natural orbitals are orthogonal to each other. A GVB–SOPP calculation is thus nothing else but a special case of a truncated MCSCF calculation, with all the advantages brought by the MO-based formulation. In addition, the transformed GVB–SOPP wave function, using Equation 9.4, has the VB advantage of interpretability as a collection of two electron bonds, in a manner close to the chemist's conception of molecules. Of course, it is straightforward to include a "core" of doubly occupied orthonormal orbitals, like the set of σ MOs for conjugated molecules.

As long as the molecule being considered consists of clearly separated local bonds and is in its equilibrium geometry, the perfect-pairing and strong-orthogonality constraints do not lead to great loss of accuracy, and at the same time result in considerable computer-time saving. On the other hand, it is clear that these restrictions would be inappropriate for delocalized electronic systems, such as benzene, whose description requires VB applications beyond the PP approximation, that is, by inclusion of all possible pairing schemes.

Since the GVB wave function takes care of the left–right electron correlation for each local bond, it incorporates a good deal of nondynamical correlation and can serve as a basis for calculations of dynamic correlation. This has been done by Carter and Goddard in their Correlation–Consistent Configuration Interaction (CCCI) method (15). The method starts from a GVB–SOPP wave function and subsequently incorporates a small set of single and double excitations that are chosen so as to include all the electronic correlation involving the orbitals that change significantly during bond breakage or formation. The method that was aimed at providing accurate bond dissociation energies, using simple wave functions, proved to be successful for describing the dissociation of single and double bonds (15). To date, GVB is the most accessible and user-friendly VB method available in most standard program packages.

9.2.2 The Spin-Coupled Valence Bond Method

The spin-coupled (SC) method, developed by Gerratt and co-workers (16–18), differs from the GVB–SOPP method in the sense that it removes the orthogonality and perfect-pairing restrictions. The method still relies on a single set of orbitals (i.e., a single-configuration type), but all the modes of spin pairing are included in the wave function, and the orbitals are allowed to overlap with each other without restrictions. Both orbitals and coefficients of the various spin-pairing modes are optimized simultaneously. Thus the method imposes no constraints, neither on the spin-coupling schemes, nor on the shapes of the orbitals; these features are determined by the variational principle alone. As such, the SC method represents the ultimate level of accuracy compatible with the orbital approximation that describes the molecule as a single configuration with fixed-orbital occupancies.

The SC wave function provides a correlated and nevertheless a lucid description of bonding in molecules. The shapes of the SC orbitals, and their variations with nuclear geometry, provide insight into the spatial arrangements of the electronic clouds, the hybridization of the atoms, and the processes of bond making and breaking throughout the course of a reaction. The spin-coupling coefficients constitute quantitative descriptors of the relative importance of the different spin-coupling modes. It is clear that the degree of delocalization of the SC orbitals is a crucial parameter for the achievement of an unambiguous association of a spin coupling form with a specific Lewis structure; the more localized the orbitals, the easier the interpretation of the wave function in terms of traditional chemical concepts. This is not *a priori* guaranteed. Nevertheless, a typical outcome of SC calculations is a set of OEOs mostly localized on atoms, but distorted (in other words slightly delocalized) toward all neighboring atoms, and especially so in the direction of bonds in the classical Lewis structure. The number of dominant spin-coupling modes generally remains very small; for molecules such as methane, the SC wave function is dominated by the GVB–PP structure (19), which is the chemist's classical structure, while for example, for benzene, the wave function is dominated by two major Kekulé structures and three much less important Dewar structures (20).

Like GVB, the SC method also uses a wave function that can serve as an appropriate basis for subsequent CI and inclusion of dynamic correlation. Subsequent CI on the SC wave function is performed in the spin-coupled valence bond (SCVB) method, which is an extension of the SC method. At the simplest level, the CI includes all the configurations that can possibly be generated by distributing the electrons within the set of the active orbitals that were optimized in the preliminary SC calculation; both formally "covalent"- and "ionic" type configurations are considered. A higher level of SCVB theory includes additional excitations, for example, from the orbitals of the core, if any (e.g., excitations from the σ orbitals when the SC orbitals are only those of π type), or to orbitals that are virtual in the one-configuration calculation. To preserve the VB character of the wave function, the virtual orbitals are made to be localized as much as possible by some appropriate technique (1,18). The final energy value is determined by nonorthogonal CI in the so-generated configuration space. From experience, the excited configurations generally bring very little stabilization in ground-state calculations; this is easily explained by the fact that the orbitals are optimized precisely so as to concentrate all the important physical effects in the reference SC configuration. On the other hand, excited configurations are important for satisfactory state ordering and electronic transition energies to excited states (21). SC calculations can be performed with the TURTLE module, now implemented in GAMESS-UK, or with the XMVB package (see below).

9.2.3 The CASVB Method

It is known that, in the MO framework, the nondynamical electron correlation is accounted for by means of a so-called CASSCF calculation, which is nothing else than a full CI in a given space of orbitals and electrons, in which the orbitals and the coefficients of the configurations are optimized simultaneously. If the active space includes all the valence orbitals and electrons, then the totality of the nondynamical correlation of the valence electrons is accounted for. In the VB framework, an equivalent VB calculation, defined with pure AOs or purely localized hybrid atomic orbitals (HAOs), would involve all the covalent and ionic structures that may possibly be generated for the molecule at hand. Note that the resulting covalent–ionic VB wave function would have the same dimension as the valence–CASSCF one (e.g., 1764 VB structures for methane, and 1764 MO SCF configurations in the CASSCF framework).

The GVB and SC methods provide wave functions that are, of course, much more compact than the corresponding valence–CASSCF one (e.g., only 14 spin-coupling modes for methane with the SC method, and a single one with the GVB method). Owing to this difference in size, the GVB and SC methods cannot be expected to include the totality of the nondynamical correlation, even if these two methods treat well, by definition, the left–right correlation for each bond of the molecule. Physically, this is because the various local ionic situations are not interconnected with these methods. For example, the two ionic situations **1** and **2** in Scheme 9.1 are found to have equal weights at the GVB or SC levels, while CASSCF would differentiate their weights, **2** being more important than **1**, as expected on a physical basis.

Scheme 9.1

Despite these restrictions, the GVB and SC methods generally provide energies that are much closer in quality to CASSCF than to Hartree–Fock (19), and wave functions that are close to the CASSCF wave function having the same number of electrons and orbitals in the active space. This property has been used to devise a fast method to get approximate SC wave functions.

Since a CASSCF calculation is faster than a direct SC calculation, owing to the advantages associated with orbital orthogonality in CASSCF, it is practical to extract an approximate SC wave function (or another type of VB function, e.g., a multiconfigurational one) from a CASSCF wave function. The conversion from one wave function to the other relies on the fact that a CASSCF wave function is invariant under linear transformations of the active orbitals. Based on this invariance principle, two different procedures were developed and both share the same name "CASVB". Thus, CASVB is not a straightforward VB method, but rather a projection method that bridges between CASSCF and VB wave functions.

In the CASVB method of Thorsteinsson et al. (22,23), one transforms the canonical CASSCF orbitals so that the wave function (which we recall, is kept unchanged in this process) involves a dominant component of a VB-type wave function, Ψ_{VB}, which is chosen in advance and may be single- or multi-configuration, as in Equation 9.5:

$$\Psi_{CAS} = S_{VB}\Psi_{VB} + \left(1 - S_{VB}^2\right)\Psi_{VB}^{\perp} \tag{9.5}$$

Here $\Psi_{VB}^{\perp}$ is the orthogonal complement of Ψ_{VB} to the CASSCF wave function, and S_{VB} is the overlap between Ψ_{CAS} and Ψ_{VB}. To ensure that the so-obtained VB function is as close as possible to the starting CASSCF wave function, the procedure transforms the orbitals in a manner that maximizes the overlap S_{VB}. An alternative procedure is one that minimizes the energy of the VB function Ψ_{VB}. This latter procedure is, however, more expensive than the first one. As the two methods generally yield similar sets of orbitals, the method of S_{VB} maximization is generally preferred (23). The S_{VB} based CASVB method is implemented in the MOLPRO and MOLCAS packages.

The CASVB method developed by Hirao et al. (24) differs from the previous one in the requirement that the final CASVB wave function, obtained after the transformation of the CASSCF canonical orbitals, is strictly equivalent to the starting CASSCF wave function. The price of this strict equivalence is that the orbitals that are used to construct the "VB structures" remain more or less delocalized, however, these VB-type delocalized orbitals are localized as much as possible following various localization procedures (1,24,25).

9.2.4 The Generalized Resonating Valence Bond Method

Multiconfigurational extensions of the GVB method have been specifically devised for delocalized electronic systems, like benzene, which require the use of two, or more, resonance structures, and for which the one-structure VB representation (e.g., as GVB–PP) is not appropriate. This situation is widespread and includes a wide variety of open-shell electronic states, as, for example, allyl radical and its analogues, pentadienyl anion and its analogues, transition states of chemical reactions, core-ionized diatomic molecules, $n-\pi^*$ excited molecules containing two equivalent carbonyl groups, n-ionized molecules having equivalent but remote lone pairs, and so on. In the general case, one-configuration-based methods will yield poor energies for a delocalized system that requires at least two structures (e.g., transition state of a reaction) relative to parts of the potential surface that are well described by a single VB structure (e.g., reactants or products). In some of these species, for example, core-ionized diatomic molecules, the use of a single-configuration method leads to the   so-called "symmetry dilemma" (26). The symmetry dilemma was analyzed in VB terms by McLean (27), and shown to arise from a competition between two effects. One is the familiar resonance effect by which a mixture of two resonance structures is lower in energy than either one taken separately. The second is the so-called "orbital size effect", whereby a specific VB structure gains stabilization if it can have its own particular set of optimal orbitals. The two effects cannot be simultaneously taken into account in any one-configuration based theory, be it of VB or MO type, because such a theory employs a single set of orbitals. In the (most frequent) case where the orbital-size effect is more important than the resonance effect, the wave function will take more or less the form of one particular VB structure leading thereby to a nonphysical symmetry breaking of the wave function.

Within the MO formalism, the symmetry-breaking phenomenon can be remedied by MCSCF-type calculations or a CI procedure within the set of symmetry-broken solutions. Thus, Jackels and Davidson (28) eliminated the problem in the NO_2 radical by using a symmetry-adapted combination of two symmetry-broken Hartree–Fock wave functions, by means of a 2×2 nonorthogonal CI. Similarly, within the VB framework, elimination of artificial symmetry-breaking can be achieved by allowing different orbitals for different VB structures in the course of the orbital optimization. Voter and Goddard employed this idea in the GVB method (29–31). Considering an electronic state as a superposition of two possible VB structures **A** and **B** (the generalization to multiple terms is trivial), the total wave function is written as a linear combination of two symmetry-broken subwave functions Φ_A and Φ_B of the GVB type, one for each VB structure:

$$\Psi = C_A \Phi_A + C_B \Phi_B \tag{9.6}$$

Of course, Φ_A and Φ_B are nonorthogonal wave functions.

In a preliminary version of the method (29), called "Resonating-Generalized Valence Bond" (RGVB), each symmetry-broken sub-wavefunction is optimized separately, and the orbitals are not reoptimized in the presence of resonance between the wave functions. This may lead to underestimation of the resonance energy since the orbital optimization only takes care of the orbital-size effect, while not optimizing the resonance energy. To remove this defect, Voter and Goddard subsequently improved their method by allowing the sub-wavefunctions to be optimized in the presence of each other, leading to the final GRVB method (30,31). The quality of such a simple wave function and the usefulness of the method can be appreciated by the results of the difficult test case of the $HF + D \rightarrow H + FD$ exchange reaction. In a basis set of double-zeta + polarization quality, the GRVB method yields a barrier of 47.7 kcal/mol, in very good agreement with higher level calculations using the same basis set. In contrast, a simple one-configuration GVB calculation leads to a much higher barrier of 69.5 kcal/mol (30).

9.2.5   Multiconfiguration Valence Bond Methods with Optimized Orbitals

The Valence Bond Self-Consistent Field (VBSCF) method has been devised by Balint-Kurti and van Lenthe (32), and was further modified by Verbeek (6,33) who also developed an efficient implementation in a package called TURTLE (11). Basically, the VBSCF method is a multiconfiguration SCF procedure that allows the use of nonorthogonal orbitals of any type. The wave function is given as a linear combination of VB structures, Φ_K (Eq. 9.7).

$$\Psi = \sum_K C_K \Phi_K \tag{9.7}$$

Unlike the SC method that has a single configuration, the VBSCF method has no restrictions upon the number of configurations and the orbital occupancy. In addition, the VBSCF wave function can use an arbitrary number of spin-coupling modes. Both the orbitals and coefficients of VB structures are optimized simultaneously, by means of a Super CI algorithm (34). The bottleneck of this algorithm is the evaluation of the nonorthogonal matrix elements, however, a number of improvements for speeding up this evaluation have been implemented (33), and can be found in the TURTLE package.

The VBSCF method permits complete flexibility in the definition of the orbitals used for constructing the VB structures, Φ_K. The orbitals can be allowed to delocalize freely during the orbital optimization (resulting in

OEOs), thereby performing GVB or SCVB calculations. The orbitals can also be defined by pairs that are allowed to delocalize on only two centers (BDOs), or they can be defined as strictly localized on a single center or fragment (see below). The VBSCF method is implemented in the TURTLE module (now being a part of GAMESS–UK) and in the XMVB package.

Other multiconfiguration VB methods have also been devised, like the biorthogonal valence bond method of McDouall (35,36) or the spin-free approach of McWeeny (37). For an overview of these methods, the reader is advised to consult a recent review (1).

9.3   VALENCE BOND METHODS BASED ON LOCALIZED ORBITALS

Quantitites such as resonance energies, covalency and ionicity, which are native to VB theory, are not treated properly by all the above methods that rely on semilocalized orbitals. The calculation of these quantities requires "pure" VB methods with orbitals that are strictly localized on a single atom or fragment. However, as these orbitals are allowed to hybridize from the AOs of a given atomic center, it is customary to use the descriptive term "hybrid atomic orbitals", to qualify these local orbitals. The advantage of the "pure" VB method is that it allows a clear distinction between covalent structure (where the fragments A and B each bear a singly occupied orbital) and ionic ones (one fragment orbital is doubly occupied, the other is empty). In this framework, the accurate description of the A–B bond requires taking into account the covalent and ionic VB structures, as has been seen in Chapter 3 with the example of the H_2 molecule.

In the case where the fragments A and B are polyatomic (e.g., $A = B = H_3C$, in $H_3C–CH_3$) and only the A–B bond is being described in a VB manner, the orbitals that are used to construct the determinants may be FOs that have tails on the adjacent atoms (e.g., the H atoms of a given CH_3 group), but remain localized in the sense that an orbital of fragment A has no coefficient on B, and vice versa. In such a case, the description of the A–B bond will still require explicit consideration of ionic and covalent components, and the so-defined FOs will be considered throughout the text as a particular case of localized orbitals.

9.3.1   Valence Bond Self-Consistent Field Method with Localized Orbitals

The flexibility of the valence bond self-consistent field (VBSCF) method can be exploited to calculate VB wave functions based on orbitals that are purely localized on a single atom or fragment. In such a case, the VBSCF wave function takes a classical VB form, which has the advantage of giving a very detailed description of an electronic system, as, for example, the interplay between the various covalent and ionic structures in a reaction. On the other hand, since covalent and ionic structures have to be explicitly considered for each bond, the number of VB structures that have to be taken into account grows rapidly with the number of bonds. Therefore, this method would not be practical for the VB description of large electronic systems, not only because of the large number of structures, but primarily because the interpretability of the wave function becomes somewhat cumbersome. Fortunately, the "reacting part" of a molecular system is often small in most elementary reactions, and this allows the definition of an active and a spectator part in the electronic system, which can be treated at different levels. Practically, the usual way of using VBSCF is to define an active part involving only those orbitals and electrons that undergo significant changes during the process, like bond breaking and/or formation; the remaining orbitals and electrons form the spectator part. While the active system is subject to a detailed VB treatment involving the complete set of chemically relevant Lewis structures, the spectator part is treated, at the MO level, as a set of doubly occupied MOs. Of course, all orbitals, including those of the spectator part, are optimized during the VBSCF procedure. As such, the VBSCF method is a close analogue of the CASSCF procedure in MO-based theory.

Consider, for example, the dissociation of an A–B bond, where A and B are two polyelectronic fragments. Including the two HAOs that are involved in the bond in the active space, and the adjacent orbitals and electrons in the spectator space, the VBSCF wave function reads as follows:

$$\Psi_{VBSCF} = C_1(|\psi\bar{\psi}\chi_a\bar{\chi}_b| - |\psi\bar{\psi}\bar{\chi}_a\chi_b|) + C_2|\psi\bar{\psi}\chi_a\bar{\chi}_a| + C_3|\psi\bar{\psi}\chi_b\bar{\chi}_b| \qquad (9.8)$$

where χ_a and χ_b are the active orbitals, common to all the structures, and ψ is a generic term that represents the product of spectator orbitals.

For a typical elementary reaction, such as S_N2, consisting, for example, of the breaking of a C–F bond followed by the formation of a new C–Cl bond (Eq. 9.9),

$$Cl^- + CH_3-F \rightarrow [Cl\ ...CH_3\ ...\ F]^- \rightarrow Cl-CH_3 + F^- \qquad (9.9)$$

the active system is made of the four electrons and three orbitals involved in the C–F bond and in the attacking lone pair of Cl^-. At any part of the potential surface, this electronic system is rigorously described at the VB level by a set of six VB structures (**3–8**), depicted in Scheme 9.2. On the other hand, three lone pairs of fluorine, three other lone pairs of chlorine, and three C–H bonds of carbon remain unchanged during the reaction and will form the "inactive" system. The active electrons are thus explicitly correlated (by Coulomb correlation), while the inactive electrons are not. One expects that the lack of Coulomb correlation in the inactive subsystem will result in a constant error throughout the potential surface, and therefore just uniformly shift the calculated energies relative to the fully correlated surface. Note that in this model all the occupied orbitals, active and inactive, are affected by the progress of

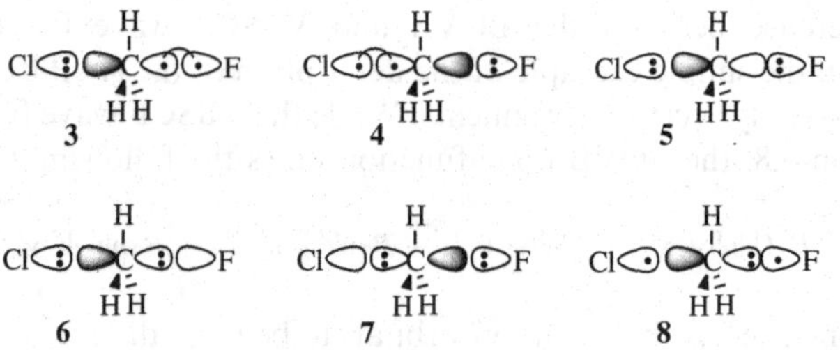

Scheme 9.2

the reaction, and thereby rearrange and adapt themselves to the local charges of the VB structures and to their mixing at all points of the reaction coordinate.

The above definitions of active–inactive subsystems are of course not restricted to the study of reactions, but can be generalized to all species whose qualitative description can be made in terms of resonating Lewis structures, such as conjugated molecules and mixed-valence compounds. Note that while all the VB structures Φ_K in Equation 9.8 differ from one another by their orbital occupancies, they still share the same set of orbitals. If a complete set of VB structures (covalent and ionic) is generated for a given electronic system, the resulting VBSCF wave function would be equivalent to a full-valence CASSCF wave function, and thereby accounting for all the nondynamic electronic correlation. Of course, as in CASSCF, here too the dynamic correlation (the definition of which will become clear in the next section) is still missing at this level, with the consequence that the calculation of bonding energies or reaction barriers is sometimes only semiquantitative. Due to the rapid progress in method development, together with the current capabilities of the XMVB package, one can carry out calculations of fairly large systems, for example, the spin states of an iron–oxo complex, $(NH_3)_5Fe{=}O^{2+}$. Thus, VBSCF calculations are becoming practical in areas that were completely off-limits a few years ago.

9.3.2 The Breathing-Orbital Valence Bond Method

The breathing-orbital valence bond (BOVB) method (38–40) was devised as an improvement of the VBSCF method, by providing the VB structures the very best possible orbitals that minimize the energy of the final multistructure state. This improvement is achieved by introducing an extra degree of freedom in the orbital optimization process: Instead of dealing with a common set of orbitals for all the VB structures, as done in VBSCF, a key specificity of the BOVB method is that the orbitals are variationally optimized with the freedom to be different for different VB structures. It is important to note that during this optimization process, *the different VB structures are not optimized separately, but rather in the presence of each other,* so that the orbital optimization not only lowers the energies of individual VB structures, but also maximizes the resonance energy resulting from the VB mixing.

The difference between the BOVB and VBSCF wave function can be illustrated on the simple example of the description of the A–B bond, where A and B are two polyelectronic fragments. While the VBSCF wave function reads like Equation 9.8, the BOVB wave function takes the following form:

$$\Psi_{\mathrm{BOVB}} = B_1(|\psi\bar\psi\chi_a\bar\chi_b| - |\psi\bar\psi\bar\chi_a\chi_b|) + B_2|\psi'\bar\psi'\chi_a'\bar\chi_a'| + B_3|\psi''\bar\psi''\chi_b''\bar\chi_b''| \qquad (9.10)$$

Logically, one expects the χ_a' and χ_b'' orbitals to be more diffuse than χ_a and χ_b, since the former are doubly occupied while the latter are only singly occupied. Similarly, the spectator orbitals in the different structures should have different sizes and shapes depending on whether they reside on cationic, neutral, or ionic fragments. This qualitative difference between VBSCF and BOVB wave functions is illustrated in Scheme 9.3 in a qualitative manner for the example of the F_2 molecule. Here, the VBSCF wave function is seen to have identical orbitals in all three VB structures (**9–11**). By contrast, the orbitals of the ionic structures **13** and **14** in the BOVB wave function are represented as bigger or smaller than those of the neutral structure **12**, depending on whether they are attached to an F^+ or F^- fragment. Thus, all the orbitals, active or spectators, are allowed to breath and adapt themselves to the local fields of the VB structures, as well as to the VB mixing of these structures, hence the name Breathing-Orbital. This instantaneous adaptation, which is missing at the VBSCF, GVB, SC, and CASSCF levels, results in better accuracy, as shown in benchmark calculations of bond dissociation energies and reaction barriers (39,41).

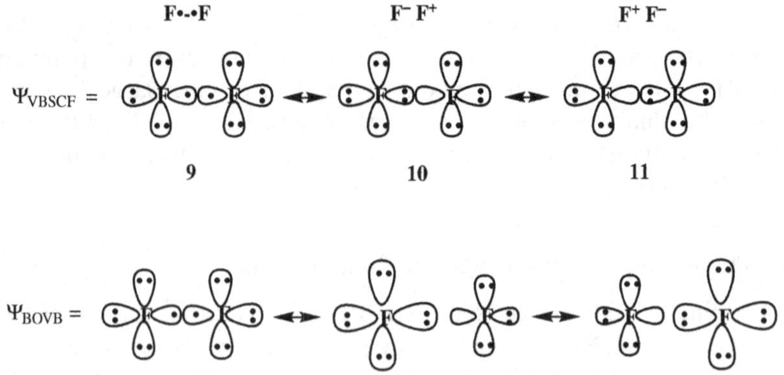

Scheme 9.3

The BOVB method has several levels of accuracy. At the most basic level, referred to as L-BOVB, all orbitals are strictly localized on their respective fragments. One way of improving the energetics is to increase the number of degrees of freedom by permitting the inactive orbitals to be delocalized. This option, which does not alter the interpretability of the wave function, accounts better for the nonbonding interactions between the fragments and is referred to

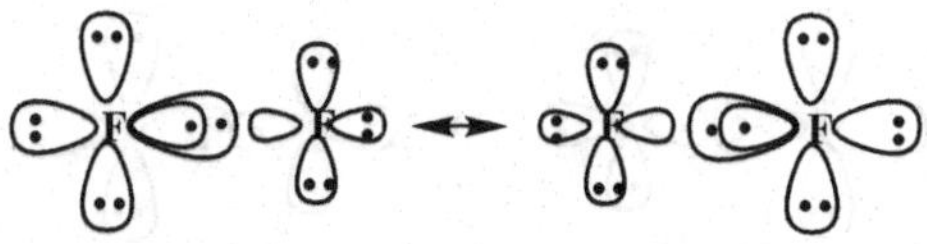

15   16

Scheme 9.4

as D-BOVB. Another improvement can be achieved by incorporating radial electron correlation in the active orbitals of the ionic structures, by allowing the doubly orbitals to split into two singly occupied ones that are spin paired. This splitting is pictorially represented in **15** and **16** (Scheme 9.4), which improve the descriptions of **13** and **14**. This option carries the label "S" (for split), leading to the SL-BOVB and SD-BOVB levels of calculation, the latter being the most accurate one.

The BOVB method has been successfully tested for its ability to reproduce dissociation energies and/or dissociation energy curves, close to the results (or estimated ones) of full CI or to other highly accurate calculations performed with the same basis sets. A variety of two-electron and odd-electron bonds, including difficult test cases as F_2, FH, and F_2^- (38,42), and the H_3M-Cl series (M = C, Si, Ge, Sn, Pb) (39,43,44) were investigated.

It is interesting to interpret the improvement of BOVB with respect to a CASSCF, GVB, or SC calculation of a two-electron bond. These four methods account for all the nondynamic correlation associated with the formation of the bond from separate fragments, but differ in the content of their dynamic correlation. With a medium-sized basis set, say 6–31+G(d), CASSCF, GVB, and SC account for less than one-half of the bonding energy of F_2, 14–16 kcal/mol (38). On the other hand, an SD-BOVB calculation yields a bonding energy of 36.2 kcal/mol[‡] versus an experimental value of 38.2 kcal/mol (39). Thus, the BOVB method brings in just what is missing in the CASSCF calculation, that is, dynamical correlation, not all of it but just *that part that is associated with the bond that is being broken or formed*. In other words, the BOVB method takes care of the differential dynamic correlation.

The importance and physical nature of dynamic correlation is even better appreciated in the case of 3e bonds, a type of bond in which the electron correlation is entirely dynamic, since there is no left-right correlation associated with odd-electron bonds. As noted earlier, the Hartree–Fock and simple VB functions for 3e bonds (hence, GVB, SC, or VBSCF) are nearly equivalent and yield about similar bonding energies. Taking the F_2^- radical anion as an example, it turns out that, compared to the experimental bonding energy of

---

[‡] Note that this bonding energy is overestimated with respect to a full CI calculation in the same basis set, which is estimated as close to 30 kcal/mol. The SD-BOVB systematically overestimates bonding energies relative to full CI, and yields values that are intermediate between full CI and experiment, a "fortunate" systematic error that compensates for basis set insufficiency.

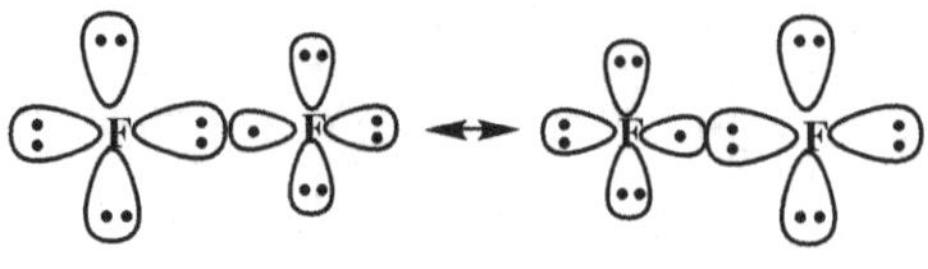

17   18

Scheme 9.5

30.3 kcal/mol, the calculated value is exceedingly small (only 4 kcal/mol), at the spin-unrestricted Hartree–Fock (UHF) level, and even worse at the restricted open-shell Hartree–Fock (ROHF) level. In contrast, the SD-BOVB calculation, which involves only two VB structures (**17, 18**) with breathing orbitals, as in Scheme 9.5, yields an excellent bonding energy of 28.0 kcal/mol.

The BOVB method is implemented in the XMVB package and also in the TURTLE module in GAMESS–UK.

9.3.3 The Valence Bond Configuration Interaction Method

In the same spirit as BOVB, the valence bond configuration interaction (VBCI) method (45,46) aims at retaining the conceptual clarity of the classical VBSCF method, while improving the energetic aspect by introducing further electron correlation. This accuracy-compactness combination is achieved by augmenting the VBSCF calculation with subsequent configuration interaction and then condensing the final CI space into a minimal number of VB structures.

In a first step, all the fundamental VB structures are involved, and a VBSCF wave function is calculated and taken as a zeroth-order wave function.

$$\Psi^{\text{VBSCF}} = \sum_K C_K^{\text{SCF}} \Phi_K^0 \qquad (9.11)$$

As in the original VBSCF method, the occupied orbitals that constitute the Φ_K^0 VB structures may be constrained to be block-localized. The blocks can be defined as atoms, fragments, a pair of atoms, and so on, depending on whether HAOs, BDOs, or other type of orbitals are required for the particular application. Then, for each occupied orbital in the VBSCF wave function, a projector is used to define a set of virtual orbitals, each being strictly localized on precisely the same block as the corresponding occupied orbital. These virtual orbitals are used to create excited VB structures in the following way: Given a fundamental VB structure Φ_K^0, an excited VB structure Φ_K^i is built by replacing one or more occupied orbital(s) by the same number of virtual orbitals located on the same block. This way, the excited VB structure Φ_K^i retains the same electron-pairing pattern and charge distribution as Φ_K^0. In other words, both Φ_K^0 and Φ_K^i describe the same "classical" VB structure. A subsequent VBCI calculation will involve all the fundamental and excited

VB structures in Equation 9.12:

$$\Psi^{\text{VBCI}} = \sum_K \sum_i C_{Ki} \Phi_K^i \qquad (9.12)$$

where the coefficients C_{Ki} and the final energy are determined by solving the secular equations without further orbital optimization.

The CI space can be truncated following the usual CI methodology. The starting point always involves single excitations, that is, VBCIS. This can be followed by VBCISD, VBCISDT, and so on, where D stands for double and T for triple excitations. Practical experience with the method shows that going beyond double excitation is usually not necessary (41). The method was also augmented by second-order perturbation theoretic (PT2) treatment, so that the role of higher excitation can be taken into account; for example, VBCISPT2 accounts for doubles using PT2, and so on.

Since the virtual orbitals conserve the nature of the fundamental VB structures, the entire VBCI wave function can be rewritten in a compact form,

$$\Psi^{\text{VBCI}} = \sum_K C_K^{\text{CI}} \Phi_K^{\text{CI}} \qquad (9.13)$$

where the VB structure Φ_K^{CI} is of the form of Equation 9.14, which collects all the VB functions that belong to the same structure in terms of spin pairing and charge distribution:

$$\Phi_K^{\text{CI}} = \sum_i C'_{Ki} \Phi_K^i \qquad (9.14)$$

VBCI is implemented in the XMVB package.

9.4 METHODS FOR GETTING VALENCE BOND QUANTITIES FROM MOLECULAR ORBITAL-BASED PROCEDURES

This section describes methods that do not follow a straightforward VB manner, but still yield VB-type information.

9.4.1 Using Standard Molecular Orbital Software to Compute Single Valence Bond Structures or Determinants

This technique utilizes an option, offered by most *ab initio* standard programs, to compute the energy of any guess function even if the latter is based on nonorthogonal orbitals. The technique orthogonalizes the orbitals without changing the Slater determinant, and then computes the expectation value of the energy by use of Slater's rules. This expectation value appears as the energy at iteration zero in the course of the SCF procedure of optimizing the Hartree–Fock orbitals. If the guess determinant is made of localized bonding orbitals that correspond to a specific VB structure, then the expectation energy of this wave function at iteration zero defines the energy of this VB structure at the Hartree-Fock level. Practically, the localized bonding orbitals that are used to construct the guess determinant can be determined by any convenient means. A method of this type has been used by Kollmar to construct a reference Kekulé structure for conjugated molecules (47). For example, a Kekulé structure of benzene displays a set of three two-centered π-bonding MOs, which can arise from the Hartree–Fock calculation of an ethylene molecule (47,48). In a VBSCD calculation, the energy of the crossing point will be the energy of a guess function made of the orbitals of the reactants, but in the geometry of the transition state, without further orbital optimization. The zero-iteration technique has also been used to estimate the energy of spin-alternant determinants [the quasiclassical (QC) determinant] (49,50). A sample of inputs for getting the energy of the guess for a Kekulé structure of benzene and its QC determinant is given at the end of this chapter.

9.4.2 The Block-Localized Wave Function and Related Methods

The Block-Localized Wave Function (BLW) method is primarily aimed at evaluating the electronic delocalization and charge-transfer effects in molecules, for the computational cost of a simple Hartree–Fock calculation. It is related to the preceding technique, but unlike the latter, the BLW method performs the calculations to self-consistency, while keeping the orbitals block localized. As this *ab initio* method deals with nonorthogonal orbitals, it can be considered as a VB-type method. The basic principle consists of partitioning the full basis set of orbitals into subsets each centered on a given fragment, which can be an atom or a group of atoms. The MOs are then optimized in a Hartree–Fock way, with the restriction that each orbital is expanded only on its own fragment. The MOs of a given fragment are orthogonal among themselves, but the orbitals of different fragments have finite overlaps. Thus, the block-localized wave function can be constructed so as to represent a diabatic state that can serve as a localized reference against which the energy of the fully delocalized wave function can be compared.

Although the theory behind BLW is more general, a typical application of the method is the energy calculation of a specific resonance structure in the context of resonance theory. As a resonance structure is, by definition, composed of local bonds plus core and lone pairs, a bond between atoms A and B will be represented as a bonding MO strictly localized on the A and B centers, a lone pair will be an AO localized on a single center, and so on. With these restrictions on orbital extension, the SCF solution can be decomposed to coupled Roothaan-like equations, each of which corresponds to a block. The final block-localized wave function is optimized at the constrained Hartree–Fock level and is expressed by a Slater determinant. Consequently, the energy difference between the Hartree–Fock wave function, where all electrons are free to delocalize in the whole system, and the block-localized wave function, where electrons are confined to specific zones of the system, is defined generally as the electron delocalization energy. Further extensions of the BLW method allow us to calculate the electronic coupling energy resulting from the mixing of two, or more, diabatic states (1). The BLW method is implemented in the XMVB and GAMESS–US packages.

The BLW method can be considered as an extension of the orbital deletion procedure (ODP) (51,52), a simpler method that can only be applied to carbocations (52) and boranes (51). The ODP consists of representing a resonance structure displaying an electronic vacancy (Lewis acid character) by deleting the primitive basis functions corresponding to the empty site before launching the SCF calculation. As a typical example, the ODP has been applied to calculate the resonance energy of the allyl cation (52).

A related method, specifically devised for weak interactions, is the SCF–MI method (self-consistent field for molecular interactions) (53,54), which aims at avoiding the basis set superposition error (BSSE) in an *a priori* fashion. Like the preceding method, SCF–MI partitions the full basis set of orbitals into subsets each centered on a given fragment, and optimizes the fragment orbitals at the constrained Hartree–Fock level. As each orbital is expanded only on its own fragment, the calculated interaction between the fragments avoids the BSSE right at the outset. This Hartree–Fock-like zeroth-order wave function may then serve as a basis for either perturbation theory or configuration interaction approaches to account for electron correlation (1). Because of the strictly partitioned nature of the basis set of orbitals, charge transfer is not formally included in the above approach. Therefore, the SCF–MI method is more appropriate for weak interactions rather than for charge- or resonance-assisted hydrogen bonds.

9.5 A VALENCE BOND METHOD WITH POLARIZABLE CONTINUUM MODEL

The valence bond method with polarizable continuum model (VBPCM) method (55) includes solute–solvent interactions in the VB calculations. It uses the same continuum solvation model as the standard PCM model implemented in current *ab initio* quantum chemistry packages, where the solvent is represented as a homogeneous medium, characterized by a dielectric constant, and is polarizable by the charge distribution of the solute. The interaction between the solute charges and the polarized electric field of the solvent is taken into account through an interaction potential that is embedded in the

Hamiltonian and determined by a self-consistent reaction field (SCRF) procedure.

In its actual implementation, the VBPCM method is based on the VBSCF method (see above). Thus, the wave function is expressed in the usual manner as a linear combination of VB structures, Equation 9.8, but now these VB structures are optimized and interacting with one another in the presence of a polarizing field of the solvent, by a self-consistent procedure. Within this model, the interaction between solute and solvent is represented by an interaction potential, V_R, which is treated as a perturbation to the Hamiltonian H° of the solute molecule in vacuum. The Schrödinger equation for the VB wave function now reads

$$(H^\circ + V_R)\Psi^{VBPCM} = E\Psi^{VBPCM} \qquad (9.15)$$

Equation 9.15 is solved iteratively; the interaction potential V_R for the ith iteration is given as a function of electronic density of the $(i - 1)$th iteration and is expressed in the form of one-electronic matrix element that is computed by a standard PCM procedure. The detailed procedures are as follows:

1. A VBSCF procedure in a vacuum is performed (see above), and the electron density is computed.
2. Given the electron density from Step 1, effective one-electron integrals are obtained by a standard PCM subroutine.
3. A standard VBSCF calculation is carried out with the effective one-electron integrals obtained from Step 2. The electron density is computed with the newly optimized VB wave function.
4. Steps 2 and 3 are repeated until the energy difference between the two iterations reaches a given threshold.

By performing the above procedures, the solvent effect is taken into account at the VBSCF level, whereby the orbitals and structural coefficients are optimized till self-consistency is achieved. Like VBSCF, the VBPCM method is suitable for diabatic states, which are calculated with the same solvent field as the one for the adiabatic state. Thus, it has the ability to compute the energy profile of the full state as well as that of individual VB structures throughout the course of a reaction, and in so doing to reveal the individual effects of solvent on the different constituents of the wave function. In this spirit, it has been used to perform a quantitative VBSCD analysis of a reaction that exhibits a marked solvent effect, the S_N2 reaction $Cl^- + CH_3Cl \rightarrow CH_3Cl + Cl^-$ (55).

The VBPCM procedure is not, in principle, restricted to the VBSCF method; it has the potential ability to be implemented to more sophisticated methods like BOVB, VBCI, or other methods. The method is implemented in XMVB.

APPENDIX

9.A.1   SOME AVAILABLE VALENCE BOND PROGRAMS

Other than the GVB method that is now implemented in many packages, here we offer brief descriptions of the main VB softwares that we are aware of and with which we had some experience to varying degrees.

9.A.1.1   The TURTLE Software

TURTLE is a general program (56) that is designed to perform multi-structure VB calculations. It can execute either nonorthogonal CI, or nonorthogonal MCSCF calculations with simultaneous optimization of orbitals and coefficients of VB structures. Complete freedom is given to the user to deal with HAOs, BDOs, or OEOs, so that calculations of the VBSCF, SCVB, BLW, or BOVB can be performed. Currently, TURTLE involves analytical gradients to optimize the energies of individual VB structures or multistructure electronic states with respect to the nuclear coordinates (57). A parallel version has been developed and implemented using the message-passing interface (MPI), for the sake of making the software portable. Because of the structure of the Super-CI optimization method, 99 % of the program could be parallelized (58).

9.A.1.2   The XMVB Program

The XMVB software (59) enables us to perform the VBSCF, SCVB, BLW, BOVB, and VBCI calculations with optimized orbitals defined in any form according to requirements. Particularly, the four "pure" VB methods, VBSCF, BOVB, VBPCM and VBCI, described in Section 9.3 are implemented as standard methods in the package and can be easily performed by specifying appropriate keywords in the input file. In this sense, XMVB is a fairly user-friendly package applicable to a variety of chemical problems, even to small organometallic complexes. The XMVB is a stand-alone program, but for flexibility, it can be interfaced to most QM softwares, for example, GAUSSIAN and GAMESS–US. In addition, it is also feasible to combine XMVB with *ab initio* MO packages to perform hybrid VB method calculations, such as VB-DFT and VBPCM. The parallel version of XMVB, based on the Message Passing Interface, is also available (60). Most recently XMVB has been incorporated also into GAMESS–US (see Chapter 10 for input file for the interfaced XMVB).

9.A.1.3   The CRUNCH Software

The CRUNCH (Computational Resource for Understanding Chemistry) has been written originally in Fortran by Gallup (61), and recently translated into

C (62). This program can perform multiconfiguration VB calculations with fixed orbitals, plus a number of MO-based calculations like RHF, ROHF, UHF (followed by MP2), Orthogonal CI, and MCSCF.

9.A.1.4 The VB2000 Software

VB2000 (63) is an *ab initio* VB package that can be used for performing nonorthogonal CI, multistructure VB with optimized orbitals, as well as SCVB, GVB, and CASVB calculations in the spirit of Hirao's method (24) (see Section 9.2.3).

9.A.2 IMPLEMENTATIONS OF VALENCE BOND METHODS IN STANDARD *AB INITIO* PACKAGES

- XMVB can be used as a stand-alone program that is freely available from the author (email: weiwu@xmu.edu.cn; website: http://ctc.xmu.edu.cn/xmvb/). It can also be used as a plug-in module in GAMESS(US) (64).
- VB2000 can be used as a plug-in module for GAMESS(US) (64) and Gaussian98/03 (65) so that some of the functionalities of GAMESS and Gaussian can be used for calculating VB wave functions. GAMESS also provides interface (option) for the access of VB2000 module. The Windows version of GAMESS (WinGAMESS) has VB2000 module compiled in, that is WinGAMESS is VB2000 ready, although it requires a license to use VB2000 legally.
- TURTLE used to be freely available as a stand-alone program. But now it is implemented in the GAMESS–UK program (66), which has to be purchased and licensed.
- The CASVB method of Thorsteinsson et al. (22) is incorporated in the MOLPRO (67) and in the MOLCAS (68) packages. In addition to the features of the original CASVB method, the CASVB code also permits fully variational VB calculations, which can be single- or multiconfigurational in nature.
- The BLW and SCF–MI methods are implemented in the GAMESS–US package.

REFERENCES

1. P. C. Hiberty, S. Shaik, *J. Comput. Chem.* **28**, 137 (2007). A Survey of Recent Developments of Ab Initio Valence Bond Theory.
2. F. Prosser, S. Hagstrom, *Int. J. Quant. Chem.* **2**, 89 (1968). On the Rapid Computation of Matrix Elements.

3. J. Gerratt, *Adv. At. Mol. Phys.* **7**, 141 (1971). General Theory of Spin-Coupled Wave Functions for Atoms and Molecules.

4. M. Raimondi, W. Campion, M. Karplus, *Mol. Phys.* **34**, 1483 (1977). Convergence of the Valence Bond Calculation for Methane.

5. G. A. Gallup, R. L. Vance, J. R. Collins, J. M. Norbeck, *Adv. Quant. Chem.* **16**, 229 (1982). Practical Valence Bond Calculations.

6. J. Verbeek, J. H. van Lenthe, *J. Mol. Struct. (Theochem)* **229**, 115 (1991). On the Evaluation of Nonorthogonal Matrix Elements.

7. X. Li, Q. Zhang, *Int. J. Quant. Chem.* **36**, 599 (1989). Bonded Tableau Unitary Group Approach to the Many-electron Correlation Problem.

8. R. McWeeny, *Int. J. Quant. Chem.* **34**, 25 (1988). A Spin-Free Form of Valence Bond Theory.

9. J. Li, W. Wu, *Theor. Chim. Acta* **89**, 105 (1994). New Algorithm for Nonorthogonal *Ab Initio* Valence Bond Calculations.1. New Strategy and Basic Expressions.

10. W. Wu, A. Wu, Y. Mo, M. Lin, Q. Zhang, *Int. J. Quant. Chem.* **67**, 287 (1998). Efficient Algorithm for the Spin-Free Valence Bond Theory. I. New Strategy and Primary Expressions.

11. J. Verbeek, Nonorthogonal Orbitals in *Ab Initio* Many-Electron Wavefunctions, PhD Thesis, Utrecht University, 1990.

12. C. A. Coulson, I. Fischer, *Philos. Mag.* **40**, 386 (1949). Notes on the Molecular Orbital Treatment of the Hydrogen Molecule.

13. F. B. Bobrowicz, W. A. Goddard, III, in H. F. Schaefer, Ed., *Methods of Electronic Structure Theory*, Plenum Press, New York, 1977, pp. 79–127.

14. W. A. Goddard, III, T. H. Dunning, Jr., W. J. Hunt, P. J. Hay, *Acc. Chem. Res.* **6**, 368 (1973). Generalized Valence Bond Description of Bonding in Low-Lying States of Molecules.

15. E. A. Carter, W. A. Goddard, *J. Chem. Phys.* **88**, 3132 (1988). Correlation-Consistent Configuration Interaction: Accurate Bond Dissociation Energies from Simple Wave Functions.

16. D. L. Cooper, J. Gerratt, M. Raimondi, in D. J. Klein, N. Trinajstic, eds., *Valence Bond Theory and Chemical Structure*, Elsevier, New York, 1990, pp. 287–349.

17. D. L. Cooper, J. Gerratt, M. Raimondi, in I. Gutman, S. J. Cyvin, Eds., *Top. Curr. Chem.* **153**, 41 (1990). Advances in the Theory of Benzenoid Hydrocarbons.

18. M. Sironi, M. Raimondi, R. Martinazzo, F. A. Gianturco, in *Valence Bond Theory*, D. L. Cooper, Ed., Elsevier, Amsterdam, The Netherlands, 2002, pp. 261–277. Recent Developments of the SCVB Method.

19. F. Penotti, J. Gerratt, D. L. Cooper, M. Raimondi, *J. Mol. Struct. (Theochem)* **169**, 421 (1988). The *Ab Initio* Spin-Coupled Description of Methane: Hybridization Without Preconceptions.

20. D. L. Cooper, J. Gerratt, M. Raimondi, *Nature (London)* **323**, 699 (1986). The Electronic Structure of the Benzene Molecule.

21. E. C. da Silva, J. Gerratt, D. L. Cooper, M. Raimondi, *J. Chem. Phys.* **101**, 3866 (1994). Study of the Electronic States of the Benzene Molecule Using Spin-Coupled Valence Bond Theory.

22. D. L. Cooper, T. Thorsteinsson, J. Gerratt, *Adv. Quant. Chem.* **32**, 51 (1998). Modern VB Representation of CASSCF Wave Functions and the Fully-Variational Optimization of Modern VB Wave Functions Using the CASVB Strategy.

23. D. L. Cooper, P. B. Karadakov, T. Thorsteinsson, in *Valence Bond Theory*, D. L. Cooper, Ed., Elsevier, Amsterdam, The Netherlands, 2002, pp. 41–53.

24. K. Hirao, H. Nakano, K. Nakayama, *Int. J. Quant. Chem.* **66**, 157 (1998). Theoretical Study of the $\pi \rightarrow \pi^*$ Excited States of Linear Polyenes: The Energy Gap Between $1^1B_u^+$ and $2^1A_g^-$ States and their Character.

25. K. Ruedenberg, M. W. Schmidt, M. M. Gilbert, S. T. Elbert, *Chem. Phys.* **71**, 41 (1982). Are Atoms Intrinsic to Molecular Electronic Wavefunctions? I. The FORS Model.

26. P. Lykos, G. W. Pratt, *Rev. Mod. Phys.* **35**, 496 (1963). Discussion on the Hartree-Fock Approximation.

27. A. D. McLean, B. H. Lengsfield, III, J. Pacansky, Y. Ellinger, *J. Chem. Phys.* **83**, 3567 (1985). Symmetry Breaking in Molecular Calculations and the Reliable Prediction of Equilibrium Geometries. The Formyloxyl Radical as an Example.

28. C. F. Jackels, E. R. Davidson, *J. Chem. Phys.* **64**, 2908 (1976). The Two Lowest Energy $^2A'$ States of NO_2.

29. A. F. Voter, W. A. Goddard, III, *Chem. Phys.* **57**, 253 (1981). A Method for Describing Resonance between Generalized Valence Bond Wavefunctions.

30. A. F. Voter, W. A. Goddard, III, *J. Chem. Phys.* **75**, 3638 (1981). The Generalized Resonating Valence Bond Method: Barrier Heights in the HF + D and HCl + D Exchange Reactions.

31. A. F. Voter, W. A. Goddard, III, *J. Am. Chem. Soc.* **108**, 2830 (1986). The Generalized Resonating Valence Bond Description of Cyclobutadiene.

32. J. H. van Lenthe, G. G. Balint-Kurti, *J. Chem. Phys.* **78**, 5699 (1983). The Valence-Bond Self-Consistent Field method (VB–SCF): Theory and Test Calculations.

33. J. H. van Lenthe, J. Verbeek, P. Pulay, *Mol. Phys.* **73**, 1159 (1991). Convergence and Efficiency of the Valence Bond Self-Consistent Field Method.

34. A. Banerjee, F. Grein, *Int. J. Quant. Chem.* **10**, 123 (1976). Convergence Behavior of Some Multiconfiguration Methods.

35. N. O. J. Malcom, J. J. W. McDouall, *J. Comput. Chem.* **15**, 1357 (1994). A Variational Biorthogonal Valence Bond Method.

36. J. J. W. McDouall, in *Valence Bond Theory*, D. L. Cooper, Ed., Elsevier, Amsterdam, 2002, pp. 227–260. The Biorthogonal Valence Bond Method.

37. R. McWeeny, *Int. J. Quant. Chem.* **74**, 87 (1999). An *Ab Initio* Form of Classical Valence Bond Theory.

38. P. C. Hiberty, S. Humbel, C. P. Byrman, J. H. van Lenthe, *J. Chem. Phys.* **101**, 5969 (1994). Compact Valence Bond Functions with Breathing Orbitals: Application to the Bond Dissociation Energies of F_2 and FH.

39. P. C. Hiberty, S. Shaik, *Theor. Chem. Acc.* **108**, 255 (2002). BOVB—A Modern Valence Bond Method That Includes Dynamic Correlation.

40. P. C. Hiberty, in *Modern Electronic Structure Theory and Applications in Organic Chemistry*, E. R. Davidson, Ed., World Scientific, River Edge, NJ, 1997, pp. 289–367. The Breathing Orbital Valence Bond Method.

41. L. Song, W. Wu, P. C. Hiberty, D. Danovich, S. Shaik, *Chem. Eur. J.* **9**, 4540 (2003). An Accurate Barrier for the Hydrogen Exchange Reaction from Valence Bond Theory: Is this Theory Coming of Age?

42. P. C. Hiberty, S. Humbel, P. Archirel, *J. Phys. Chem.* **98**, 11697 (1994). Nature of the Differential Electron Correlation in Three-Electron Bond Dissociations. Efficiency of a Simple Two-Configuration Valence Bond Method with Breathing Orbitals.

43. A. Shurki, P. C. Hiberty, S. Shaik, *J. Am. Chem. Soc.* **121**, 822 (1999). Charge-Shift Bond in Group IVB Halides: A Valence Bond Study of MH_3–Cl (M = C, Si, Ge, Sn, Pb) Molecules.

44. D. Lauvergnat, P. C. Hiberty, D. Danovich, S. Shaik, *J. Phys. Chem.* **100**, 5715 (1996). Comparison of C–Cl and Si–Cl Bonds. A Valence Bond Study.

45. W. Wu, L. Song, Z. Cao, Q. Zhang, S. Shaik, *J. Phys. Chem. A* **106**, 2721 (2002). A Practical Valence Bond Method Incorporating Dynamic Correlation.

46. L. Song, W. Wu, Q. Zhang, S. Shaik, *J. Comput. Chem.* **25**, 472 (2004). A Practical Valence Bond Method: A Configuration Interaction Method Approach with Perturbation Theoretic Facility.

47. H. Kollmar, *J. Am. Chem. Soc.* **101**, 4832 (1979). Direct Calculation of Resonance Energies of Conjugated Hydrocarbons with *ab initio* MO Methods.

48. S. S. Shaik, P. C. Hiberty, J.-M. Lefour, G. Ohanessian, *J. Am. Chem. Soc.* **109**, 363 (1987). Is Delocalization a Driving Force in Chemistry ? Benzene, Allyl Radical, Cyclobutadiene and their Isoelectronic Species.

49. R. Méreau, M. T. Rayez, J. C. Rayez, P. C. Hiberty, *Phys. Chem. Chem. Phys.* **3**, 3650 (2001). Alkoxy Radical Decomposition Explained by a Valence-Bond Model.

50. P. C. Hiberty, D. Danovich, A. Shurki, S. Shaik, *J. Am. Chem. Soc.* **117**, 7760 (1995). Why Does Benzene Possess a D_{6h} Symmetry? A Quasiclassical State Approach for Probing π-Bonding and Delocalization Energies.

51. Y. Mo, J. H. Jiao, P.v.R. Schleyer, *J. Org. Chem.* **69**, 3493 (2004). Hyperconjugation Effect in Substituted Methyl Boranes: An Orbital Deletion Procedure Analysis.

52. Y. Mo, *J. Org. Chem.* **69**, 5563 (2004). Resonance Effect in the Allyl Cation and Anion: A Revisit.

53. A. Famulari, E. Gianinetti, M. Raimondi, M. Sironi, *Int. J. Quant. Chem.* **69**, 151 (1998). Implementation of Gradient-Optimization Algorithms and Force Constant Computations in BSSE-Free Direct and Conventional SCF Approaches.

54. A. Famulari, E. Gianinetti, M. Raimondi, M. Sironi, *Theor. Chem. Acc.* **99**, 358 (1998). Modification of Guest and Saunders Open Shell SCF Equations to Exclude BSSE from Molecular Interaction Calculations.

55. L. Song, W. Wu, Q. Zhang, S. Shaik, *J. Phys. Chem. A* **108**, 6017 (2004). VBPCM: A Valence Bond Method that Incorporates a Polarizable Continuum Model.

56. J. Verbeek, J. H. Langenberg, C. P. Byrman, F. Dijkstra, J. H. van Lenthe, TURTLE—A gradient VB/VBSCF program (1998–2004), Theoretical Chemistry Group, Utrecht University, Utrecht. See van Lenthe, J. H.; Dijkstra, F.; Havenith, R. W. A. in *Valence Bond Theory*, D. L. Cooper, Ed., Elsevier, Amsterdam, The Netherlands, 2002, pp. 79–116.

57. F. Dijkstra, J. H. van Lenthe, *J. Chem. Phys.* **113**, 2100 (2000). Gradients in Valence Bond Theory.

58. F. Dijkstra, J. H. van Lenthe, *J. Comput. Chem.* **22**, 665 (2001). Software News and Updates. Parallel Valence Bond.

59. L. Song, W. Wu, Y. Mo, Q. Zhang, XMVB-01: An *Ab Initio* Non-orthogonal Valence Bond Program, Xiamen University, Xiamen 361005, China, 2003. See the website: http://ctc.xmu.edu.cn/xmvb/

60. L. Song, Y. Mo, Q. Zhang, W. Wu, *J. Comput. Chem.* **26**, 514 (2005). XMVB: A Program for Ab Initio Nonorthogonal Valence Bond Computations.

61. G. A. Gallup, *Valence Bond Methods*, Cambridge University Press, Cambridge, 2002.

62. See the website: http://phy-ggallup.unl.edu/crunch/

63. J. Li, B. Duke, R. McWeeny, VB2000 Version 1.8, SciNet Technologies, San Diego, CA, 2005. For details, see the website: http//www.scinetec.com/.

64. M. W. Schmidt, K. K. Baldridge, J. A. Boatz, S. T. Elbert, M. S. Gordon, J. J. Jensen, S. Koseki, N. Matsunaga, K. A. Nguyen, S. Su, T. L. Windus, M. Dupuis, J. A. Montgomery, *J. Comput. Chem.* **14**, 1347 (1993). See the website of GAMESS. http://www.msg.ameslab.gov/GAMESS/GAMESS.html

65. Gaussian 03, Revision C.02, Frisch, M. J. et al., Gaussian, Inc., Wallingford CT, 2004. See the website: http://www.gaussian.com/

66. GAMESS–UK is a package of ab initio programs written by M. F. Guest, J. H. van Lenthe, J. Kendrick, K. Schöffel, P. Sherwood, R. J. Harrison, with contributions from R. D. Amos, R. J. Buenker, M. Dupuis, N. C. Handy, I. H. Hillier, P. J. Knowles, V. Bonacic-Koutecky, W. von Niessen, V. R. Saunders, A. Stone. The package is derived from the original GAMESS code due to M. Dupuis, D. Spangler, J. Wendoloski, NRCC Software Catalog, Vol. 1, Program No. QG01 (GAMESS), 1980. See M. F. Guest, I. J. Bush, H. J. J. van Dam, P. Sherwood, J. M. H. Thomas, J. H. van Lenthe, R. W. A. Havenith, J. Kendrick, *Mol. Phys.* **103**, 719 (2005). For the latest version, see the website: http://www.cfs.dl.ac.uk/gamess-uk/index.shtml

67. MOLPRO is a package of *ab initio* programs written by H.-J. Werner, P. J. Knowles, M. Schütz, R. Lindh, P. Celani, T. Korona, G. Rauhut, F. R. Manby, R. D. Amos, A. Bernhardsson, A. Berning, D. L. Cooper, M. J. O. Deegan, A. J. Dobbyn, F. Eckert, C. Hampel, G. Hetzer, A. W. Lloyd, S. J. McNicholas, W. Meyer, M. E. Mura, A. Nicklaß, P. Palmieri, R. Pitzer, U. Schumann, H. Stoll, A. J. Stone, R. Tarroni, T. Thorsteinsson. See the website: http://www.molpro.net/

68. MOLCAS Version 5.4, K. Andersson, M. Barysz, A. Bernhardsson, M. R. A. Blomberg, D. L. Cooper, M. P. Fülscher, C. de Graaf, B. A. Hess, G. Karlström, R. Lindh, P.-Å. Malmqvist, T. Nakajima, P. Neogrády, J. Olsen, B. O. Roos, B. Schimmelpfennig, M. Schütz, L. Seijo, L. Serrano-Andrés, P. E. M. Siegbahn, J. Stålring, T. Thorsteinsson, V. Veryazov, P.-O. Widmark, Lund University, Sweden (2002).

Gaussian Input 9.1
These two inputs will yield the energy of a Kekulé structure of benzene.

The first input performs a standard RHF calculation of the ground state of benzene, and saves the MOs in the %chk file:

```
$RunGauss
%chk=benz
# rhf/6-31G test punch=mo
scf=tight

 benzene

 0 1
 C
 C,1,cc
 C,2,cc,1,ccc
 C,3,cc,2,ccc,1,0.,0
 C,4,cc,3,ccc,2,0.,0
 C,5,cc,4,ccc,3,0.,0
 H,2,hc,1,hcc,3,180.,0
 H,3,hc,2,hcc,4,180.,0
 H,4,hc,3,hcc,5,180.,0
 H,5,hc,4,hcc,3,180.,0
 H,6,hc,5,hcc,4,180.,0
 H,1,hc,6,hcc,5,180.,0

 cc=1.400
 ccc=120.
 hc=1.0857
 hcc=120.
```

The second input uses the MOs saved in the %chk file as a guess function, while the π MOs of this guess function are modified so as to represent a Kekulé structure. The new π MOs arise from a separate calculation of ethylene at the same C–C distance as in benzene and are given at the end of the input.

```
$RunGauss
%chk=benz
#p rhf/6-31G test iop(5/22=1) scfcyc=1
iop(4/6=2) iop(5/13=1) guess=(read,cards)

 A Kekule structure of benzene

 0 1
 C
 C,1,cc
 C,2,cc,1,ccc
 C,3,cc,2,ccc,1,0.,0
 C,4,cc,3,ccc,2,0.,0
 C,5,cc,4,ccc,3,0.,0
 H,2,hc,1,hcc,3,180.,0
```

```
H,3,hc,2,hcc,4,180.,0
H,4,hc,3,hcc,5,180.,0
H,5,hc,4,hcc,3,180.,0
H,6,hc,5,hcc,4,180.,0
H,1,hc,6,hcc,5,180.,0


cc=1.400
ccc=120.
hc=1.0857
hcc=120.

(5D15.8)
   17 Alpha MO OE=-0.49679058D+00
0.00000000D+00 0.00000000D+00 0.00000000D+00 0.00000000D+00
0.36986000D+00
0.00000000D+00 0.00000000D+00 0.00000000D+00 0.32135000D+00
0.00000000D+00
0.00000000D+00 0.00000000D+00 0.00000000D+00 0.00000000D+00
0.00000000D+00
0.00000000D+00 0.00000000D+00 0.00000000D+00 0.00000000D+00
0.00000000D+00
0.00000000D+00 0.00000000D+00 0.00000000D+00 0.00000000D+00
0.00000000D+00
0.00000000D+00 0.00000000D+00 0.00000000D+00 0.00000000D+00
0.00000000D+00
0.00000000D+00 0.00000000D+00 0.00000000D+00 0.00000000D+00
0.00000000D+00
0.00000000D+00 0.00000000D+00 0.00000000D+00 0.00000000D+00
0.00000000D+00
0.00000000D+00 0.00000000D+00 0.00000000D+00 0.00000000D+00
0.00000000D+00
0.00000000D+00 0.00000000D+00 0.00000000D+00 0.00000000D+00
0.36986000D+00
0.00000000D+00 0.00000000D+00 0.00000000D+00 0.32135000D+00
0.00000000D+00
0.00000000D+00 0.00000000D+00 0.00000000D+00 0.00000000D+00
0.00000000D+00
0.00000000D+00 0.00000000D+00 0.00000000D+00 0.00000000D+00
0.00000000D+00
0.00000000D+00
   20 Alpha MO OE=-0.33224126D+00
0.00000000D+00 0.00000000D+00 0.00000000D+00 0.00000000D+00
0.00000000D+00
0.00000000D+00 0.00000000D+00 0.00000000D+00 0.00000000D+00
0.00000000D+00
0.00000000D+00 0.00000000D+00 0.00000000D+00 0.36986000D+00
0.00000000D+00
0.00000000D+00 0.00000000D+00 0.32135000D+00 0.00000000D+00
0.00000000D+00
0.00000000D+00 0.00000000D+00 0.36986000D+00 0.00000000D+00
0.00000000D+00
0.00000000D+00 0.32135000D+00 0.00000000D+00 0.00000000D+00
0.00000000D+00
```

```
 0.00000000D+00  0.00000000D+00  0.00000000D+00  0.00000000D+00
 0.00000000D+00
 0.00000000D+00  0.00000000D+00  0.00000000D+00  0.00000000D+00
 0.00000000D+00
 0.00000000D+00  0.00000000D+00  0.00000000D+00  0.00000000D+00
 0.00000000D+00
 0.00000000D+00  0.00000000D+00  0.00000000D+00  0.00000000D+00
 0.00000000D+00
 0.00000000D+00  0.00000000D+00  0.00000000D+00  0.00000000D+00
 0.00000000D+00
 0.00000000D+00  0.00000000D+00  0.00000000D+00  0.00000000D+00
 0.00000000D+00
 0.00000000D+00  0.00000000D+00  0.00000000D+00  0.00000000D+00
 0.00000000D+00
 0.00000000D+00
21 Alpha MO OE=-0.33224126D+00
 0.00000000D+00  0.00000000D+00  0.00000000D+00  0.00000000D+00
 0.00000000D+00
 0.00000000D+00  0.00000000D+00  0.00000000D+00  0.00000000D+00
 0.00000000D+00
 0.00000000D+00  0.00000000D+00  0.00000000D+00  0.00000000D+00
 0.00000000D+00
 0.00000000D+00  0.00000000D+00  0.00000000D+00  0.00000000D+00
 0.00000000D+00
 0.00000000D+00  0.00000000D+00  0.00000000D+00  0.00000000D+00
 0.00000000D+00
 0.00000000D+00  0.00000000D+00  0.00000000D+00  0.00000000D+00
 0.00000000D+00
 0.00000000D+00  0.36986000D+00  0.00000000D+00  0.00000000D+00
 0.00000000D+00
 0.32135000D+00  0.00000000D+00  0.00000000D+00  0.00000000D+00
 0.00000000D+00
 0.36986000D+00  0.00000000D+00  0.00000000D+00  0.00000000D+00
 0.32135000D+00
 0.00000000D+00  0.00000000D+00  0.00000000D+00  0.00000000D+00
 0.00000000D+00
 0.00000000D+00  0.00000000D+00  0.00000000D+00  0.00000000D+00
 0.00000000D+00
 0.00000000D+00  0.00000000D+00  0.00000000D+00  0.00000000D+00
 0.00000000D+00
 0.00000000D+00  0.00000000D+00  0.00000000D+00  0.00000000D+00
 0.00000000D+00
 0.00000000D+00
```

After this second calculation is completed, the energy of the Kekulé structure is read in the output as the RHF energy at iteration zero.

Gaussian Input 9.2

These two inputs will yield the energy of a spin-alternant determinant of benzene.

The first input makes a standard UHF calculation of the ground state of benzene, and saves the $\alpha-$ and $\beta-$ molecular spin-orbitals in the %chk file:

```
$RunGauss
%chk=benz
# uhf/6-31G test punch=mo
scf=tight

benzene

   0 1
C
C,1,cc
C,2,cc,1,ccc
C,3,cc,2,ccc,1,0.,0
C,4,cc,3,ccc,2,0.,0
C,5,cc,4,ccc,3,0.,0
H,2,hc,1,hcc,3,180.,0
H,3,hc,2,hcc,4,180.,0
H,4,hc,3,hcc,5,180.,0
H,5,hc,4,hcc,3,180.,0
H,6,hc,5,hcc,4,180.,0
H,1,hc,6,hcc,5,180.,0

cc=1.400
ccc=120.
hc=1.0857
hcc=120.
```

The second input uses the spin-orbitals saved in the %chk file as a guess function, but the π MOs of this guess function are replaced by one-centered AOs of alternate spins so as to represent a spin-alternant determinant. These AOs are all the same and can be taken, as done here, as the constituting AOs of an ethylenic π MO. Alternatively, the π AOs can arise from a separate calculation of planar CH_3.

```
$RunGauss
%chk=benz
# uhf/6-31G test iop(5/22=1) scfcyc=1
 iop(4/6=2) iop(5/13=1) guess=(read,cards)

Spin-alternant determinant of benzene

   0 1
C
C,1,cc
C,2,cc,1,ccc
C,3,cc,2,ccc,1,0.,0
C,4,cc,3,ccc,2,0.,0
```

Gaussian Intput 9.2, *continued*

```
C,5,cc,4,ccc,3,0.,0
H,2,hc,1,hcc,3,180.,0
H,3,hc,2,hcc,4,180.,0
H,4,hc,3,hcc,5,180.,0
H,5,hc,4,hcc,3,180.,0
H,6,hc,5,hcc,4,180.,0
H,1,hc,6,hcc,5,180.,0

cc=1.400
ccc=120.
hc=1.0857
hcc=120.

(5D15.8)
   17 Alpha MO OE=-0.49679058D+00
 0.00000000D+00  0.00000000D+00  0.00000000D+00  0.00000000D+00
0.36986000D+00
 0.00000000D+00  0.00000000D+00  0.00000000D+00  0.32135000D+00
0.00000000D+00
 0.00000000D+00  0.00000000D+00  0.00000000D+00  0.00000000D+00
0.00000000D+00
 0.00000000D+00  0.00000000D+00  0.00000000D+00  0.00000000D+00
0.00000000D+00
 0.00000000D+00  0.00000000D+00  0.00000000D+00  0.00000000D+00
0.00000000D+00
 0.00000000D+00  0.00000000D+00  0.00000000D+00  0.00000000D+00
0.00000000D+00
 0.00000000D+00  0.00000000D+00  0.00000000D+00  0.00000000D+00
0.00000000D+00
 0.00000000D+00  0.00000000D+00  0.00000000D+00  0.00000000D+00
0.00000000D+00
 0.00000000D+00  0.00000000D+00  0.00000000D+00  0.00000000D+00
0.00000000D+00
 0.00000000D+00  0.00000000D+00  0.00000000D+00  0.00000000D+00
0.00000000D+00
 0.00000000D+00  0.00000000D+00  0.00000000D+00  0.00000000D+00
0.00000000D+00
 0.00000000D+00  0.00000000D+00  0.00000000D+00  0.00000000D+00
0.00000000D+00
 0.00000000D+00  0.00000000D+00  0.00000000D+00  0.00000000D+00
0.00000000D+00
 0.00000000D+00  0.00000000D+00  0.00000000D+00  0.00000000D+00
0.00000000D+00
 0.00000000D+00
   20 Alpha MO OE=-0.33224126D+00
 0.00000000D+00  0.00000000D+00  0.00000000D+00  0.00000000D+00
0.00000000D+00
 0.00000000D+00  0.00000000D+00  0.00000000D+00  0.00000000D+00
0.00000000D+00
 0.00000000D+00  0.00000000D+00  0.00000000D+00  0.00000000D+00
0.00000000D+00
 0.00000000D+00  0.00000000D+00  0.00000000D+00  0.00000000D+00
0.00000000D+00
 0.00000000D+00  0.00000000D+00  0.36986000D+00  0.00000000D+00
0.00000000D+00
 0.00000000D+00  0.32135000D+00  0.00000000D+00  0.00000000D+00
0.00000000D+00
```

```
   0.00000000D+00  0.00000000D+00  0.00000000D+00  0.00000000D+00
0.00000000D+00
   0.00000000D+00  0.00000000D+00  0.00000000D+00  0.00000000D+00
0.00000000D+00
   0.00000000D+00  0.00000000D+00  0.00000000D+00  0.00000000D+00
0.00000000D+00
   0.00000000D+00  0.00000000D+00  0.00000000D+00  0.00000000D+00
0.00000000D+00
   0.00000000D+00  0.00000000D+00  0.00000000D+00  0.00000000D+00
0.00000000D+00
   0.00000000D+00  0.00000000D+00  0.00000000D+00  0.00000000D+00
0.00000000D+00
   0.00000000D+00  0.00000000D+00  0.00000000D+00  0.00000000D+00
0.00000000D+00
   0.00000000D+00
   21 Alpha MO OE=-0.33224126D+00
   0.00000000D+00  0.00000000D+00  0.00000000D+00  0.00000000D+00
0.00000000D+00
   0.00000000D+00  0.00000000D+00  0.00000000D+00  0.00000000D+00
0.00000000D+00
   0.00000000D+00  0.00000000D+00  0.00000000D+00  0.00000000D+00
0.00000000D+00
   0.00000000D+00  0.00000000D+00  0.00000000D+00  0.00000000D+00
0.00000000D+00
   0.00000000D+00  0.00000000D+00  0.00000000D+00  0.00000000D+00
0.00000000D+00
   0.00000000D+00  0.00000000D+00  0.00000000D+00  0.00000000D+00
0.00000000D+00
   0.00000000D+00  0.00000000D+00  0.00000000D+00  0.00000000D+00
0.00000000D+00
   0.00000000D+00  0.00000000D+00  0.00000000D+00  0.00000000D+00
0.00000000D+00
   0.36986000D+00  0.00000000D+00  0.00000000D+00  0.00000000D+00
0.32135000D+00
   0.00000000D+00  0.00000000D+00  0.00000000D+00  0.00000000D+00
0.00000000D+00
   0.00000000D+00  0.00000000D+00  0.00000000D+00  0.00000008D+00
0.00000000D+00
   0.00000000D+00  0.00000000D+00  0.00000000D+00  0.00000000D+00
0.00000000D+00
   0.00000000D+00  0.00000000D+00  0.00000000D+00  0.00000000D+00
0.00000000D+00
   0.00000000D+00


   17 Beta MO OE=-0.49679058D+00
   0.00000000D+00  0.00000000D+00  0.00000000D+00  0.00000000D+00
0.00000000D+00
   0.00000000D+00  0.00000000D+00  0.00000000D+00  0.00000000D+00
0.00000000D+00
   0.00000000D+00  0.00000000D+00  0.00000000D+00  0.36986000D+00
0.00000000D+00
   0.00000000D+00  0.00000000D+00  0.32135000D+00  0.00000000D+00
0.00000000D+00
   0.00000000D+00  0.00000000D+00  0.00000000D+00  0.00000000D+00
0.00000000D+00
```

Gaussian Intput 9.2, *continued*

```
 0.00000000D+00  0.00000000D+00  0.00000000D+00  0.00000000D+00
 0.00000000D+00
 0.00000000D+00  0.00000000D+00  0.00000000D+00  0.00000000D+00
 0.00000000D+00
 0.00000000D+00  0.00000000D+00  0.00000000D+00  0.00000000D+00
 0.00000000D+00
 0.00000000D+00  0.00000000D+00  0.00000000D+00  0.00000000D+00
 0.00000000D+00
 0.00000000D+00  0.00000000D+00  0.00000000D+00  0.00000000D+00
 0.00000000D+00
 0.00000000D+00  0.00000000D+00  0.00000000D+00  0.00000000D+00
 0.00000000D+00
 0.00000000D+00  0.00000000D+00  0.00000000D+00  0.00000000D+00
 0.00000000D+00
 0.00000000D+00  0.00000000D+00  0.00000000D+00  0.00000000D+00
 0.00000000D+00
 0.00000000D+00
    20 Beta MO OE=-0.33224126D+00
 0.00000000D+00  0.00000000D+00  0.00000000D+00  0.00000000D+00
 0.00000000D+00
 0.00000000D+00  0.00000000D+00  0.00000000D+00  0.00000000D+00
 0.00000000D+00
 0.00000000D+00  0.00000000D+00  0.00000000D+00  0.00000000D+00
 0.00000000D+00
 0.00000000D+00  0.00000000D+00  0.00000000D+00  0.00000000D+00
 0.00000000D+00
 0.00000000D+00  0.00000000D+00  0.00000000D+00  0.00000000D+00
 0.00000000D+00
 0.00000000D+00  0.00000000D+00  0.00000000D+00  0.00000000D+00
 0.00000000D+00
 0.00000000D+00  0.36986000D+00  0.00000000D+00  0.00000000D+00
 0.00000000D+00
 0.32135000D+00  0.00000000D+00  0.00000000D+00  0.00000000D+00
 0.00000000D+00
 0.00000000D+00  0.00000000D+00  0.00000000D+00  0.00000000D+00
 0.00000000D+00
 0.00000000D+00  0.00000000D+00  0.00000000D+00  0.00000000D+00
 0.00000000D+00
 0.00000000D+00  0.00000000D+00  0.00000000D+00  0.00000000D+00
 0.00000000D+00
 0.00000000D+00  0.00000000D+00  0.00000000D+00  0.00000000D+00
 0.00000000D+00
 0.00000000D+00  0.00000000D+00  0.00000000D+00  0.00000000D+00
 0.00000000D+00
 0.00000000D+00
    21 Beta MO OE=-0.33224126D+00
 0.00000000D+00  0.00000000D+00  0.00000000D+00  0.00000000D+00
 0.00000000D+00
 0.00000000D+00  0.00000000D+00  0.00000000D+00  0.00000000D+00
 0.00000000D+00
 0.00000000D+00  0.00000000D+00  0.00000000D+00  0.00000000D+00
 0.00000000D+00
 0.00000000D+00  0.00000000D+00  0.00000000D+00  0.00000000D+00
 0.00000000D+00
 0.00000000D+00  0.00000000D+00  0.00000000D+00  0.00000000D+00
 0.00000000D+00
```

```
0.00000000D+00  0.00000000D+00  0.00000000D+00  0.00000000D+00
0.00000000D+00
0.00000000D+00  0.00000000D+00  0.00000000D+00  0.00000000D+00
0.00000000D+00
0.00000000D+00  0.00000000D+00  0.00000000D+00  0.00000000D+00
0.00000000D+00
0.00000000D+00  0.00000000D+00  0.00000000D+00  0.00000000D+00
0.00000000D+00
0.00000000D+00  0.00000000D+00  0.00000000D+00  0.00000000D+00
0.36986000D+00
0.00000000D+00  0.00000000D+00  0.00000000D+00  0.32135000D+00
0.00000000D+00
0.00000000D+00  0.00000000D+00  0.00000000D+00  0.00000000D+00
0.00000000D+00
0.00000000D+00  0.00000000D+00  0.00000000D+00  0.00000000D+00
0.00000000D+00
0.00000000D+00
```

After this second calculation is completed, the energy of the spin-alternant determinant is read in the output as the UHF energy at iteration zero.

Reference to GAUSSIAN 98: Frisch, M. J.; Trucks, G. W.; Schlegel, H. B.; Scuseria, G. E.; Robb, M. A.; Cheeseman, J. R.; Zakrzewski, V. G.; Montgomery, J. A. Jr.; Stratmann, R. E.; Burant, J. C.; Dapprich, S.; Millam, J. M.; Daniels, A. D.; Kudin, K. N.; Strain, M. C.; Farkas, O.; Tomasi, J.; Barone, V.; Cossi, M.; Cammi, R.; Mennucci, B.; Pomelli, C.; Adamo, C.; Clifford, S.; Ochterski, J.; Petersson, G. A.; Ayala, P. Y.; Cui, Q.; Morokuma, K.; Malick, D. K.; Rabuck, A. D.; Raghavachari, K.; Foresman, J. B.; Cioslowski, J.; Ortiz, J. V.; Baboul, A. G.; Stefanov, B. B.; Liu, G.; Liashenko, A.; Piskorz, P.; Komaromi, I.; Gomperts, R.; Martin, R. L.; Fox, D. J.; Keith, T.; Al-Laham, M. A.; Peng, C. Y.; Nanayakkara, A.; Gonzalez, C.; Challacombe, M.; Gill, P. M. W.; Johnson, B. G.; Chen, W.; Wong, M. W.; Andres, J. L.; Head-Gordon, M.; Replogle, E. S.; Pople, J. A., Gaussian, Inc.: Pittsburgh, PA, 1998.

10 Do Your Own Valence Bond Calculations—A Practical Guide

10.1 INTRODUCTION

As noted in previous chapters, the reemergence of VB theory is characterized, among other things, by the development of a growing number of *ab initio* methods that can be applied to chemical problems of interest. This chapter is intended to guide the reader in the use of some of these VB methods in order to run meaningful calculations. To that aim, we plan to calculate the energy and wave function of a simple molecule, F_2, at various computational levels, we will also have a look at the shape of the orbitals and assess the accuracy of the various methods by calculating the bond dissociation energy (BDE) of the molecule. The reason for this choice is that the BDE of F_2 is known to be a difficult test case; the Hartree–Fock method finds the molecule to be unbound (negative BDE), many of the higher level methods are quantitatively inaccurate and dynamic electron correlation appears to be particularly important. The know-how established here for this example may be useful for more complex tasks like generation of a VBSCD for chemical reactions. We will comment on such calculations as well.

10.2 WAVE FUNCTIONS AND ENERGIES FOR THE GROUND STATE OF F_2

All the calculations of F_2 are carried out with a simple basis set of double-zeta polarization type, the standard 6-31G(d) basis set, and are performed at a fixed interatomic distance of 1.44 Å, which is approximately the optimized distance for a full CI calculation in this basis set. Only the σ bond is described in a VB fashion, and the corresponding orbitals are referred to as the "active orbitals", while the orbitals representing the lone pairs, so-called "spectator orbitals", remain doubly occupied in all calculations. A common point to the various VB methods we use, except the VBCI method, is that at the dissociation limit, the methods converge to two $F^{\bullet}$ fragments at the restricted-open-shell Hartree–Fock (ROHF) level.

A Chemist's Guide to Valence Bond Theory, by Sason Shaik and Philippe C. Hiberty
Copyright © 2008 John Wiley & Sons, Inc.

10.2.1   GVB, SC, and VBSCF Methods

As discussed in Chapter 9, the GVB and SC methods describe the F−F bond with a single formally covalent structure that links two orbitals of the Coulson−Fischer type (overlap-enhanced orbitals, OEOs). On the other hand, one can use purely localized orbitals as in the VBSCF method and describe the bond by three VB structures, one covalent and two ionic $(F_a\bullet-\bullet F_b, \; F_a^- F_b^+, \; F_a^+ F_b^-)$. The VB calculations will adjust the proportions of the covalent versus ionic structures as a balance between the tendency to minimize the Coulomb repulsion between the electrons as in the covalent structure, on one hand, and on the other, to maximize the covalent−ionic resonance interaction by incorporating the ionic structures. This adjustment is called nondynamic correlation or left−right correlation. The GVB, SC, and VBSCF methods take care of much of this nondynamical correlation energy of the valence electrons. In MO-based terms, this quantity is defined as the difference between the Hartree−Fock energy and the energy of a CASSCF calculation involving all valence MOs and all valence electrons. In the case of F_2, the corresponding valence−CASSCF involves 14 electrons, 7 occupied MOs and 1 virtual MO, and is referred to as CAS(14,8) in standard program packages.

GVB Calculations of F_2   The GVB method is available in most standard *ab initio* program packages. In the case at hand, only the σ bond will be described in a GVB way, while the other valence orbitals, which represent the six lone pairs, will form a core of doubly occupied MOs. Such a calculation is exactly equivalent to a CAS(2,2) calculation involving two electrons and two MOs, namely, the σ bonding orbital (ϕ) and the corresponding antibonding one (ϕ*), also called natural orbitals (see Section 9.2.1). The calculation is fairly automatic, and one has only to ascertain that the correct occupied and virtual MOs are placed in the active space. This is usually done using an appropriate choice of the guess function, which is illustrated in Input 10.1. Thus, as can be seen, the guess function arising from a former Hartree−Fock calculation is rearranged by switching the MOs 5 and 9, so that the σ-bonding and σ-antibonding valence MOs become the frontier orbitals. Then, the keyword "CAS(2,2)" would automatically select the highest lying MO and the lowest vacant one to construct the active space.

The GVB results are normally presented in the form of an MO−CI wave function (recall Section 9.2.1), that is a linear combination of MO-based configurations constructed with natural orbitals (NOs), as in Equation 10.1:

$$\Psi_{GVB} = 0.96142|\psi\bar{\psi}\phi\bar{\phi}| - 0.27507|\psi\bar{\psi}\phi^*\bar{\phi^*}| \qquad (10.1)$$

Here, ψ represents the core of doubly occupied MOs, while ϕ and ϕ* are the NOs, that are displayed in the two right-hand columns in Table 10.1, which is extracted from the corresponding output.

TABLE 10.1   Active Orbitals for the GVB Description of the σ Bond of F_2, in the Natural Orbital (NO) and in the GVB Pair Representations

Basis Functions[a]	Natural Orbitals		GVB Pair	
	ϕ	ϕ^*	φ_a	φ_b
F_a 2S	0.03461	−0.08824	−0.0111	0.07214
2PX	0.00000	0.00000	0.00000	0.00000
2PY	0.00000	0.00000	0.00000	0.00000
2PZ	0.45756	0.60997	0.69116	0.11577
3S	0.08514	−0.15729	0.00089	0.14926
3PX	0.00000	0.00000	0.00000	0.00000
3PY	0.00000	0.00000	0.00000	0.00000
3PZ	0.28298	0.30254	0.39222	0.10683
4D0	−0.02891	0.01722	−0.01737	−0.03361
4D+1	0.00000	0.00000	0.00000	0.00000
4D−1	0.00000	0.00000	0.00000	0.00000
4D+2	0.00000	0.00000	0.00000	0.00000
4D−2	0.00000	0.00000	0.00000	0.00000
F_b 2S	0.03461	0.08824	0.07214	−0.0111
2PX	0.00000	0.00000	0.00000	0.00000
2PY	0.00000	0.00000	0.00000	0.00000
2PZ	−0.45756	0.60997	−0.11577	−0.69116
3S	0.08514	0.15729	0.14926	0.00089
3PX	0.00000	0.00000	0.00000	0.00000
3PY	0.00000	0.00000	0.00000	0.00000
3PZ	−0.28298	0.30254	−0.10683	−0.39222
4D0	−0.02891	−0.01722	−0.03361	−0.01737
4D+1	0.00000	0.00000	0.00000	0.00000
4D−1	0.00000	0.00000	0.00000	0.00000
4D+2	0.00000	0.00000	0.00000	0.00000
4D−2	0.00000	0.00000	0.00000	0.00000

[a]The 2S and 2P basis functions are more compact than the 3S and 3P ones.

The GVB wave function can also be expressed in its VB form, Equation 10.2:

$$\Psi_{GVB} = \left|\psi\bar{\psi}\varphi_a\overline{\varphi_b}\right| - \left|\psi\bar{\psi}\overline{\varphi_a}\varphi_b\right| \tag{10.2a}$$

$$\varphi_a = z_a + \varepsilon z_b' \tag{10.2b}$$

$$\varphi_a = z_b + \varepsilon z_a' \tag{10.2c}$$

The orbitals φ_a and φ_b are the OEOs, which are seen in the equation to be coupled in a singlet manner and are called a "GVB pair". The z_a, z_a', and z_b, z_b' are the pure AOs located, respectively, on atoms F_a and F_b. These φ_a and φ_b orbitals are given in the last two columns in Table 10.1. It can be seen that each OEO is mostly concentrated on a single center, with the delocalization on the other center being minor, albeit nonnegligible.

This "GVB pair" is provided by some program packages, but not in others. When the GVB pair is not provided, the transformation from NOs to GVB pairs is readily done through Equation 9.4. Note that in Equation 10.2, the primed atomic orbitals z_a' and z_b' are not exactly equivalent to the unprimed ones, z_a and z_b, in terms of diffuseness and hybridization. This nonequivalence comes from our use of a nonminimal basis set (being of double-zeta + polarization type), and is related to the shape of the NOs. Indeed, it can be seen in Table 10.1 that the AOs that constitute the bonding combination ϕ are less hybridized and more diffuse than the AOs of ϕ^*. This latter feature can be appreciated by looking at the 3PZ/2PZ ratio, 0.62 in the bonding combination versus only 0.49 in the antibonding one. This difference in AO shapes will be shown below to have a slight energy lowering effect.

As long as the lone pairs are kept in the form of a core of doubly occupied orbitals, the SC wave function of F_2 is practically similar to the GVB one. The only difference is that the OEOs that describe the σ bond are not restricted any more to be orthogonal to the other orbitals. Nevertheless, the energies of the GVB and SC wave functions are identical up to the fifth digit.

The BDEs provided by the various VB methods are displayed in Table 10.2. Inspection of the first four entries shows that the HF method has a large negative BDE, while the CAS and GVB methods lead to positive BDEs. The full valence-CASSCF energy, in the second entry, is lower than

TABLE 10.2 Total Energies and Bond Dissociation Energies (BDEs) for $F_2{}^a$

Entry	Method	Total Energy[b]	BDE[c]
1	HF	−198.66642	−33.9
2	CAS(14,8)	−198.75016	18.6
3	CAS(2,2)	−198.74601	16.0
4	GVB	−198.74601	16.0
5	L-VBSCF	−198.73599	9.7
6	D-VBSCF	−198.74423	14.9
7	L-BOVB	−198.76763	29.6
8	SL-BOVB	−198.77088	31.6
9	π-D-BOVB	−198.77092	31.6
10	π-SD-BOVB	−198.77685	35.3
11	L-VBCISD[d]	−198.93918	31.3
12	π-D-VBCISD[d]	−198.94316	33.8

[a] A fixed interatomic distance of 1.44 Å is taken for F_2. The 6-31G(d) basis set is used throughout.
[b] Absolute energy in hartrees. $E[F\bullet$ (ROHF) $\times 2] = -198.72052$.
[c] Bond dissociation energy, in kcal/mol.
[d] Energies including Davidson's correction. The VBCISD reference energy for the atoms separated by 20 Å is -198.88925 hartrees.

the GVB energy of F_2, as expected since the CASSCF wave function involves 10 symmetry-adapted configurations, nine of which represent excitations from the lone pairs to the antibonding σ MO. However, these configurations have a rather weak stabilizing effect, so that the BDE calculated with GVB energy is only 2.6 kcal/mol smaller than the CASSCF value. Still, in both methods the BDE is no more than 18.6 kcal/mol. Thus, although posing a great improvement compared with HF, still both the CASSCF and the GVB methods greatly underestimate the BDE.

VBSCF Calculations of F_2 A VBSCF calculation can easily be carried out with the TURTLE or XMVB programs, by simply specifying the VB structures that have to compose the wave function. The degree of localization–delocalization is specified by providing, for each orbital of the VB structures, a list of basis functions on which this orbital is allowed to expand. If all the orbitals, including the active ones (σ bond) or the lone pairs, are defined as single-centered hybrid atomic orbitals (HAOs), the respective calculation is referred to as L-VBSCF (L standing for "localized"). An example of a complete procedure for running an L-VBSCF calculation is shown in Input 10.2, which is taken from the stand-alone XMVB package. In order not to confuse the reader hereafter we show only XMVB inputs, since this program is freely available to users and is user-friendly. As can be seen (from the third step of the input) L-VBSCF involves three VB structures, one covalent and two ionics, and these structures are constructed from a unique set of valence HAOs, labeled 1–8. The latter are in turn defined over the basis functions of a single atom.

In principle, an L-VBSCF calculation for F_2, explicitly involving one covalent and two ionic VB structures, should yield an energy and a BDE close to those of a GVB wave function, since the same amount of electron correlation is taken into account. However, while the two methods would indeed be exactly equivalent in minimal basis set, here it can be seen in Table 10.2 (Entries 4 vs. 5) that L-VBSCF leads to a higher energy than GVB and to a poorer BDE. One reason for this difference is the freedom given to the GVB method to delocalize the spectator orbitals, here the lone pairs. It is, however, very easy to apply the same freedom with the VBSCF method, by letting the spectator orbitals delocalize while keeping the active orbital localized, as shown in Input 10.3. This extra degree of freedom given to the wave function does not alter the interpretation in terms of VB structures, and generally results in some energy lowering, here 5.2 kcal/mol. As seen in Table 10.2, this level of calculation, referred to as D-VBSCF, yields a total energy value very close to the GVB one and a very close BDE value. The remaining difference is due to a subtle effect that appears if one expands the GVB wave function Ψ_{GVB} (Eq. 10.2) in AO-based determinants, as in Equation 10.3:

$$\Psi_{GVB} = |\psi\bar{\psi}(z_a\bar{z}_b - \bar{z}_a z_b)| + \varepsilon^2|\psi\bar{\psi}(z_a'\bar{z}_b' - \bar{z}_a'z_b')|$$
$$+\varepsilon|\psi\bar{\psi}(z_a\bar{z}_a' - \bar{z}_a'z_a)| + \varepsilon|\psi\bar{\psi}(z_b\bar{z}_b' - \bar{z}_b'z_b)| \tag{10.3}$$

The last two terms represent the ionic structures, in which the two active electrons are placed in singly occupied orbitals ($z_a \neq z_a'$ and $z_b \neq z_b'$) that are coupled to a singlet. In contrast, the ionic structures in the VBSCF calculation are closed shell, and as can be seen in Equation 10.4 the expression for Ψ_{VBSCF} is nothing else but Equation 10.3 with $z_a = z_a'$ and $z_b = z_b'$.

$$\Psi_{\mathrm{VBSCF}} = C_1 \left| \psi \overline{\psi} (z_a \overline{z_b} - \overline{z_a} z_b) \right| + C_2 \left| \psi \overline{\psi} (z_a \overline{z_a} + z_b \overline{z_b}) \right| \qquad (10.4)$$

It follows that the GVB calculation takes into account some radial correlation, not present at the VBSCF level, hence the higher energy of the latter. Of course, this is not a limitation of the VBSCF method, because one could have started with ionic structures that match the ones embedded in the GVB wave function.

10.2.2   The BOVB Method

To run a BOVB calculation smoothly, it is advisable to start from an appropriate guess function, which may be, for example, a preliminary VBSCF wave function. In the XMVB program, the BOVB procedure sets up automatically by coding the keyword "bovb" in the input. As with the VBSCF method, the spectator orbitals in the BOVB method may be defined as either localized or delocalized, resulting in the L- and D-BOVB methods, respectively.

L-BOVB calculations of F$_2$ First, let us consider an L-BOVB calculation of the F_2 molecule using the familiar 3-structure VB wave function, where all orbitals are single-centered HAOs. However, now each VB structure has its own specific set of orbitals, as in Equation 10.5:

$$\Psi_{\mathrm{L-BOVB}} = C_1 \left| \psi \overline{\psi} (z_a^\bullet \overline{z_b^\bullet} - \overline{z_a^\bullet} z_b^\bullet) \right| + C_2 \left| \psi' \overline{\psi'} z_a^- \overline{z_a^-} \right| + C_2 \left| \psi'' \overline{\psi''} z_b^- \overline{z_b^-} \right| \qquad (10.5)$$

where ψ, ψ', and ψ'' stand for the cores of doubly occupied spectator orbitals, which are different for the three structures. The active orbitals are labeled with the superscripts, which signify the local charge of the fragment, such that $\bullet$, $+$, and $-$, respectively, signify a neutral, cationic, or anionic fluorine atom. Practically, the most straightforward way to do an L-BOVB calculation is to modify Input 10.2 by simply replacing the keyword "vbscf" by "bovb". This user-friendly option generates, in fact, a hidden input that is shown as Input 10.4, which illustrates how one sets up the calculations and requires different orbitals for different structures. Thus, orbitals 1–8 belong to the covalent structure $F\bullet{-}\bullet F$, 9–15 belong to F^-F^+, and 16–22 to F^+F^-. The different numbering of the orbitals ensures that the resulting VB wave function will have different orbitals for different structures. Of course, the L-BOVB calculation could also be performed by using Input 10.4 directly. A fragment of the output of the L-BOVB calculation is Output 10.1, showing the Hamiltonian matrix, the

TABLE 10.3 Some Optimized L-BOVB Orbitals for the Ground State of F_2

Basis Functions	Active orbitals[a]		π Spectator orbitals[a]		
	$z_a^{\bullet}$	z_a^{-}	$x_a^{\bullet}$	x_a^{-}	x_a^{+}
F_a 2S	0.05396	0.06635	0.00000	0.00000	0.00000
2PX	0.00000	0.00000	0.68144	0.62039	0.76344
2PY	0.00000	0.00000	0.00000	0.00000	0.00000
2PZ	0.73371	0.68133	0.00000	0.00000	0.00000
3S	−0.01888	−0.03624	0.00000	0.00000	0.00000
3PX	0.00000	0.00000	0.46671	0.53281	0.36863
3PY	0.00000	0.00000	0.00000	0.00000	0.00000
3PZ	0.40379	0.46105	0.00000	0.00000	0.00000
4D0	0.02530	0.08397	0.00000	0.00000	0.00000
4D+1	0.00000	0.00000	0.00978	0.03158	−0.01083
4D−1	0.00000	0.00000	0.00000	0.00000	0.00000
4D+2	0.00000	0.00000	0.00000	0.00000	0.00000
4D−2	0.00000	0.00000	0.00000	0.00000	0.00000

[a]The superscripts designate the nuclear charge of the fragment that bears the orbital: neutral ($\bullet$), anionic ($-$) or cationic ($+$).

overlap matrix between the VB structures, as well as their coefficients and weights. It can be seen that we are still dealing with 3×3 matrices and a 3-structure wave function, despite the use of 22 VB orbitals. As such, the VB information at the L-BOVB level is simple as in L-VBSCF.

How different are these BOVB orbitals from each other? Table 10.3 displays some of these orbitals, chosen from the active set ($z_a^{\bullet}$, z_a^{-}) or from the spectator set (e.g., $x_a^{\bullet}$, x_a^{+} and x_a^{-}, that represent lone pairs of π symmetries in the xz plane). It can be seen (e.g., from the coefficients of the 3P vs. 2P basis functions) that all the orbitals that are associated with an anionic fragment (z_a^{-}, x_a^{-}) are more diffuse than those belonging to a neutral fragment ($z_a^{\bullet}$, $x_a^{\bullet}$). Similarly, the spectator-orbital x_a^{+} of a cationic fragment is more contracted than the other spectator orbitals. These variations in the orbital size make good chemical sense, thereby demonstrating that BOVB describes the ionic VB structures in a balanced way, relative to the covalent structure, by contrast to the VBSCF level that forces the ionic structures to populate orbitals that are optimized for an average neutral situation. This better description is reflected in the significantly lower L-BOVB energy compared with the valence-CASSCF limit (cf. Entries 2 and 7 in Table 10.2). Since these two computational levels converge to the same dissociation limit, that is, twice the ROHF energy of an $F^{\bullet}$ radical, the L-BOVB value of the BDE of F_2 is highly improved compared with the full valence−CASSCF value.

SL-BOVB Calculations of F_2   The L-BOVB wave function can be further improved by incorporating radial electron correlation for the active electrons of the ionic structures. This is achieved by allowing the doubly occupied active orbitals to split into two singly occupied ones that are spin-paired, much as in the ionic structures in the GVB wave function as expanded in Equation 10.3 (see also Section 9.3.2). The corresponding calculation is referred to as SL-BOVB (S for split, L for localized orbitals).

The improved electron correlation allows the active electrons of the ionic structures to better avoid one another by locating into orbitals of different sizes, that is, one electron in a compact orbital z_a^c and the other in a more diffuse one z_a^d, and vice versa. The wave function now expresses as in Equation 10.6:

$$\Psi_{SL-BOVB} = C_1 \left| \psi\overline{\psi} \left(z_a^{\cdot}\overline{z_b^{\cdot}} - \overline{z_a^{\cdot}}z_b^{\cdot} \right) \right| + C_2 \left| \psi'\overline{\psi'} \left(z_a^c\overline{z_a^d} - \overline{z_a^c}z_a^d \right) \right|$$
$$+ C_2 \left| \psi''\overline{\psi''} \left(z_b^c\overline{z_b^d} - \overline{z_b^c}z_b^d \right) \right| \tag{10.6}$$

As seen from Table 10.2 (Entries 8 vs. 7), the improvement of SL-BOVB compared with L-BOVB amounts to 2 kcal/mol. Practically, the SL-BOVB wave function can be calculated directly in the form of Equation 10.6, but an alternative way may be preferred, for the sake of a faster convergence. Thus, instead of using the open-shell formulation of the orbital splitting, the calculations describe each ionic structure as a combination of two closed-shell structures, as in Equation 10.7.

$$\Psi_{SL-BOVB} = C_1 \left| \psi\overline{\psi} \left(z_a^{\cdot}\overline{z_b^{\cdot}} - \overline{z_a^{\cdot}}z_b^{\cdot} \right) \right| + C_2^0 \left| \psi'\overline{\psi'} z_a^0\overline{z_a^0} \right| + C_2^* \left| \psi'\overline{\psi'} z_a^*\overline{z_a^*} \right|$$
$$+ C_2^0 \left| \psi''\overline{\psi''} z_b^0\overline{z_b^0} \right| + C_2^* \left| \psi''\overline{\psi''} z_b^*\overline{z_b^*} \right| \tag{10.7}$$

After orbital optimization, the orbitals z_a^* and z_b^* display a node, while the z_a^0 and z_b^0 orbitals are nodeless, like z_a^- and z_b^- in Equation 10.5. Note that the two sets of spectator orbitals, ψ' or ψ'', are common to two ionic structures. The SL-BOVB calculation is described in Input 10.5, which differs from Input 10.4 by having two supplementary ionic structures (see "nstruct" keyword) and two more orbitals. The active orbitals 23 and 24 correspond to z_a^* and z_b^*, and it is noteworthy that the ionic structures 2 and 4 (3 and 5) share the same set of spectator orbitals, respectively, 9–14 (16–21).

The wave functions in Equations 10.6 and 10.7 are exactly equivalent, and the transformation from one to the other is achieved with the simple formulas below, which are reminiscent of the transformation between the two forms of a GVB wave function (GVB pair or natural orbitals):

$$z_a^c = N(z_a^0 + \lambda z_a^*), \qquad z_a^d = N'(z_a^0 - \lambda z_a^*), \qquad \lambda^2 = -\frac{C_2^*}{C_2^0} \tag{10.8}$$

where N and N' are normalization constants, which are of no importance since the program renormalizes the orbitals automatically. The practical advantage of the formulation in Equation 10.7 is that the guess function is very easy to construct from the orbitals calculated at the L-BOVB level. To start the calculation, z_a^0 can be taken as z_a^-, while z_a^* will be the same orbital with a node between the two most important basis functions. By experience, the convergence of the 5-structure formulation (Eq. 10.7) is always very fast, while formulation (Eq. 10.6) may be slower.

D-BOVB and SD-BOVB Calculations of F$_2$ Like VBSCF, the BOVB method is improved by allowing the spectator orbitals to be delocalized in the so-called D-BOVB or SD-BOVB levels. As long as the spectator orbitals can be distinguished from the active ones by symmetry, which is the case of the lone pairs of π symmetry in the F$_2$ molecule, delocalizing these orbitals is easily done: The user has only to specify which basis functions the spectator orbitals are allowed to be made of, much the same as in VBSCF. This is shown in Input 10.6 for a calculation of D-BOVB type, in which the π spectator orbitals are allowed to delocalize.

The results of these calculations are shown in Entries 9 and 10 in Table 10.2, where the methods are also signified with the letter π to stress that only the π-type lone-pair orbitals undergo delocalization. It is seen that the delocalization improves the BDE by 2 kcal/mol. The final BDE value of 35.3 kcal/mol is close to experimental value (38.3 kcal/mol) and at par with the estimated value for a full CI calculation with the same basis set.

An additional 1 kcal/mol or so could be further gained by letting the 2s lone pairs delocalize, but currently this cannot be done in the BOVB framework as easily as with the VBSCF method, for symmetry reasons. As will be explained in detail below, the localized nature of the active orbital *is a requirement* for a valid BOVB calculation. Thus allowing the 2s spectator orbitals (having the same symmetry as the bond orbitals) to delocalize in the course of orbital optimization might lead to a switch between the 2s lone pairs and the active orbitals. This might result in an effective delocalization of the active orbitals. The solution to avoid this problem and still delocalize the 2s orbitals is a two-step procedure that is detailed in the appendix. This refinement is, however, not really necessary, since already at the π-SD-BOVB level, the calculated BDE of 35.3 kcal/mol constitutes a significant improvement with respect to the valence–CASSCF level, while keeping the simplicity of a 3-structure VB wave function.

Avoidance of Nonphysical Results in BOVB Calculations   Of course, it would be tempting to further lower the BOVB energies by allowing the active orbitals to delocalize freely over the whole molecule. This, however, *should never be done,* as the procedure would end with severely biased results. The following cases are important to keep in mind:

1. ***VBSCF with Covalent and Ionic Structures and Delocalized Active Orbitals:*** We recall that delocalizing the active orbitals in a unique formally covalent structure is one way of introducing the ionic structures in an implicit way, as done in GVB. On the other hand, explicitly introducing the ionic structures

defined with localized orbitals, as is done in BOVB, is another way of accounting for bond ionicity effects. Therefore, explicitly introducing the ionic structures and letting the orbitals delocalize necessarily causes redundancy (double counting effect). This redundancy of information has no serious consequences at the VBSCF level. For example, a 1-structure VBSCF calculation that uses delocalized active orbitals (OEOs) in a single formally covalent VB structure (much in the GVB way) yields exactly the same energy as a 3-structure calculation that uses the same type of orbitals and includes one covalent and two ionic structures. However, while the energy is not affected, the *3-structure wave function based on delocalized orbitals is rather unreadable and, moreover, the structural weights are arbitrary, making this type of calculation senseless.*

2. *BOVB with Covalent and Ionic Structures and Delocalized Active Orbitals:* A BOVB calculation that includes covalent and ionic structures explicitly and uses delocalized active orbitals is not only useless, but also nonphysical and misleading. In such a calculation, all the three structures would assume delocalized orbitals, and each of the structures would resemble some kind of a GVB wave function. In such a case, the freedom of each VB structure to have orbitals different from those of the other structures is used by the variational process to correlate the electrons of the spectator orbitals, while this extra correlation is not present in the separate fragments. As such, the calculated BDE increases steeply due to the imbalanced treatment of the molecule and its fragments, rather than to a variational improvement of the method. To demonstrate this imbalance, we performed a computational test on the BDE of F_2. Starting from an L-BOVB wave function and letting the active orbitals delocalize leads to a BDE jump from 29.6 to a nonphysical value of 51.5 kcal/mol, way too large relative to experiment. Worse, the artifact associated with the delocalization of the active orbitals is even more apparent if one starts from the SD-BOVB level and allows the active orbitals to be delocalized. Now the BDE reaches a value of 102 kcal/mol, more than 60 kcal/mol higher than the experimental BDE (38.3 kcal/mol). Thus, *using strictly localized active orbitals is a fundamental condition of validity for BOVB calculations that includes the spectator orbitals in the breathing orbital set.*

10.2.3 The VBCI Method

As discussed in Chapter 9, the VBCI method provides results that are at par with the BOVB method, the difference being that the electrons of the spectator orbitals are correlated too in the VBCI method. The wave function starts from a VBSCF wave function and augments it with subsequent local configuration interaction that can be restricted to single excitations (VBCIS level), or single and double excitations (VBCISD), or higher excitations. Here, we will consider only the VBCISD level, which is a good compromise between accuracy and cost efficiency.

The VBCI method is implemented in the XMVB program and is fairly straightforward. This is done in exactly the same way as in a VBSCF or BOVB calculation. Thus, for a VBCISD calculation in which all the orbitals are requested to be localized on their respective fragment, the input will simply be Input 10.2 in which the keyword "vbscf" is replaced by "vbcisd". This calculation will be referred to as L-VBCISD. Of course, it is also possible to delocalize the π-spectator orbitals as has been done above in the BOVB framework, which is accomplished by replacing "bovb" by "vbcisd" in Input 10.6. This latter level is referred to as "π-D-VBCISD" in Table 10.2.

The VBCI calculations require a guess function; in each case this guess is the corresponding VBSCF wave function, obtained in a preliminary calculation. For example, for L-VBCISD one uses an L-VBSCF guess, while the guess function for a π-D-VBCISD calculation must come from a VBSCF calculation of π-D-VBSCF type, and so on. The updated XMVB version also automatically calculates the Davidson correction for the VBCISD wave function. The results of some VBCISD calculations are displayed in Table 10.2, Entries 11 and 12. It is apparent that the absolute energies are lower than those of the BOVB levels, as expected since all electrons are correlated. Note that the reference for the calculation of the bonding energy, that is, the energy of the two fluorine atoms at infinite separation, cannot be taken as twice the energy of a single fluorine atom as calculated at the VBCISD level, since it is known that CISD methods, in general, are not size consistent. Note, however, that unlike the MO-based CISD method, the VBCISD has very small size inconsistency that is easily removed by the Davidson correction. Nevertheless, to avoid any size inconsistency, the VBCISD energy of the reference structure is calculated here using an elongated F..F system (e.g., with a distance of 20 Å). It can be seen that the VBCISD BDEs are close to the BOVB ones, and that in both cases, delocalizing the π spectator orbitals increases the BDE by $\sim$2 kcal/mol.

In the above examples, the VBCI calculations include 549 and 1089 configurations, respectively, at the L-VBCISD and π-D-VBCISD levels. However, note that the VBCI outputs also provide the VB information in condensed form, in terms of the three fundamental VB structures. This is shown in Output 10.2 for the L-VBCISD calculation, which displays a 3×3 Hamiltonian matrix, the corresponding overlap matrix between the fundamental VB structures (which are defined according to Eq. 9.14) and the weights of these structures.

10.3 VALENCE BOND CALCULATIONS OF DIABATIC STATES AND RESONANCE ENERGIES

One of the most valuable features of theoretical methods based on classical VB structures is their ability to calculate the energy of a diabatic state. Contrary to adiabatic states, a diabatic state is not an eigenfunction of the Hamiltonian. Such a state can be a single VB structure, separate VB curves of covalent and ionic structures, or a Lewis–VB curve in the VBSCD of a chemical reaction (see Chapter 6). Diabatic states have many applications (e.g., in dynamics), but their major impact is conceptual, since their use allows one to quantify important concepts of organic chemistry, such as resonance energy, covalency and ionicity of a bond. The following section defines diabatic states and provides some guidelines for their calculations by means of VB theory.

10.3.1   Definition of Diabatic States

The concept of a diabatic state has different definitions. Strictly speaking, a basis of diabatic states (ϑ, ϑ'...) should be such that Equation 10.9 is satisfied for any variation ∂Q of the geometrical coordinates (Q).

$$< \vartheta|\partial/\partial Q|\vartheta' >= 0 \qquad (10.9)$$

However, it is impossible to fulfill this condition in the general case with more than one geometric degree of freedom. Therefore, one has to search for a compromise in the form of a function whose physical meaning remains as constant as possible along a reaction coordinate. In this sense, a single VB structure, that keeps the same bonding scheme irrespective of the geometry of the system, is the choice definition for a general diabatic state (1). For example, if we consider the F_2 molecule in the VBSCF or BOVB framework, the ground state (made of three VB structures) will be an eigenfunction of the Hamiltonian (i.e., an adiabatic state), while the three VB structures, respectively $F\bullet–\bullet F$, $F^+ F^-$, and $F^- F^+$, will be the diabatic states. Generally, however, a diabatic state can also be a mixture of VB structures that represent a given bonding scheme. For example, in the radical displacement reaction $A^\bullet + B{-}C \rightarrow A{-}B + C^\bullet$, one diabatic state could be the bonding scheme of the reactants, $A^\bullet B{-}C$, while the other would represent the products, $A{-}B\ C^\bullet$. In this case, each diabatic state would be made of three VB structures, respectively $A^\bullet B\bullet{-}\bullet C$, $A^\bullet B^+ C^-$, and $A^\bullet B^- C^+$ for the reactant-like diabatic state (corresponding to the covalent and two ionic components of B–C bond), and $A\bullet{-}\bullet B\ C^\bullet$, $A^+ B^- C^\bullet$, and $A^- B^+ C^\bullet$ for the product-like one. Such diabatic states constitute the crossing curves of the VB state correlation diagrams (VBSCDs).

10.3.2   Calculations of Meaningful Diabatic States

Having defined a diabatic state as a unique VB structure, or more generally as a linear combination of *a subset* of VB structures leading to a specific bonding scheme, the question is now: How do we calculate such a state in a meaningful way?

The Nonvariational Method (Method I)   An initial possibility is to keep the same orbitals that optimize the adiabatic state for the diabatic state; something that seems simple and appealing. In practice, this would be done as follows:

F•––•F	F⁻ F⁺	F⁺ F⁻
–65.717367	24.257687	24.258857
24.257687	–65.278968	–5.907975
24.258857	–5.907975	–65.278966

Scheme 10.1

Once the orbitals have been determined at the end of a VBSCF or BOVB orbital optimization process, the program constructs a Hamiltonian matrix in the space of the VB structures. The adiabatic energies are calculated by diagonalization of the Hamiltonian matrix (see, e.g., Outputs 10.1 and 10.2). The energies of the diabatic states are just the respective diagonal matrix elements of the Hamiltonian matrix. An example of a Hamiltonian matrix, corresponding to an L-BOVB calculation of F_2, is displayed in Scheme 10.1. Thus, were we interested in the energy of the covalent structure alone, we would have taken the first diagonal element of this matrix. In turn, to get a value of the resonance energy between this covalent structure and the two ionic structures, we would simply take the energy difference between the diabatic (the diagonal covalent structure) and adiabatic states. This sounds straightforward, but there is a caveat.

The main problem with this diabatization procedure is that this does not guarantee the best possible orbitals for the diabatic states, except for cases where one uses a minimal basis set. Indeed, the BOVB orbitals are optimized so as to minimize the energy of the multistructure ground state, and are therefore the best compromise between the need to lower the energies of the individual VB structures and to maximize the resonance energy between all the VB structures. This latter requirement implies that the final orbitals are not the best possible orbitals to minimize each of the individual VB structures taken separately. It follows that the diabatic states calculated in this way will very often possess very high energy and are not recommended.

The Quasi-variational method (Method II)   An alternative approach, which we recommend, consists of optimizing each diabatic state separately, in an independent calculation. Consequently, the resulting orbitals of the diabatic states are different from those of the adiabatic states, and each diabatic state possesses its best possible set of orbitals. The diabatic energies are obviously lower compared with those obtained by the previous method, and are therefore quasi-variational. The diabatic energies of the covalent and ionic structures of F_2, calculated with Methods I and II in the L-BOVB framework, are shown in Table 10.4. It is seen that the ionic structures have much lower energies in the quasi-variational procedure, and as such, the procedure can serve a basis for deriving quantities such as resonance energies (see below).

It might be argued that, in the limit of an infinite basis set, there would be so many and so diverse polarization functions that the optimized orbitals could not be considered to be localized anymore. Thus the diabatic state in the

TABLE 10.4   L-BOVB/6-31G(d) Calculated Energies of the Ground State and Diabatic States of the F_2 Molecule[a]

Entry	State	Absolute Energy[b]	Relative Energy[c]
1	Ground state	−198.76763	0
2	F•−•F (Method I)[d]	−198.65609	70.0
3	F•−•F (Method II)[e]	−198.67608	57.4
4	F^+ F^- (Method I)[d]	−198.21769	345.1
5	F^+ F^- (Method II)[e]	−198.27027	312.1

[a]Calculated in 6−31G(d) basis set, with an F−F distance of 1.44 Å.
[b]Absolute energy in hartrees.
[c]Energy relative to the ground state, in kcal/mol.
[d]Diagonal element of the 3-structure Hamiltonian matrix.
[e]Calculated as an optimized single VB structure, independently of the ground state.

quasi-variational procedure would converge to the ground state rather than to a specific VB structure. In practice, however, one finds that the desired quantities, such as resonance energies, are much less prone to basis set dependencies in the quasi-variational procedure (2). Thus, for example, a few tests on several molecules (protonated formyl fluoride, protonated formic acid and protonated formamide) showed that, as long as standard basis sets are used, for example, 6-31G(d) to 6-311G(2d,f), basis set dependency of diabatic state remains marginal (3).

10.3.3   Resonance Energies

Many molecules are represented as a set of resonating structures. This is also the case for all transition states of elementary reactions, which are represented in a VBSCD as a superposition of two VB structures, one corresponding to the reactants, the other to the products (see Chapter 6). The resonance energy (the *B* parameter in a VBSCD) characterizes the stabilization arising from the mixing of the two structures. Another resonance energy of interest is the energy resulting from the mixing between the covalent VB structure and the ionic ones in a single bond, so-called "covalent−ionic resonance energy" (RE_{CS}).

Calculations of RE$_{CS}$ Although the chemical community is accustomed to the idea of covalent−ionic superposition in the description of two-electron bonding, the corresponding RE_{CS} quantities have never been determined. An exception was Pauling's attempt to define the RE_{CS} of heteronuclear bonds. Thus, for an A−X bond, Pauling used the average of the BDEs of the homonuclear bonds (A−A and X−X) to quantify the "covalent bond energy" of the A−X bond. The difference compared with the actual bond energy of A−X was defined by Pauling as the covalent−ionic resonance energy of the heteronuclear bond. In Pauling's scheme, all homonuclear bonds have $RE_{CS} = 0$ per definition, which is incorrect as will be seen immediately.

Thus, an experimentally based method for determining the RE_{CS} is still highly desirable. Recently, it was shown that RE_{CS} for hydrogen halides can be quantified from the differences of the reaction barriers for H-atom transfer between halogens vis-à-vis halogen atom transfer between hydrogen atoms (4). Therefore, there is much interest to quantify these resonance energies for a variety of bonds, including homonuclear ones (e.g., in F_2).

In the general case, the resonance energy for a molecule is defined as the difference between the energy of the ground state and the energy of its most stable VB structure (the reference structure). In the F_2 case, the covalent–ionic resonance energy is defined as in Equation 10.10:

$$RE_{CS} = E(F\bullet-\bullet F) - E(F-F) \tag{10.10}$$

where the last term of the equation corresponds to the energy of the 3-structure ground state, and the first one is the energy of the covalent structure. As just discussed, here we have a choice between two methods (Methods I and II) to calculate the energy of the covalent structure. It can be seen from Entries 2 vs. 1 and Entries 3 vs. 1 in Table 10.4 that the LBOVB-calculated covalent–ionic resonance energy of F_2 amounts to 70.0 kcal/mol with Method I, and to 57.4 kcal/mol with Method II. As expected, Method II provides a smaller value than Method I. Recall, however, that the RE_{CS} quantity has a very clear meaning; it characterizes the contribution of the two ionic structures to the bond energy. As such, the only unique determination of RE_{CS} is by reference to a quasi-variational covalent structure with the *best possible* orbitals in its wave function, precisely as provided by Method II.

Thus, considering that the BDE is only 38 kcal/mol, the RE_{CS} value of 57.4 kcal/mol for F_2 may appear as surprisingly large. In fact, $F-F$ is an extreme case for a homonuclear bond, and is the prototype of a specific category of bonds that have been termed "charge-shift bonds" where the bonding is not contributed by either the covalent or the ionic structures, but rather by the resonance mixing of the two structures. There are many CS bonds and their properties have a variety of chemical consequences (2,4). Traditional covalent bonds exist too, but their RE_{CS} quantities are small (e.g., 11.7 kcal/mol for the $H-H$ bond), whereas the major bonding energy is contributed by the covalent structure itself (2). Clearly, Pauling's assumption that $RE_{CS} = 0$ for homonuclear bonds is far from being true. Another issue concerning the calculations of RE_{CS} is the temptation to allow the covalent structure to delocalize the active orbitals. As repeatedly shown throughout this book, delocalization of the active orbitals in a formally covalent wave function (as in GVB) implicitly introduces the ionic structures (see, e.g., Eq. 10.3). Therefore, in such a case the calculation of RE_{CS} becomes meaningless. Accordingly, the general method that we recommend for calculating RE_{CS} quantities in the VBSCF or BOVB frameworks consists of separate optimizations of the ground state and of the major VB structure (the one that has the largest weight in the wave function), while using strictly localized orbitals.

Comments on the Calculations of RE of Conjugated Molecules   The resonance energies (RE) of molecules like benzene are extremely important and continue to preoccupy chemical thought in parts of the chemical community, and to generate controversies. The RE of a molecule like benzene is well defined as the difference between the energy of a single Kekulé structure and the fully delocalized state of benzene. If we focus on the vertical RE (VRE), this quantity is calculated as the energy difference between benzene and its Kekulé structure in a given geometry that corresponds to the equilibrium structure of benzene, with identical C-C bond lengths. If we are interested in the adiabatic RE (ARE), we should calculate the energy difference between benzene and a Kekulé structure in its own optimized geometry (alternating C−C and C=C bonds of different lengths). Similar considerations apply to the calculations of REs for any conjugated molecule.

The main problem is the meaningful calculation of the Kekulé structure, since in addition to the covalent structures there are plenty of ionic structures that contribute to the double bond in the Kekulé structure. Of course, one can carry out a quasi-variational calculation (Method II type) of the covalent structure combined with the subset of ionic structures that contribute to the Kekulé structure. The resulting energy can be compared to that of the ground state involving the complete set of VB structures; this, however, is tedious. A more economical and much easier way consists of calculating the RE by using orbitals of Coulson−Fischer (CF) type. Thus, a single Kekulé structure will display three bonds, which will be described by means of bond distorted orbitals (BDOs) coupled in a pairwise manner. Such BDOs are orbitals of the CF type, which restrict the delocalization tails of the orbitals to the two atoms that are bonded in the Kekulé structure (see Input 10.7). As shown repeatedly, this type of local GVB wave function implicitly involves the two ionic structures that contribute to the bond. In a second calculation, one can calculate benzene by including only covalent VB structures and use CF orbitals that can have delocalization tails *on all the carbon atoms*. This unrestricted calculation is equivalent to the use of the five covalent structures (two Kekulé and three Dewar type) and the 170 ionic structures of benzene. In fact, it can be verified in Table 10.5 that including the three Dewar structures is not even necessary, as their effect can very well be retrieved by the effect of orbital delocalization if one only includes the two Kekulé structures (cf. Entries 2 and 3 in Table 10.5). The input for the two-structure calculation of the ground state of benzene is displayed in Input 10.8.

One can apply this method to other conjugated systems as well. In this manner, the resonance energy is the difference between the variational energies of the full state and the reference VB structure. As such, the resonance energy itself is variational. Tests of these variational resonance energies show that they reproduce experimentally determined values, for example, for benzene and cyclobutadiene (5).

TABLE 10.5 Energies of a Single Kekulé Structure and of the Ground State of Benzene[a]

Entry	State	Absolute Energy[b]	Relative Energy[c]
1	Single Kekulé structure	−230.65837	0
2	2-Structure ground state[d]	−230.76765	−68.6[d]
3	5-Structure ground state[e]	−230.76796	−68.8[e]

[a]Calculated by a GVB-type method in 6-31G(d) basis set, with a uniform R(C-C) distance of 1.40 Å.
[b]Absolute energy in hartrees.
[c]Energy relative to the Kekulé structure, in kcal/mol.
[d]Wave function involving the two Kekulé structures. RE = 68.6 kcal/mol.
[e]Wave function involving the two Kekulé structures and the three Dewar structures. RE = 68.8 kcal/mol.

10.4 COMMENTS ON CALCULATIONS OF VBSCDS AND VBCMDS

One of key developments of modern VB methods is the ability to compute barriers for elementary reactions (6,7), sometimes with high accuracy (6). Equally important is the current capability to analyze these barriers using the VB diagrams, VBSCD or VBCMD, described in Chapter 6, and to compute reactivity quantities like the promotion gap, G, and the resonance energy of the TS, B. These are multi-layered calculations in which both the adiabatic and diabatic curves are calculated variationally.

VBSCD Calculations Let us exemplify a VBSCD calculation for the hydrogen-exchange reaction by reference to Fig. 10.1. As long as one uses a moderate basis set, for example, up to 6-31++G**, one can simply calculate each of the curves in the VBSCD using a variational procedure within the subset of VB structures that define reactants or products. Thus, Ψ_R is computed using a variational energy calculation of the covalent and two ionic VB structures (**1–3**), describing the bonding in H• + H−H', while Ψ_P is that variational curve made from the covalent and ionic structures (**4–6**) describing the bonding in the product, H−H/•H'. A more economical calculation will generate the two curves using a single VB structure for each diabatic curve, with BDOs that are allowed to delocalize only across the two bonded atoms.

Once the curves are calculated, one has the promotion energy, G, and the height of the crossing point, ΔE_c, which defines the f factor, $f = \Delta E_c/G$. In a separate calculation, one uses the entire VB structure set, (**1–8**), and calculates the adiabatic state, which is the curve in bold in Fig. 10.1. These calculations provide the energy barrier and the resonance energy of the TS. For this reaction, it was shown that the VBCISD//cc-pVTZ leads to an accurate barrier (10 kcal/mol), and B and G values that match semiempirical estimates.

There are two caveats in these calculations, but none of them affects the calculations of the crossing point or of B, but both concern the ability to calculate G in fully variational procedures.

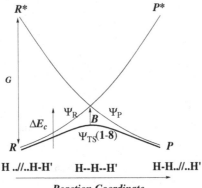

FIGURE 10.1 A VBSCD for the exchange reaction H• + H−H' → H−H + H•'. The parameter Ψ_R is a variational combination of VB structures (**1−3**), while Ψ_P is made from (**4−6**). The adiabatic bold curve is a variational combination of all structures, (**1−8**).

The first caveat concerns variational collapse of the excited states in the VBSCD. Thus, since the excited state (e.g., R^* in Fig. 10.1), for H-abstraction is not a spectroscopic state, then using large basis sets would tend to mimic the corresponding ground state, and R^* will tend to collapse to R. To avoid this, one has to freeze the orbital of the unbound fragment, while allowing the orbitals of the bound fragment to be optimized (6). Thus, for example, in the calculations of the Ψ_R curve, this entails keeping the orbital of the left-hand side H• as in the free fragment while allowing the orbitals in the H−H' fragment (the bonded one) to be optimized during the quasi-variational procedure. For the Ψ_P curve, the H•' is kept with frozen orbital while the H−H fragment is allowed to undergo orbital optimization. In this manner, one generates the entire $R \rightarrow R^*$ and $P \rightarrow P^*$ curves, as well as the corresponding G value in good agreement with its estimated value (e.g., $G = 0.75\Delta E_{ST}$).

The second caveat concerns the limitation of the quasi-variational procedure to produce an entire diabatic curve when there are low lying excited states that cut through it. For example, in the case of H-abstraction by an electronegative atom, or in S_N2 reactions (7), the ionic structure of the bond lies below the R^* and P^* image states in the VBSCD. Therefore, past the crossing point of the curves, instead of correlating to the image states, the curve will collapse to the ionic structure, which is lower lying. Thus, in an S_N2 reaction, for example, $Cl^- + CH_3-Cl' \rightarrow Cl-CH_3 + Cl'^-$, the reactant curve made from the covalent $(Cl^-/H_3C\bullet-\bullet Cl')$ and ionic $(Cl^-\ H_3C^+\ Cl'^-)$ structures that describe the right-hand side C–Cl' bond will initially contain a mixture of the two structures as it should, but at a very long distance between the two bonded fragments it will collapse to the triple-ionic structure, instead of correlating to the corresponding charge-transfer state. There is no good remedy for this limitation of the quasi-variational procedure. In such a case, one simply calculates the curves up to their crossing point, and then finds G by directly calculating the excited state of the VBSCD [in this case, $Cl'^\bullet//(H_3C \therefore Cl')^-$].

VBCMD Calculations In many cases, the most lucid insight into reactivity is provided by calculating the many-curve VB diagram, called VBCMD. For example, the hydrogen-abstraction reaction from hydrogen-halide $(X\bullet + H-X)$ has a much lower barrier than the corresponding halogen abstraction $(H\bullet + X-H)$, even though in both reactions one breaks and makes the exact same bond $(H-X)$ (4). The qualitative VBCMD, in Fig. 10.2, gives a straightforward answer to this trend. Thus, it is apparent that in the halogen abstraction reaction, the ionic structures are destabilized in the region of the TS and thereby raise the barrier.

The calculation of the VBCMD is straightforward. One separately computes the covalent structures, then the ionic structures; all in variational procedures that optimize the orbitals for the individual structures. In this manner, the answer is clear and is usually unique. Of course, with very large basis sets, the VBCMD will suffer from the caveats mentioned above for single VB structures.

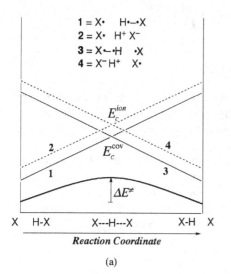

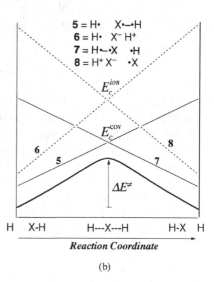

(a)   (b)

FIGURE 10.2 Qualitative VBCMDs for $F\bullet + H-F$ (a), and $H\bullet + F-H$ (b).

For a given basis set, one has to ascertain that the VBCMD energy ordering preserves the ordering of the structures, as reflected from their relative weights in the adiabatic calculations.

APPENDIX

10.A.1   CALCULATING AT THE SD–BOVB LEVEL IN LOW SYMMETRY CASES

This chapter considered molecules with high enough symmetry that can assist the distinction between active and inactive orbitals. Such facility is not always present in the general case, and this poses a danger that during the BOVB orbital optimization there will occur some flipping between the sets of active and inactive orbitals. This, however, depends on the BOVB level.

The simplest level, L-BOVB, presents no particular practical problem and is an automatic process in all cases. Using the more accurate SL-BOVB level, merely requires one to check that the orbital being split (in an ionic structure) is indeed an active orbital, and does not end up belonging to the inactive space after the optimization process. While this condition is generally met by the choice of an appropriate guess function in high symmetry cases (e.g., F_2), there is no guarantee that, in the general case, the unwanted exchange between the active and inactive spaces will not actually take place. For example, in a low symmetry molecule, such as hydrazine (Scheme 10.2), the orbital optimization may lead to **1** instead of the correct structure **2**.

1   **2**

Scheme 10.2

To circumvent this difficulty, a general procedure was developed. After the L-BOVB step, the orbitals are initially subject to localization[1] using any standard localization procedure, and then the active orbitals are split while the inactive ones are kept frozen during the optimization process.

Delocalization of the inactive orbitals (D-BOVB or SD-BOVB) is important for getting accurate energetics. Once again, it is important to make sure that the orbitals that are delocalized are the inactive ones, while the active set remains purely localized, which is the basic tenet of the BOVB method. To avoid a spurious exchange between the active and inactive spaces during the

---

[1]*This requires prior orthogonalization of the orbitals within each fragment.*

orbital optimization process, it is possible to start from an L-BOVB or SL-BOVB wave function, then allowing delocalization of the inactive orbitals while, this time, freezing, the *active* orbitals during the subsequent optimization process that leads to the D-BOVB or SD-BOVB levels, respectively.

REFERENCES

1. T. F. O'Malley, *Adv. At. Mol. Phys.* **7**, 223 (1971). Diabatic States of Molecules—Quasistationary Electronic States.
2. S. Shaik, D. Danovich, B. Silvi, D. Lauvergnat, P. C. Hiberty, *Chem. Eur. J.* **11**, 6358 (2005). Charge-Shift Bonding—A Class of Electron-Pair Bonds that Emerges from Valence Bond Theory and Is Supported by the Electron Localization Function Approach.
3. B. Braïda, *Faraday Discuss.* **135**, 367 (2007). Chemical Concepts from Quantum Mechanics.
4. P. C. Hiberty, C. Mégret, L. Song, W. Wu, S. Shaik, *J. Am. Chem. Soc.* **128**, 2836 (2006). Barriers of Hydrogen vs. Halogen Exchange: An Experimental Manifestation of Charge-Shift Bonding.
5. S. Shaik, A. Shurki, D. Danovich, P. C. Hiberty, *Chem. Rev.* **101**, 1501 (2001). A Different Story of π-Delocalization—The Distortivity of π-Electrons and Its Chemical Manifestations.
6. L. Song, W. Wu, P. C. Hiberty, D. Danovich, S. Shaik, *Chem. Eur. J.* **9**, 4540 (2003). An Accurate Barrier for the Hydrogen Exchange Reaction from Valence Bond Theory: Is this Theory Coming of Age?
7. L. Song, W. Wu, P. C. Hiberty, S. Shaik, *Chem. Eur. J.* **12**, 7458 (2006). The Identity S_N2 Reaction $X^- + CH_3X \rightarrow XCH_3 + X^-$ in Vacuum and in Aqueous Solution: A Valence Bond Study.

Gaussian Input 10.1. GVB Calculations of F_2

This input[a] will yield the CAS(2,2) or GVB energy and wave function of F_2. The calculation uses a guess function from a previous Hartree–Fock calculation that has been saved in the %chk file. In this guess, the σ-bonding and σ-antibonding MOs are, respectively, the fifth and tenth ones. This guess is altered in the following input by the keyword "alter", and the alteration consists of switching the MOs 9 and 5, so that the two MOs that are to be included in the active space are moved to become the frontier orbitals.

```
$RunGauss
%chk=f2-cas2
#P CAS(2,2)/6-31G(d) 5d test gfprint gfinput  pop=full
scfcyc=1000 scf=tight punch=mo guess=(check,alter)

f2

0 1
 f
 f 1 r

 r  1.44

 5 9
```

[a]Gaussian 03, Revision C.02, Frisch, M. J. et al., Gaussian, Inc., Wallingford CT, 2004. See the website: http://www.gaussian.com/

XMVB[b] Input 10.2. L-VBSCF and L-BOVB Calculations of F_2

This is the input for the stand-alone version of XMVB. The new release of GAMESS–US will include XMVB and the input instructions will vary accordingly, as shown in the end of the inputs section.

The present input describes a complete procedure to perform an L-VBSCF calculation on F_2 (or L-BVOB by simply changing the keyword VBSCF to BOVB–see in the end of the input). The calculation involves three steps that constitute a continuous stream of instructions. Here the steps are separated for the sake of clarity.The first step is a simple Hartree–Fock calculation (herein by means of Gaussian98), which is necessary to get the atomic integrals.

```
g98<< $$
%chk=f2
%int=f2
%rwf=f2
# hf/6-31G* test 5d nosym noraff  scf=nodirect scfcyc=100
 gfinput iop(6/7=3)

F2

0 1
f
f     1 1.44

$$
```

The second step specifies which orbitals are frozen and which ones will be optimized in the VB calculation. After the file specification (three first lines), the first line shows that there are 28 basis functions. The second line specifies that 2 MOs arising from the Hartree–Fock calculation will be frozen during the VB calculation, and that among the 28 basis functions, only 26 will be kept as the basis on which the VB orbitals will be expanded. The third and fourth lines indicate, respectively, the MOs that are frozen, and the basis functions that are kept for the variational procedure. In the present case, the MOs that are frozen correspond to the 1s core, and the 1s basis functions are eliminated from the VB orbitals.

```
echo "'f2.rwf'
'f2.int'
'f2.chk'
28
2   26
1 2
 2 3 4 5 6 7 8 9 10 11 12 13 14 16 17 18 19 20 21 22 23 24 25 26 27 28
" >ff.bfi
g98int ff.bfi
intprep ff.bfi
```

The third step is the VB calculation itself. Hereafter, we only consider the 14 valence electrons and the 26 basis functions that are kept for the definition of the VB orbitals, in the order 2S, 2PX, 2PY, 2PZ, 3S, 3PX, and so on, as in the gaussian program (see Table 10.1). There are eight VB orbitals, the definitions of which are specified in the section "$orb". In this section, the first line indicates the number of basis functions over which each VB orbital is being expanded. These basis functions are specified in the following lines, for each VB orbital. The VB orbital 1 can be expanded on the basis functions 1, 4, 5, 8, 9, that is, the 2S, 2PZ, 3S, 3PZ, and D0 basis functions of the F_a atom. The second VB orbital can span the basis functions 2PX, 3PX and D+1 on F_a, it is therefore an orbital of π type, and so on. Note that the numbering of the basis functions, in this step, corresponds to the *basis functions that are kept* for the VB calculations. This numbering is therefore different from the one that appears in the Hartree−Fock output.

The VB structures are specified in section "$struct". In the first line, the VB orbitals 1−6 are doubly occupied, while 7 and 8 are singly occupied and singlet coupled. As 7 and 8 are VB orbitals of σ type, each localized on their respective fragments, this line designates the covalent VB structure F•−•F. The two following lines correspond to the ionic structures F^-F^+ and F^+F^-, respectively.

The keyword "vbscf" means that a calculation of VBSCF type will be performed. As the VB orbitals are all localized, this will be an L-VBSCF calculation. A guess function is automatically provided at the end of the input. However, no guess function is required for this calculation, which converges very fast after a few iterations.

```
echo 'F2      6-31G*
$ctrl
nbasis=26 norb=8  nelectron=14 nstruct=3 nmul=1
iopt=1   iout=1 itmax=200 vbscf
$end
$struct
  1  1  2  2  3  3  4  4  5  5  6  6  7  8
  1  1  2  2  3  3  4  4  5  5  6  6  7  7
  1  1  2  2  3  3  4  4  5  5  6  6  8  8
$end
$orb
  5 3 3 5 3 3 5 5
   1 4 5 8 9
   2 6 10
   3 7 11
   14 17 18 21 22
   15 19 23
   16 20 24
   4 1 5 8 9
   17 14 18 21 22
 $end
 ' > ff.xmi
xmvb ff.xmi
```

Replacing the keyword "vbscf" by "bovb" would do automatically an L-BOVB calculation. In such a case, doing a preliminary L-VBSCF calculation and using its orbitals as a guess is recommended, albeit not compulsory.

Using Input 10.2 with the keyword «bovb» is the simplest way of doing an L-BOVB calculation. This procedure generates a hidden extended input that is shown below as Input 10.4. Of course, Input 10.4 could also be used directly and would lead to the same results.

XMVB Input 10.3. D-VBSCF Calculations of F_2

This input describes the VB part of a D-VBSCF calculation. The preceding steps are analogous to Input 10.2. The difference with the L-VBSCF input is in the definition of the spectator orbitals $1-6$, which are now allowed to delocalize over both atoms. On the other hand, the active orbitals 7 and 8 remain localized.

```
$ctrl
nbasis=26 norb=8  nelectron=14 nstruct=3 nmul=1
iopt=1    iout=1 itmax=200 vbscf
$end
$struct
  1  1  2  2  3  3  4  4  5  5  6  6  7  8
  1  1  2  2  3  3  4  4  5  5  6  6  7  7
  1  1  2  2  3  3  4  4  5  5  6  6  8  8
$end
$orb
  10 6 6 10 6 6 5 5
   1 4 5 8 9 14 17 18 21 22
   2 6 10 15 19 23
   3 7 11 16 20 24
  14 17 18 21 22 1 4 5 8 9
  15 19 23 2 6 10
  16 20 24 3 7 11
   4 1 5 8 9
  17 14 18 21 22
$end
' > ff.xmi
xmvb ff.xmi
```

XMVB Input 10.4. Extended Input for an L-BOVB Calculations of F_2

Although the simplest way to do an L-BOVB calculation is to use Input 10.2 with the keyword "bovb" instead of "vbscf" (see above), one can also directly use the extended input that is shown below. Doing this can be a useful check before running an SL-BOVB calculation as shown later.

This input describes the VB part of an L-BOVB calculation, and shows how different orbitals are to be required for different structures. Orbitals 1–8 correspond to the covalent structure, 9–15 to F^-F^+, and 16–22 to F^+F^-. Of course, both ways of doing an L-BOVB calculation lead to the same results and provide a guess function that contains the 22 VB orbitals.

```
echo 'F2      6-31G*
$ctrl
nbasis=26 norb=22 nelectron=14 nstruct=3 nmul=1
iopt=1    iout=1 itmax=20 vbscf
$end
$struct
 1  1  2  2  3  3  4  4  5  5  6  6  7  8
 9  9 10 10 11 11 12 12 13 13 14 14 15 15
16 16 17 17 18 18 19 19 20 20 21 21 22 22
$end
$orb
 5 3 3 5 3 3 5 5 5 3 3 5 3 3 3 5 5 3 3 5 3 3 5
 1 4 5 8 9
 2 6 10
 3 7 11
14 17 18 21 22
15 19 23
16 20 24
 4 1 5 8 9
17 14 18 21 22
 1 4 5 8 9
 2 6 10
 3 7 11
14 17 18 21 22
15 19 23
16 20 24
 4 1 5 8 9
 1 4 5 8 9
 2 6 10
 3 7 11
14 17 18 21 22
15 19 23
16 20 24
17 14 18 21 22
$end
' > ff.xmi
xmvb ff.xmi
```

XMVB Input 10.5. SL-BOVB Calculations of F_2

This input describes the VB part of an SL-BOVB calculation. Orbitals 15, 23 and 22, 24 are the active orbitals of the ionic structures **4** and **5**, respectively. In the guess function, Orbital 23 (resp. 24) is derived from 15 (resp. 22) by creating a node, for example, $\lambda(2PZ) + \nu(3PZ) \rightarrow \lambda(2PZ) - \nu(3PZ)$.

```
echo 'F2      6-31G*
$ctrl
nbasis=26 norb=24 nelectron=14 nstruct=5 nmul=1
iopt=1    iout=1 itmax=20 vbscf
$end
$struct
  1  1  2  2  3  3  4  4  5  5  6  6  7  8
  9  9 10 10 11 11 12 12 13 13 14 14 15 15
 16 16 17 17 18 18 19 19 20 20 21 21 22 22
  9  9 10 10 11 11 12 12 13 13 14 14 23 23
 16 16 17 17 18 18 19 19 20 20 21 21 24 24
$end
$orb
 5 3   3 5   3 3 5 5 5 3   3 5   3 3 5 5 3   3 5   3 3 5 5 5
 1 4  5 8 9
 2 6 10
 3 7 11
14 17 18 21 22
15 19 23
16 20 24
 4 1 5 8 9
17 14 18 21 22
 1 4 5 8 9
 2 6 10
 3 7 11
14 17 18 21 22
15 19 23
16 20 24
 4 1 5 8 9
 1 4 5 8 9
 2 6 10
 3 7 11
14 17 18 21 22
15 19 23
16 20 24
17 14 18 21 22
 4 1 5 8 9
17 14 18 21 22
$end
' > ff.xmi
xmvb ff.xmi
```

XMVB Input 10.6. π- D-BOVB Calculation of F_2

This input describes the VB part of a π- D-BOVB calculation. Here, only the π spectator orbitals (2, 3, 5, 6) are allowed to delocalize. The same extension of the π orbitals, applied to Input 10.5, would yield a π-SD-BOVB calculation.

```
echo 'F2    6-31G*
$ctrl
nbasis=26 norb=8  nelectron=14 nstruct=3 nmul=1
iopt=1   iout=1 itmax=20 bovb
$end
$struct
 1  1  2  2  3  3  4  4  5  5  6  6  7  8
 1  1  2  2  3  3  4  4  5  5  6  6  7  7
 1  1  2  2  3  3  4  4  5  5  6  6  8  8
$end
$orb
 5  6 6 5  6 6 5 5
  1 4 5 8 9
  2 6 10 15 19 23
  3 7 11 16 20 24
  14 17 18 21 22
  15 19 23 2 6 10
  16 20 24 3 7 11
  4 1 5 8 9
  17 14 18 21 22
$end
' > ff.xmi
xmvb ff.xmi
```

XMVB Input 10.7. Calculations of the Energy of a Single Kekulé Structure of Benzene

This input describes the complete procedure to calculate the energy of a single Kekulé structure of benzene in a GVB way, by using bond-distorted orbitals (BDOs). The atomic integrals are calculated at the Hartree−Fock step. In the second step, all the σ-MOs are frozen, and only the π-type basis functions are retained to form the VB orbitals, which are labeled from 1 to 6. Basis functions 1−4 belong to the first atom, 5−8 to the second one, and so on. In the VB state, the Kekulé structure is specified as a configuration in which all VB orbitals are singly occupied. By convention, the spin couplings are 1-2, 3-4, and 5-6.

```
g98<< $$
%chk=benzene
%int=benzene
%rwf=benzene

gaussian input for benzene, in 6-31 G(d) basis set

$$

echo "'benzene.rwf'
'benzene.int'
'benzene.chk'
96
18 24
1 2 3 4 5 6 7 8 9 10 11 12 13 14 15 16 18 19
  4 8 12 14 18 22 26 28 32 36 40 42 46 50 54 56 60 64 68 70 74 78 82 84
```

```
" >ff.bfi

g98int ff.bfi
intprep ff.bfi

echo 'benzene      6-31G*
$ctrl
nbasis=24 norb=6  nelectron=6  nstruct=1 nmul=1
iopt=1    iout=1 itmax=200 vbscf
$end
$struct
 1 2 3 4 5 6
$end
$orb
 8  8  8  8  8  8
 1  2  3  4  5  6  7  8
 5  6  7  8  1  2  3  4
 9 10 11 12 13 14 15 16
13 14 15 16  9 10 11 12
17 18 19 20 21 22 23 24
21 22 23 24  17 18 19 20
$end
' > ff.xmi
xmvb ff.xmi
```

It is seen in the $orb section that the two first VB orbitals are located on atoms 1 and 2, the two following ones on atoms 3 and 4, and so on.

XMVB Input 10.8. Calculations of Benzene with two Kekulé Structures

This input describes the VB part of the calculation of the ground state of benzene. The wave function consists of two Kekulé structures, and the VB orbitals are all allowed to delocalize freely over the whole molecule, leading to CF or OEO atomic orbitals.

```
echo 'benzene      6-31G*
$ctrl
nbasis=24 norb=6  nelectron=6  nstruct=2 nmul=1
iopt=1    iout=1 itmax=20 vbscf
$end
$struct
 1 2 3 4 5 6
 1 6 2 3 4 5
$end
$orb
24 24 24 24 24 24
 1  2  3  4  5  6  7  8  9 10 11 12 13 14 15 16 17 18 19 20 21 22 23 24
 5  6  7  8  9 10 11 12 13 14 15 16 17 18 19 20 21 22 23 24  1  2  3  4
 9 10 11 12 13 14 15 16 17 18 19 20 21 22 23 24  1  2  3  4  5  6  7  8
13 14 15 16 17 18 19 20 21 22 23 24  1  2  3  4  5  6  7  8  9 10 11 12
17 18 19 20 21 22 23 24  1  2  3  4  5  6  7  8  9 10 11 12 13 14 15 16
21 22 23 24 1  2  3  4  5  6  7  8  9 10 11 12 13 14 15 16 17 18 19 20
$end
' > ff.xmi
xmvb ff.xmi
```

XMVB Input 10.9. L-VBSCF Calculations of F_2 Using the XMVB-GAMESS[c] Link

```
$SCF DIRSCF=.FALSE. FDIFF=.FALSE. $END
$XMVB INPUT=CUSTOM VBMETH=VBSCF $END
$SYSTEM MWORDS=15 $END
$VBDAT
#CTRL
NSTR= 3
#END
#STRU
 1  1  2  2  3  3  4  4  5  5  6  6  7  8
 1  1  2  2  3  3  4  4  5  5  6  6  7  7
 1  1  2  2  3  3  4  4  5  5  6  6  8  8
#END
#ORB
 14 14 14 14 14 14 14 14
 1-14
 1-14
 1-14
 15-28
 15-28
 15-28
 1-14
 15-28
#END
$END
$BASIS   GBASIS=N31 NGAUSS=6 NDFUNC=1 $END
$GUESS   GUESS=HUCKEL $END
$DATA
  F2 RHF/6-31G*
  CN 1

   F
   F           1          1.44
$END
```

XMVB Output 10.1. L-BOVB of F$_2$

This output displays part of the information that is given by the XMVB program at the end of an L-BOVB calculation on F$_2$. The energies arising from the diagonalization of the Hamiltonian matrix must be shifted by the "Nuclear Repulsion Energy" indicated below (this quantity also includes the electronic energy of the frozen electrons, if any).

```
F2      6-31G*

OPTIMIZATION METHOD:  DFP-BFS WITHOUT GRADIENT

Number of Structures:    3

The following structures are used in calculation:

  1 *****    1  1  2  2  3  3  4  4  5  5  6  6  7  8
  2 *****    9  9 10 10 11 11 12 12 13 13 14 14 15 15
  3 *****   16 16 17 17 18 18 19 19 20 20 21 21 22 22

Nuclear Repulsion Energy:    -132.938723

Total Energy:    -198.767632


            ****** MATRIX  OF  HAMILTONIAN ******

         1          2          3
  1  -65.717367  24.257687  24.258857
  2   24.257687 -65.278968  -5.907975
  3   24.258857  -5.907975 -65.278966


            ****** MATRIX  OF  OVERLAP  ******

1        2          3
  1   1.000000  -0.366057  -0.366075
  2  -0.366057   1.000000   0.088416
  3  -0.366075   0.088416   1.000000


******  COEFFICIENTS OF STRUCTURES ******

  1    0.75138  ******    1  1  2  2  3  3  4  4  5  5  6  6  7  8

  2   -0.26099  ******    9  9 10 10 11 11 12 12 13 13 14 14 15 15

  3   -0.26099  ******   16 16 17 17 18 18 19 19 20 20 21 21 22 22


            ****** WEIGHTS OF STRUCTURES ******

  1    0.70815  ******    1  1  2  2  3  3  4  4  5  5  6  6  7  8

  2    0.14592  ******    9  9 10 10 11 11 12 12 13 13 14 14 15 15

  3    0.14592  ******   16 16 17 17 18 18 19 19 20 20 21 21 22 22
```

XMVB Output 10.2. L-VBCISD of F_2

This output displays part of the information that is given by the XMVB program at the end of an L-VBCISD calculation on F_2. Each "fundamental structure" is a linear combination of VB functions that possess the same nature in terms of spin-pairing and charge distributions. The coefficients of these VB functions are extracted from the multistructure ground-state wave function, and are renormalized (see Eqs. 9.13-9.14).

```
NO ORBITAL OPTIMIZATION

Number of Structures:    3

The following structures are used in calculation:

    1 *****    1  1  2  2  3  3  4  4  5  5  6  6  7  8
    2 *****    1  1  2  2  3  3  4  4  5  5  6  6  7  7
    3 *****    1  1  2  2  3  3  4  4  5  5  6  6  8  8

Nuclear Repulsion Energy:      -132.938725

        549  CI Structures

Max valence electrons:              10

VBPP is applied

VBSCF ENERGY:    -198.735995021256

VBCI ENERGY:     -198.931121537566
Total Energy:    -198.931122

Davidson corrected energy:    -198.939178403990


            ******  COEFFICIENTS OF STRUCTURES  ******

    1      0.76692  ******    1  1  2  2  3  3  4  4  5  5  6  6  7  8

    2     -0.22658  ******    1  1  2  2  3  3  4  4  5  5  6  6  7  7

    3     -0.22658  ******    1  1  2  2  3  3  4  4  5  5  6  6  8  8

    4     -0.00012  ******    1  9  2  2  3  3  4  4  5  5  6  6  7  8

    5     -0.00253  ******    1 10  2  2  3  3  4  4  5  5  6  6  7  8

    6      0.00021  ******    1 11  2  2  3  3  4  4  5  5  6  6  7  8

    7     -0.00016  ******    1  1  2 12  3  3  4  4  5  5  6  6  7  8

    8      0.00256  ******    1  1  2 13  3  3  4  4  5  5  6  6  7  8

    9      0.00012  ******    1  1  2 14  3  3  4  4  5  5  6  6  7  8

   10     -0.00221  ******    1  1  2  2  3 15  4  4  5  5  6  6  7  8

. . . . . . . . . . . .  etc.  . . . . . . . . . . . .
```

```
****** WEIGHTS OF FUNDAMENTAL STRUCTURES ******


1      0.74484  ******    1  1  2  2  3  3  4  4  5  5  6  6  7  8

2      0.12758  ******    1  1  2  2  3  3  4  4  5  5  6  6  7  7

3      0.12758  ******    1  1  2  2  3  3  4  4  5  5  6  6  8  8
```

```
         ****** OVERLAP OF FUNDAMENTAL ******


        1          2          3
1    1.000000   0.353629   0.353628
2    0.353629   1.000000   0.041312
3    0.353628   0.041312   1.000000
```

```
         ****** HAMILTONIAN OF FUNDAMENTAL ******


         1          2          3
1   -65.892994 -23.497663 -23.497606
2   -23.497663 -65.376260  -2.821991
3   -23.497606  -2.821991 -65.376261
```

Epilogue

Our intention in writing this book was to teach a theory that has been largely abandoned in the mainstream community of theoretical chemistry, but is nevertheless still being used every day by chemists in their thinking about molecules and reactions through the Lewis representation of molecules and Pauling's resonance theory ideas. Indeed, VB theory is a chemical theory, and its neglect undermines the cultural heritage of our science. Therefore, for the sake of showing the versatility of VB theory, we have taken our reader along through 10 chapters, which describe different aspects of the theory, from its history and its rivalry with MO theory, through the development of a Hückel-type VB theory, to applications to chemical reactivity, photochemistry, and electronic structure of polyradical species, all the way to more quantitative aspects of the method; the existing computational methods, and their domains, and the available program packages. In the end, we also provided some know-how on the nature of the input and output of VB calculations, and how to run meaningful VB calculations and obtain desirable quantities, such as resonance energies. *Throughout this book we made an effort to create bridges between VB and MO theories*, and showed the advantage of fusing the special insight of these two theories. Ultimately, researchers, students, and teachers using this book will determine the success of our effort. As we wrote in the preface: Any feedback will be most welcome!

While the 10 chapters may seem plenty for one course, one must remember that there are many aspects we did not even cover. For example, we did not treat chemical bonding in detail, nor did we deal with new ideas in chemical bonding (1). In fact, VB theory predicts that alongside the traditional covalent and ionic electron-pair bonds, there exists a class where the covalent–ionic resonance energy is the root cause of the bonding and sometimes of the existence of the molecule, (e.g., as in F_2). All these bonds are typified by depleted electron density in the bonding region, and are distinctly different than the classical covalent bonds. Another form of bonding is no-pair bonding that arises in monovalent clusters with maximum spin, (e.g., $^{n+1}Cu_n$). This form maintains very strong bonding despite the lack of any electron pairs. This

---

A Chemist's Guide to Valence Bond Theory, by Sason Shaik and Philippe C. Hiberty
Copyright © 2008 John Wiley & Sons, Inc.

kind of bonding is buttressed by the resonance energy between the covalent triplet pair and the triplet ionic forms (1). Another topic, which was not treated, concerns the applications of VB theory to species and reactions of metalloenzymes (2,3), and to enzymatic reactions in general (4). Likewise, the description of TM complexes by VB theory was barely touched (5) and the application to molecular dynamics was merely mentioned (6). Nevertheless, we are confident that studying the material in this book will form an incentive to consider VB applications to many other topics, not even mentioned here.

While we argued throughout the book that current VB programs are capable today of treating a variety of problems, even some chemical reactions, we recognize that VB theory may never become a standard computational method as MO or DFT methods. However, we hope that this book shows that *the use of VB theory is all about insight*, and the ability of one to think, reason, and predict chemical patterns. This word insight brings to mind the Coulson admonition: "Give me insight not numbers". This book tries to do that, and whatever measure of success that is achieved is something about which we have to thank the great teachers of quantum chemistry. We thus dedicate this monograph to the inspirational teachers whom we were fortunate to meet during our careers: Nick Epiotis, Roald Hoffmann, Lionel Salem, and the Late Edgar Heilbronner, whose writing, teachings, and admonitions continue to guide us.

REFERENCES

1. S. Ritter, *Chem. Eng. News* **85**, 37 (2007). The Chemical Bond.
2. F. Ogliaro, S. Cohen, S. P. de Visser, S. Shaik, *J. Am. Chem. Soc.* **122**, 12892 (2000). Medium Polarization and Hydrogen Bonding Effects on Compound I of Cytochrome P450: What Kind of a Radical Is It Really?
3. S. Shaik, S. Cohen, S. P. de Visser, P. K. Sharma, D. Kumar, S. Kozuch, F. Ogliaro, D. Danovich, *Eur. J. Inorg. Chem.* 207 (2004). The "Rebound Controversy": An Overview and Theoretical Modeling of the Robound Step in C–H Hydroxylation by Cytochrome P450.
4. A. Warshel, R. M. Weiss, *J. Am. Chem. Soc.* **102**, 6218 (1980). An Empirical Valence Bond Approach for Comparing Reactions in Solutions and in Enzymes.
5. T. K. Firman, C. R. Landis, *J. Am. Chem. Soc.* **123**, 11728 (2001). Valence Bond Concepts Applied to the Molecular Mechanics Description of Molecular Shapes. 4. Transition Metals with π-Bonds.
6. D. G. Truhlar, *J. Comput. Chem.* **28**, 73 (2007). Valence Bond Theory for Chemical Dynamics.

Glossary

1. ABBREVIATIONS, TERMS, AND ACRONYMS

Active Orbitals: The set of orbitals that are treated in a valence bond fashion (see VBT). All other orbitals are called "inactive" or "spectator".

AO: Atomic orbital.

BDO: Bond distorted orbital. An orbital that is localized on one atom with a small delocalization tail on an atom with which it shares a bond.

Bond diagram: A simple mnemonics used to describe a generalized two-electron bond (see Fig. 3.2), composed of two singlet-paired electrons in two fragment orbitals of any type.

BEBO: Bond energy–bond order. A semiempirical relationship that relates the energy of a bond to its bond order. This relationship is used for semiempirical construction of potential energy surfaces.

BOVB: Breathing orbital valence bond. A VB computational method. The BOVB wave function is a linear combination of VB structures that simultaneously optimizes the structural coefficients and the orbitals of the structures and allows different orbitals for different structures. The BOVB method must be used with strictly localized active orbitals (see HAOs). When all the orbitals are localized, the method is referred to as L-BOVB. There are other BOVB levels, which use delocalized MO-type inactive orbitals, if the latter have different symmetry than the active orbitals. (See Chapters 9 and 10.)

CASSCF: Complete active space self-consistent field. A method that calculates electronic structure by using all the configurations that arise by distributing all the valence electrons within a given window of molecular orbitals. The procedure optimizes the coefficients and orbitals of the so-selected set of configurations. The CASSCF calculations belong to the general class of calculations, so-called, multiconfiguration SCF (MCSCF) calculations, where one uses more than the Hartree–Fock configuration to describe the electronic structure.

A Chemist's Guide to Valence Bond Theory, by Sason Shaik and Philippe C. Hiberty
Copyright © 2008 John Wiley & Sons, Inc.

CASPT2 (CASMP2): A method of calculation involving second-order perturbation theory done after a CASSCF calculation.

CASVB: A method that converts a CASSCF wave function to the closest possible VB wave function.

CBD: Cyclobutadiene.

CF orbitals: Coulson–Fisher-type orbitals. These are semilocalized AOs, also called overlap enhanced orbitals (OEOs) which are localized on a given center, but have small delocalization tails on other centers. Special cases of CF AOs are BDOs (see above).

CI: Configuration interaction.

CIS: Configuration interaction including only single excitations from the Hartree–Fock MO determinant. A simple method for the calculations of excited states.

COT: Cyclooctatetraene.

CRUNCH: A VB program based on the symmetric group methods of Young and written by Gallup and co-workers. (See Chapter 9.)

DFT: Density functional theory.

DIM: Diatomics in molecules. A semiempricial method used to construct potential energy surfaces of polyatomic molecules from the energy of the diatomic fragments.

EVB: Empirical VB. A method used to construct the free energy profiles in enzymatic reactions.

ESR: Electron spin resonance.

HF: The Hartree–Fock MO method.

FO–VB: Fragment orbital valence bond. A VB method that uses fragment orbitals with symmetry properties. The method is used for understanding the structures of transition states and molecules.

GAUSSIAN: A general-purpose package of programs.

GAMESS: A general-purpose package of programs. The GAMESS–US version is noncommercial. The GAMESS–UK version is commercial. Both packages contain VB programs. (See Chapter 9.)

GVB: Generalized valence bond. A theory that employs CF orbitals to calculate electronic structure with wave functions in which the electrons are formally coupled in a covalent manner. The simplest level of the theory is GVB–PP (PP-perfect pairing), in which all the electrons are paired into bonds, as in the Lewis structure of the molecule.

HAO: Hybrid atomic orbitals that are strictly localized on a single atomic center. The HAOs have no delocalization tails.

Heisenberg Hamiltonian: An effective Hamiltonian imported from physics (also called spin-Hamiltonian), and used in VB to describe spin states of molecular species with an average of one electron per site (atom), for example, of polyradicals. (See Chapter 8.)

Hubbard Hamiltonian: An effective Hamiltonian, which includes on-site electron–electron repulsion and a resonance integral, and is used in VB to calculate states of molecular species with an average of one electron per site (atom), e.g., of polymers.

HL: Heitler–London. The term corresponds to the wave function used by Heitler and London to calculate the bonding energy of the H_2 molecule in 1927, and is used as a generic name to describe a covalent many-electron wave function.

HLVB Theory: A VB theory that uses only HL-type covalent wave functions. This theory was used in the early days and the main proponents were Pauling and Slater. The HLVB is also the basis for the semiempirical potentials used in molecular dynamics.

HMO theory: Hückel molecular orbital theory.

Ionic VB Structures: Valence bond structures that involve oppositely charged centers. For example, the description of an electron pair bond A–B requires a covalent HL structure, A•–•B, and two ionic structures, $A^+:B^-$ and $A:^- B^+$.

LBO: Localized bond orbital. For a given molecule, the LBOs can be obtained from the canonical MOs (CMOs) by a unitary transformation that does not change either the total energy or the total wave function. (See Chapter 3.)

LEPS: London–Evans–Polanyi–Saito. A valence bond type method based on HLVB and used to construct potential energy surfaces for molecular dynamics.

MOPLRO: A general-purpose package of programs. It contains the CASVB method.

Pauli repulsion: The repulsion of two electrons with identical spins on two centers, (e.g., A• ↑↑• B). This repulsion also appears in VB structures bearing three and four electrons, that is, A: •B, A• :B, A: :B. The Pauli repulsion is precisely the same as the "overlap repulsion" known from qualitative MO theory. (See Table 3.1.)

PES: Photoelectron spectroscopy. A method for measuring ionization energies of molecules.

Promotion gap: The energy gap between the ground and excited states in the VBSCD (see below). This factor, labeled as G, originates the barrier in chemical reactions.

RE: Resonance energy. The energy lowering due to the mixing of VB structure(s) into a reference structure. For example, the RE of benzene is the energy lowering relative to a single Kekulé structure (See Chapter 10).

RE$_{CS}$: Charge-shift resonance energy. The energy lowering due to covalent–ionic mixing. For example, the RE_{CS} of the F–F bond is the energy lowering due to the mixing of the ionic structures F^-F^+ and F^+F^- into the covalent structure F•–•F.

Rumer structures: A basis set of linearly independent VB structures. For example, for benzene there are five Rumer structures: two Kekulé structures and three Dewar structures. The Rumer structure set is normalized, but not orthogonalized. See Chapters 3 and 4.

SAD: Spin-alternant determinant. The VB determinant with one electron per site and with alternating spins. Other terms describing the same determinant are the quasiclassical (QC) state, and the antiferromagnetic (AF) state. In nonalternant hydrocarbons, where compete spin alternation is impossible, the determinant is called MSAD, namely, the maximum spin-alternating determinant. The SAD–MSAD are the leading terms in the wave function of molecules with one electron per site, for example, conjugated hydrocarbons. In radicals (e.g., allyl radical) the SAD is the root cause of spin polarization (i.e., negative spin densities flanked by positive ones). See Chapters 7 and 8.

SC: Spin-Coupled. A VB computational method that uses OEOs and calculates the wave function of an electronic system using a single orbital configuration with all the possible spin-pairing schemes. The method is similar in spirit to GVB.

SCVB: Spin-Coupled VB. The CI-augmented SC method.

TURTLE: A VB program that can perform a variety of VB calculations (e.g., VBSCF and BOVB). The program is now incorporated into the GAMESS–UK package.

Twin-States: The ground and excited states that arise from avoided crossing in the VBSCD (see below). Usually, the twin-states correspond to a transition state of a thermal reaction and an excited-state intermediate. This excited-state intermediate can be converted to a "funnel" (a conical intersection) for the products of a photochemical reaction. See Chapter 6.

VAL-BOND: An empirical method for calculating and predicting geometries of transition metal complexes.

VBSCD: Valence bond state correlation diagram. A VB diagram that views the barrier formation as a result of avoided crossing between two state curves that are anchored in the ground and two excited states of reactants and products. The VBSCD is a paradigm for the barrier in chemical reactions (see Chapter 6).

VBCMD: Valence bond configuration mixing diagram. A VB diagram that involves many VB curves, for example, the covalent and ionic curves corresponding to some transformations, or the two Lewis state curves in the VBSCD and curves that are in a secondary high-lying configuration throughout the reaction path (see Chapter 6).

VBSCF: Valence bond self-consistent field. A VB computational method. The VBSCF wave function is a linear combination of VB structures that simultaneously optimizes the structural coefficients and the orbitals of the structures. It can be used with any type of AOs: OEOs, BDOs, and HAOs. With HAOs, we refer to the method as L-VBSCF; L = localized.

VBCI: Valence bond configuration interaction. A VB computational method that starts with a VBSCF wave function, which is further improved by CI. The CI involves virtual orbitals that are localized on exactly the same regions as the respective active orbitals. There are a few VBCI levels that are denoted by the rank of excitation into the virtual orbitals, for example, VBCISD involves single and double excitations.

VBPCM: Valence bond polarized continuum model. A VB computational method that incorporates solvent effect by using the PCM solvation model. The method can be coupled with VBSCF, BOVB, and VBCI.

VB2000:   A VB program that can perform a variety of VB calculations, for example, VBSCF, GVB, and SCVB. The program is incorporated into the GAMESS–US package.

VBT: Valence bond theory.

XMVB: A VB program that can perform a variety of VB calculations, for example, VBSCF, BOVB, VBCI, and VBPCM. The program has a stand-alone version, and a version that is incorporated into the GAMESS–US package.

2.   SYMBOLS

B	Denotes the resonance energy of the transition state in the VBSCD.
G	Denotes the promotion gap in the VBSCD.
ΔE_c	Denotes the height of the crossing point in the VBSCD.
f	Denotes the ratio $\Delta E_c/G$ in the VBSCD.
Ω	Denotes a VB determinant.
Ω_{QC}	Denotes the spin-alternant quasi-classical determinant.
Φ	Denotes a VB structure.
Ψ	Denotes a state (eigenfunction of the Hamiltonian).
$a, b, c..$ or χ_1, χ_2	These letters denote AOs.
φ	Denotes an MO or FO.
ϕ	Denotes a product of orbitals (a hartree product).
$\bar{a}$	Denotes a spin–orbital in a Slater determinant with spin β, where the identity of the spin is indicated by the bar over the orbital symbol. The lack of the bar indicates a spin orbital with spin α.
β	Denotes the reduced matrix element, for example, $\beta_{ab} = h_{ab} - 0.5(h_{aa} + h_{bb})S_{ab}$. This reduced matrix element is equivalent to the resonance integral of the Hückel type.

INDEX

Page numbers in boldface indicate main discussion of the indexed item.

A Chemist's Guide to Valence Bond Theory, by Sason Shaik and Philippe C. Hiberty
Copyright © 2008 John Wiley & Sons, Inc.

Printed in the United States
By Bookmasters